Seeking Molecular Biomarkers for Schizophrenia Using ROC Analysis

Margareth Borges Coutinho Gallo

Seeking Molecular Biomarkers for Schizophrenia Using ROC Analysis

Margareth Borges Coutinho Gallo
Department of Health and Environment
Oswaldo Cruz Foundation - FIOCRUZ
Eusébio, Ceará, Brazil

ISBN 978-3-031-59027-6   ISBN 978-3-031-59028-3   (eBook)
https://doi.org/10.1007/978-3-031-59028-3

© The Editor(s) (if applicable) and The Author(s), under exclusive license to Springer Nature Switzerland AG 2024

This work is subject to copyright. All rights are solely and exclusively licensed by the Publisher, whether the whole or part of the material is concerned, specifically the rights of translation, reprinting, reuse of illustrations, recitation, broadcasting, reproduction on microfilms or in any other physical way, and transmission or information storage and retrieval, electronic adaptation, computer software, or by similar or dissimilar methodology now known or hereafter developed.

The use of general descriptive names, registered names, trademarks, service marks, etc. in this publication does not imply, even in the absence of a specific statement, that such names are exempt from the relevant protective laws and regulations and therefore free for general use.

The publisher, the authors and the editors are safe to assume that the advice and information in this book are believed to be true and accurate at the date of publication. Neither the publisher nor the authors or the editors give a warranty, expressed or implied, with respect to the material contained herein or for any errors or omissions that may have been made. The publisher remains neutral with regard to jurisdictional claims in published maps and institutional affiliations.

This Springer imprint is published by the registered company Springer Nature Switzerland AG
The registered company address is: Gewerbestrasse 11, 6330 Cham, Switzerland

Paper in this product is recyclable.

*To God, my spiritual snuggle,
To my family, my emotional support,
To all the scientists who inspired me to write this book.
"I have stood on the shoulders of giants".*
                                              Margareth Gallo

Foreword

Searching for biomarkers to psychiatric disorders has been a challenge for decades—especially biomarkers to diseases such as schizophrenia, which not only affects more than 25 million people worldwide, but also significantly affect patients' families.

The power that omic tools have brought to the stage in the past two decades have enabled this long searched task. Whilst most of the findings in the schizophrenia field are restricted to the basic science field, some of the results might have a chance to dare to jump to the clinic.

This book approaches the challenges, pros and cons in the quest for clinical biomarkers in schizophrenia, bringing an objective analysis to the reader and providing the latest update on this long-posed and necessary question.

Researcher in the field of schizophrenia for 20 years.

Professor of Biochemistry University of Campinas, Campinas, Brazil

Daniel Martins-de-Souza

Preface

Over the span of two decades, my professional journey has revolved around the development and validation of analytical methods for quality control in plants and medicines. My experience spans both academia and the pharmaceutical industry. However, the statistical knowledge acquired during my undergraduate studies proved insufficient for the intricate applications demanded by my daily work. The global COVID-19 pandemic provided an unexpected opportunity for self-improvement, as esteemed universities offered online courses, previously elusive due to work commitments. Seizing this chance, I dedicated two consecutive years to deepening my understanding of statistics and its practical applications.

In early 2022, recognizing the need for further academic growth, I decided to embark on my fourth postdoctoral endeavor. This time, the focus was on metabolomics, and I reached out to Dr. Daniel Martins-de-Souza at the State University of Campinas, São Paulo, Brazil. The project aimed to develop a diagnostic method for schizophrenia spectrum disorders (SSD) using blood metabolites. Fortuitously, Oswaldo Cruz Foundation in Ceará granted me a one-year leave of absence, enabling me to travel 4000 kilometers to Campinas.

My initial task involved conducting a comprehensive survey of the state-of-the-art diagnostic models for SSD utilizing metabolic biomarkers. Surprisingly, only around 10% of articles reported the use of receiver operating characteristic (ROC) analysis to assess the accuracy of the predictive models. This statistical tool is widely regarded as the most suitable for evaluating biomarker performance. Intrigued by this discrepancy, I delved deeper, aiming to establish connections between the potential molecular biomarkers and the hypotheses surrounding the pathophysiology of SSD.

My curiosity extended to identifying the best biomarkers defined by ROC analysis across all studies, with the intent of assembling a potential panel for validating a method capable of distinguishing patients from healthy individuals. A thorough search on PubMed, Google Scholar, and Medline, using the keywords "metabolomics/AUC/schizophrenia," revealed that the first study in this area was published in 2011, marking a research coverage period of 12 years (2011–2023).

Navigating this complex terrain was no easy feat. The elusive etiology of the disease, subjective clinical diagnostic methods, non-standardized sample collection, and varying patient stages posed significant challenges. Moreover, the multidisciplinary nature of this field demanded expertise in disease understanding, chromatography and mass spectrometry, machine learning, and, crucially, statistics. Many studies exhibited technical inadequacy in handling statistical tools, describing parameters, and interpreting results.

Motivated by these challenges and my newfound expertise, I envisioned creating a tutorial to assist newcomers in developing a binary diagnostic model (disease vs. health) using metabolic biomarkers assessed by the ROC technique on the MetaboAnalyst platform. It imparts essential statistical concepts, provides graph examples, and elucidates interpretation and description techniques. Additionally, a checklist is included to assist with experimental design, informed by observations gleaned from various studies and systematic reviews.

Returning to the core proposal, the book features tables detailing human-related metabolic biomarkers, those related to intestinal flora (metabolites and bacteria), neurotransmitters, neurotrophic factors, neuroendocrine markers, cellular and environmental stress-related markers, and immuno-inflammatory markers. Each is accompanied by statistical scores and areas under the curve (AUCs), establishing connections with symptoms and metabolic pathways. Following the compilation of all metabolites assessed through ROC analysis, subsequent enrichment analyses were conducted to explore their associations with both disease and metabolic pathways. These comprehensive analyses have led to the identification of a panel of biomarkers exhibiting substantial potential for diagnosing SSD.

I trust that this book will prove invaluable to those who consult it. Happy reading!

Eusébio, Ceará, Brazil   Margareth Borges Coutinho Gallo

Acknowledgement

The author expresses gratitude for the distinctive technical support provided by Professor Daniel Martins-de-Souza during her post-doctoral internship at the Neuroproteomics Laboratory at the State University of Campinas. Special thanks are extended to doctors Talita Vrechi and Lícia Costa for their assistance in literature mining across databases. The author is also appreciative of Professors Ricardo Roberto da Silva and Benilton de Sá Carvalho for their invaluable guidance and expertise in statistics, enhancing the quality of the research. Furthermore, heartfelt appreciation is extended to FIOCRUZ for granting a one-year work license, facilitating the post-doctoral internship.

Contents

1 Hypotheses of the Pathophysiology of Schizophrenia Spectrum Disorders 1
   1.1 Introduction 1
   1.2 The Pathophysiology of SSD 2
      1.2.1 The Role of Genes 2
      1.2.2 The Neurodevelopmental (the 2-Hits) Hypothesis 2
      1.2.3 The Neurobiological/Neurochemical Hypothesis 3
      1.2.4 The "Dysconnectivity" Hypothesis 5
      1.2.5 The Autoimmune Hypothesis 6
      1.2.6 The Role of the Inflammatory Response 6
      1.2.7 The Role of Environment Stress 6
      1.2.8 The Role of Mitochondria and Oxidative Stress 7
   References 8

2 Biomarkers 13
   2.1 Introduction 13
   2.2 Biomarker Classification 14
   2.3 Development and Validation of Machine Learning-Based Diagnostic Models 15
      2.3.1 Data Collection and Preparation 15
      2.3.2 Model Construction 20
   2.4 Statistical Analysis of Biomarkers 22
      2.4.1 Principal Component Analysis 22
      2.4.2 Discriminant Analysis 24
      2.4.3 ROC Curve Analysis 26
   References 31

3	**Biomarkers Related to the Hypotheses of the Pathophysiology of Schizophrenia Spectrum Disorders**........................	37
	3.1 Genetic, Epigenetic, and Proteomic Biomarkers...............	37
	3.2 Metabolic Biomarkers for SSD Assessed Through ROC Analysis.................................	38
	3.2.1 Discussion......................................	68
	References...	72
4	**Immune, Inflammatory and Oxidative Stress-Related Biomarkers**	77
	4.1 Introduction...	77
	4.2 Immune, Inflammatory, and Oxidative Stress-Related Biomarkers for SSD Assessed Through ROC Analysis..........	78
	4.3 Discussion..	90
	References...	91
5	**Gut Microbiota-Brain Axis and Related Biomarkers**.............	95
	5.1 Introduction...	95
	5.2 Gut Microbiota and Associated Metabolites Assessed as Biomarkers for SSD Through ROC Analysis................	97
	5.3 Discussion..	110
	References...	112
6	**Neuroendocrine Biomarkers**................................	115
	6.1 Introduction...	115
	6.2 Neuroendocrine Biomarkers for SSD Assessed Through ROC Analysis...	116
	6.3 Discussion..	121
	References...	122
7	**Neurotrophic Biomarkers**	125
	7.1 Introduction...	125
	7.2 Neurotrophic Biomarkers for SSD Assessed Through ROC Analysis...	126
	7.3 Discussion..	132
	References...	133
8	**Neurotransmitter Biomarkers**................................	135
	8.1 Introduction...	135
	8.2 Neurotransmitter Biomarkers for SSD Assessed Through ROC Analysis...	136
	8.3 Discussion..	143
	References...	143
9	**Conclusions**	147
	9.1 Final Considerations	147
	9.2 Cautions in Pursuit of Robust Biomarkers....................	148
	9.2.1 Study Design and Planning	149
	9.2.2 Sample Collection	151

		9.2.3	Sample Preparation	153
		9.2.4	Analytical Techniques	153
		9.2.5	Data Acquisition and Processing	153
		9.2.6	Statistical Analysis	153
		9.2.7	Data Interpretation and Validation	154
		9.2.8	Reporting and Documentation	154
	References			154
10	**Step-by-Step Guide to Building a Diagnostic Model Using MetaboAnalyst**			**157**
	10.1	Introduction		157
	10.2	LC-MS Spectra Processing		162
		10.2.1	Data Upload	162
		10.2.2	Data Integrity Check	163
		10.2.3	Data Processing	163
		10.2.4	Job Submission	176
		10.2.5	Results of Data Processing	176
	10.3	Statistical Analyzes (One Factor)		180
		10.3.1	Data Uploading	182
		10.3.2	Data Processing	182
		10.3.3	Normalization	186
	10.4	Chemometrics Analysis		189
		10.4.1	Principal Component Analysis	189
		10.4.2	Partial Least Squares Discriminant Analysis	195
		10.4.3	Cluster Analysis	199
		10.4.4	Classification and Feature Selection	205
	10.5	Univariate Analysis		212
		10.5.1	Fold Change Analysis	213
		10.5.2	t-Tests	214
		10.5.3	Volcano Plot	215
		10.5.4	One-Way Analysis of Variance (ANOVA)	217
		10.5.5	Pattern Hunter	219
	10.6	Enrichment Analysis		222
		10.6.1	Overrepresentation Analysis (ORA)	222
		10.6.2	Single Sample Profiling (SSP)	224
		10.6.3	Quantitative Enrichment Analysis (QEA)	227
	10.7	Power Analysis		227
	10.8	Biomarker Analysis		232
		10.8.1	Classical Univariate ROC Analysis	233
		10.8.2	Multivariate ROC Curve-Based Exploratory Analysis	235
		10.8.3	ROC Curve-Based Model Evaluation	236
	References			242
Index				**249**

Chapter 1
Hypotheses of the Pathophysiology of Schizophrenia Spectrum Disorders

1.1 Introduction

In the early 1900s, Swiss psychiatrist Paul Eugen Bleuer coined the term schizophrenia (from the Greek root *skhizein* = splitting and *phrēn* = mind) to denote the division of the different mental domains experienced by the patients and differentiate it from the hitherto accepted concept known as *dementia praecox*, by Emil Kraepelin. Since then, the disease has been redefined several times by the International Statistical Classification of Diseases and Related Health Problems (ICD) and by the Diagnostic and Statistical Manual of Mental Disorders (DSM), being recognized as Schizophrenia Spectrum Disorders (SSD) in the most recent version of DSM, effective March 2022. The last reclassification reflects the advances in neuroscience but is still limited by not delving into the roots of the mental psychopathology [1]. The Research Domain Criteria (RDoC) project, launched in 2009 by the National Institute of Mental Health (NIMH), emerged with the mission of fostering research based on experimental psychopathology and measurement, encouraging the sharing and collection of data from multiple response systems, and circumventing this limitation [2].

A two-hit hypothesis is the basis for explaining the etiology of SSD, proposing an additive effect between genetic susceptibility and environmental risk factors occurring early in life and later in adolescence or young adulthood [3], as evidenced by nearly three hundred newly described loci genes [4]. Several symptoms occur progressively in the clinical course of the disease and can vary independently in each patient. They were clustered by factor analysis into reality distortion (delusions and hallucinations), disorganization (of thoughts and behavior, and inappropriate affect), negative symptoms (deterioration in social function), and cognitive impairment (learning and memory deficits) [5].

© The Author(s), under exclusive license to Springer Nature Switzerland AG 2024
M. Borges Coutinho Gallo, *Seeking Molecular Biomarkers for Schizophrenia Using ROC Analysis*, https://doi.org/10.1007/978-3-031-59028-3_1

1.2 The Pathophysiology of SSD

Historically, the understanding of the pathophysiology of SSD was built from the serendipity discovery of the first antipsychotics [6, 7], the study of their mechanisms of action [8–12], the changes in the brain revealed by neuroimaging of *post-mortem* and *in vivo* tissues [13, 14], and the search for causality in substance-induced psychoses [15]. All the advances achieved have allowed paradigms to be broken [16, 17], and paved the way for the postulation of several hypotheses to characterize the pathophysiological basis of SSD. Furthermore, they leveraged the biomarker findings and explained the complex interaction between antipsychotics and the set of neurotransmitter receptors on which they act [9].

Evidence has shown SSD to be a complex disorder with the involvement of multiple biological systems. This is reflected in the interweaving of several of the postulated hypotheses, as can be seen in the summaries narrated below.

1.2.1 The Role of Genes

SSD heritability estimates are 60–80%, but there is a lack of disease-specific genes [18, 19]. A recent study identified common variant associations at 287 distinct genomic loci, which were primarily enriched at genes implicated in neurodevelopmental disorders and genes expressed in both excitatory and inhibitory neurons of the CNS, indicating neurons as the main site of SSD [4].

1.2.2 The Neurodevelopmental (the 2-Hits) Hypothesis

The original hypothesis reflects the interplay of genetics and environment being accounted for the onset and progress of the disease, since environmental exposures during fetal and perinatal life such as malnutrition, infections (e.g., influenza, *Toxoplasma gondii*, and herpes simplex virus), substance abuse, obstetric complications, and psychological stress, among others, may lead to a defective neurodevelopment and/or neuroinflammation. Evidence has been compiled in a review of the role of environmental factors in the susceptibility to SSD [20]. Later, this hypothesis morphed into a larger theory, including several other influencing factors such as epigenetic modifications, which affect different aspects of normal brain maturation such as myelination, plasticity, and synaptic pruning [21, 22]. Evidence for excessive synaptic pruning lies in the low levels of the presynaptic protein synaptophysin found in cingulate cortex, frontal cortex, and hippocampus and in the reduced numbers of synaptic spines present in the cortical region of SSD patients [14].

Epigenetics is among the approximately 50% of reasons remaining outside of heredity capable of explaining the causality of SSD, although we still do not clearly know the principles that determine which dose and time of exposure to the environmental alteration would be adequate to trigger the epigenetic changes [23]. A meta-analysis suggested epigenetic modifications as potential therapeutic targets for SSD [20, 24] while another indicated them as targets for biomarker-guided interventions [25].

1.2.3 The Neurobiological/Neurochemical Hypothesis

This hypothesis postulates the cause of SSD as a function of deregulation in the production of neurotransmitters. Thus, all hypotheses related to neurotransmitters could be grouped together.

1.2.3.1 The Neurotrophins Hypothesis

This hypothesis is a derivative of the maldevelopment model that posits changes in embryonic neurogenesis as the cause of SSD. Neurotransmitter deficits were thereby interpreted in light of neurotrophic factors deficacy [26]. Two meta-analyses have confirmed low levels of neurotrophins associated with medicated and drug-naïve patients [27, 28]. Through protein-protein interaction, researchers have shown the connection of neurotrophic factors (Brain-derived neurotrophic factor (BDNF), neurotrophin 4 (NT-4), and nerve growth factor (NGF)) and cell adhesion components (catenin beta protein (CTNNB1) and cadherin gene/protein (CDH1)) with the immune-inflammatory responses of first episode (FE) patients, and the massive participation of BDNF in neurogenesis, placental angiogenesis, fetal development, and synaptic plasticity. Thus, the reduced level of BDNF observed in SSD patients would explain the reduction of its immunoregulatory and neuroprotective functions against the neurotoxic effects of the inflammatory response system [29].

1.2.3.2 The Dopamine Hypothesis

The old hypothesis stated that psychotic symptoms were related to overactive dopamine (DA) transmission, while the most current one explains positive and negative symptoms due to the co-occurrence of dopamine 2 receptor (D2R) hyperactivity in subcortical and limbic brain regions, and the hypo-functionality of D1R in the prefrontal cortex [22, 30, 31]. A DA-derived hypothesis—the dopamine prediction-error—posits that a shift in the balance of inputs and outputs to the dopaminergic system can produce paradoxical shifts in learning, which involve difficulties learning from rewarding events while simultaneously "learning too much" about irrelevant or neutral information [32].

1.2.3.3 The Serotonin Hypothesis

SSD-like psychosis caused by LSD—a potent serotonin 2A receptor (5-HT2AR) agonist—has given rise to the hypothesis that serotonin hyperactivity in the cerebral cortex could explain SSD symptoms. Thus, dopaminergic hypofunction and downstream glutamate release due to the upregulation of ascending serotonergic pathways came naturally and could explain the origin of positive and negative symptoms, the relationship of SSD to stress, cortical atrophy, peripheral depletion of phospholipids, and the effectiveness of DA blockade in treating positive symptoms [33–35].

1.2.3.4 The Glutamate Hypothesis

It postulates a dysfunction in N-methyl-D-aspartate receptors (NMDAR) and an excessive glutamate release, especially in the prefrontal cortex and hippocampus regions, as the causes of SSD symptoms. NMDAR hypofunction results in decreased activity of cortical GABAergic interneurons, leading to decrease in inhibitory control of pyramidal neurons, increase in DA release, and accumulation of glutamate in the synaptic space. Ca^{2+} influx increases, triggering an excitotoxic cascade that causes mitochondrial calcium overload, mitochondrial dysfunction, and the activation of death signals like apoptosis. In neurons, excess glutamate induces the production of nitric oxide (NO), which reacts with superoxide (O_2^-) to form peroxynitrite ($ONOO^-$), inhibiting mitochondrial complexes I, II, and V. Prolonged exposure to glutamate saturates the mitochondrial matrix with Ca^{2+} phosphate complexes, which results in inhibition of ATP synthesis and ultimately leads to neuron death or dysfunction [9, 36–39]. Dysregulation of glutamate neurotransmission in the CA1 region of the hippocampus elevates neuronal activity, eliciting attenuated psychotic symptoms. As the illness progresses, the dysfunction projects to fields internal and external to the hippocampus and frontal cortex and causes an atrophic process in which the neuropil of hippocampal cells is reduced and interneurons are lost, driving to syndromic psychosis [40]. Lieberman et al. [41] showed downregulation of glutamate dehydrogenase I as evidence that excess extracellular glutamate drives CA1 hypermetabolism. Over time, the surge in extracellular glutamate leads to a functional deficit and/or loss of viability of parvalbumin-expressing GABAergic interneurons, which in turn produce further toxicity to hippocampal cells, causing shrinkage, which may contribute to the reduction in hippocampal volume observed in SSD patients.

1.2.3.5 The Norepinephrine Hypothesis

Adrenergic receptors (ARs)—α_{1A}, α_{1B}, α_{1D}, α_{2A}, α_{2B}, α_{2C}; β_1, β_2, and β_3 subtypes—are ubiquitous in neural circuitries, suggesting norepinephrine (NE) plays a role in a range of body functions, e.g., stress response, immune defense, memory consolidation, and endocrine function. Thus, dysfunction in NE signaling could result in

1.2 The Pathophysiology of SSD

some of the SSD symptoms. A systematic review provided no evidence to support the role of NE α-ARs in the pathophysiology of SSD [42]. However, evidence came with clozapine, an atypical AP with proven antagonism on α2-ARs that relieves negative and cognitive symptoms [8], and with lumateperone, a "third-generation" antipsychotic exhibiting a unique mode of action with simultaneous activation of NMDA and AMPA neurotransmission, interaction and modulation of dopaminergic and serotonergic neurotransmission, in addition to having binding affinity for off-target serotonin receptor subtype 2C (5-HT2C), α1-AR, and histaminergic receptor subtype 1 (H1) receptors [43]. Findings about the role of the locus coeruleus-norepinephrine (LC-NE) system in brain function and cognition explain cognitive and motivational deficits observed in SSD [44].

1.2.3.6 The Cannabinoid Hypothesis

The involvement of the endocannabinoid system (eCB) in the pathophysiology of SSD arose from the finding that prolonged use of *Cannabis* increases the incidence of psychotic disorders such as SSD and that its main active, tetrahydrocannabinol (THC), acts as a partial agonist on CB1 receptor of eCB system [15, 45]. Two ubiquitous cannabinoid receptors (CB1R and CB2R) and two main endogenous ligands—anandamide (N-arachidonylethanolamide) and 2-arachidonoyl glycerol (2-AG), derivatives of polyunsaturated fatty acids (PUFAs)—compose the eCB system, which mediates the crosstalk with different neurotransmitter and neurohormonal systems. CB1R attenuates the effects of chronic stress by inhibiting HPA axis hyperactivation within the amygdala when it signals the release of glutamate from synapses. CB2R participates in the regulation of the immune system. 2-AG partakes in the CB1R dependent retrograde signaling and is an intermediate metabolite for lipid synthesis, providing arachidonic acid for prostaglandin synthesis and the inflammatory cascade. CB1R activation induces psychotic symptoms, while its blockade has antipsychotic effects. The opposite is true for CB2R, whose pharmacological activation or overexpression induces improvement while blocking it causes worsening [46–49]. Increased anandamide levels in blood and cerebrospinal fluid (CSF) were correlated with psychotic symptoms in unmedicated patients with acute SSD and normalized with clinical remission [50].

1.2.4 The "Dysconnectivity" Hypothesis

It postulates that the underlying cause of the motor-sensory, thought, or cognitive "dysconnection" seen in SSD arises due to the co-existence of aberrant wiring of connections at the level of cellular processes, and impaired synaptic plasticity, regardless of whether the two causes share a common (genetic) origin or whether one caused the other. Thus, an abnormal regulation of NMDAR-dependent synaptic plasticity by the modulatory action of neurotransmitters, such as DA, acetylcholine, or serotonin, could change the strength of glutamatergic synapses by altering the functional state and/or number of α-amino-3-hydroxy-5-methyl-4-isoxazole-propionate (AMPA) receptors [51]. SSD patients showed altered effective connectivity, which was correlated with SSD symptoms and illness duration [52]. Yue et al. [53] highlighted three studies that indicate "dysconnection" as an SSD diagnostic and risk biomarker.

1.2.5 The Autoimmune Hypothesis

In 1930, Dameshek reported SSD patients with increased levels of lymphocytes and eosinophils. However, it wasn't until a decade later that the hypothesis of a humoral response targeting the body itself emerged, linking autoimmunity to the pathogenesis of SSD symptoms [54]. Currently, it is known that autoantibodies (Ab) acting on neuronal receptors and ion channels in Central Nervous System (CNS), e.g., NMDAR-Ab and NCAM1-Ab, can alter synaptic plasticity and transmission, contributing to memory impairment, behavioral abnormalities, and psychotic disorders. More than 25 new types of autoantibodies have been recently described in patients with neurological diseases, and many of them are proving to be directly pathogenic [55–57].

1.2.6 The Role of the Inflammatory Response

At first, it was suggested that the entire inflammatory arsenal would be the mediator of the prodromal, acute, and chronic phases of SSD [58]. But later, low-grade inflammation was seen in a mere 40% of SSD patients [59]. Short-term antipsychotic treatment resolves early-onset inflammation but is ineffective in chronic phase, which is marked by a pronounced metabolic-inflammatory imbalance [60]. Thus, it remains unclear whether the altered levels of cytokines seen in SSD patients are the cause or outcome of the disorder or simply immunopathogenic mechanisms underlying SSD.

1.2.7 The Role of Environment Stress

Stressful situations induce the release of corticotropin-releasing hormone (CRH), which is co-secreted with arginine vasopressin (AVP) from the hypothalamus. This chemical signaling is transmitted to the anterior lobe of the pituitary gland, where the secretion of adrenocorticotropic hormone (ACTH) takes place, which in turn leads to the release of cortisol from the adrenal cortex into the bloodstream as a defense response of the body. The activation of inflammation induces the production of C-reactive protein (CRP), the pro-inflammatory transcription factor NF-κB, and IL-6, which in turn activates the entire hypothalamic pituitary adrenal (HPA) axis. More cortisol is released to reduce the inflammatory response, stimulate gluconeogenesis, and protect the body from an excessive immune response. Chronic stress and inflammation can lead to dysregulation of the HPA axis, which secretes excess cortisol and triggers dopaminergic neurotransmission, exacerbating or causing some symptoms such as delusions and impaired response to psychosocial stressors [46]. A systematic review corroborated HPA axis hyperactivity at the onset of psychosis, despite the cortisol levels being contradictory and dependent on confounding factors [61]. Later research showed the HPA axis-inflammatory process relationship [62]. HPA axis hyperactivity causes immune cells to resist the anti-inflammatory effects of cortisol, activating neuroendocrine hormones, which in turn can trigger dysbiosis in the gut microbiota and disrupt the intestinal epithelial barrier, allowing the entry of bacterial compounds, such as lipopolysaccharide (LPS), leading to increased HPA stress response and heightened inflammation. This could be a way of explaining the co-occurrence of gastrointestinal inflammatory comorbidities, such as celiac disease, colitis, irritable bowel syndrome, and metabolic dysfunction, with SSD that end up promoting additional changes in the gut microbiota and its respective metabolites. The latter can play an important role in modulating physiological processes in the host involved in signaling, amino acid transport and degradation, and regulation of the immune system [21, 39, 46, 47, 53, 63, 64]. Two systematic reviews have corroborated the occurrence of significant differences in the β-diversity of the gut microbiota in SSD patients, which showed an enrichment of lactic acid-producing bacteria, a reduction of anti-inflammatory butyrate-producing bacteria, and high production of bacteria associated with the metabolism of glutamate and gamma-amino-butyric acid (GABA) [65, 66]. An integrative analysis showed an aberrant gut metabolome-immune network influencing the immune response by regulating host metabolic processes [67]. Important evidence for the role of gut microbiota in mental disorders came from the ability of gut microbiota transplanted from sick to healthy mice to elicit the symptoms of depression by altering PUFA metabolism related to eCB system [68] and to cause cognitive deficits, perhaps by modulating tryptophan metabolism and GABA degradation [69]. Composition changes in gut microbiota were associated with SSD pathogenesis and antipsychotic-induced metabolic perturbations [70].

1.2.8 The Role of Mitochondria and Oxidative Stress

Genetic association and gene-expression studies indicated that SSD patients have an altered ability to assemble antioxidant mechanisms, as well as an altered expression of mtDNA-encoded genes, resulting in significant decrease in mitochondrial transcripts in brain tissue [39].

Changes in mitochondria function affect systemic metabolic regulation, cellular aging, social behavior and can trigger atypical cellular energy state that may lead to gradual disturbances of astrocyte function. Impaired mitochondria influence neurodevelopment, impacting neuronal connectivity, neurotransmission, and myelination. The sex disparities seen in SSD can be partly attributed to estrogens influence on mitochondrial morphology and functions, particularly the efficiency of the mitochondrial electron transport chain (ECT) and the antioxidant defense mechanisms to fight free radicals in the brain [71, 72]. In the caudate and putamen, mitochondrial number is linked to symptom severity [73]. Decrease in mitochondrial complex I activity induces changes in ETC activity and increase in oxidative stress markers, e.g., thiobarbituric acid reactive substances (TBARS). Generation of high levels of reactive oxygen species (ROS) causes impairment of mitochondrial oxidative phosphorylation through oxidation of mitochondrial lipids, sulfhydryl groups, and iron sulfur complexes of mitochondrial respiratory enzymes. Oxidative damage to membrane lipids, proteins, and DNA impairs cell viability and function, and activates immune-inflammatory pathways. Oxidative stress can induce depressive state via mitochondrial fatty acid metabolism, down-regulating acetyl-L-carnitine, which in turn regulates the activation of the metabotropic glutamate receptor mGlu2 by epigenetic acetylation [39, 71]. But whether mitochondria dysfunction is a primary cause of toxicity or secondary response to damage remains to be found.

Considering so many formulated hypotheses and so many new ones emerging, but without leaving the physiological axis, perhaps it is time to radically change the way of interpreting what is happening to the patient in terms of their thoughts and feelings.

References

1. Kontaxakis V, Konstantakopoulos G (2015) From DSM-I to DSM-5. Psychiatrike 26:13–16
2. Cuthbert BN (2022) Research Domain Criteria (RDoC): progress and potential. Curr Dir Psychol Sci 31:107–114. https://doi.org/10.1177/09637214211051363
3. CGJ G, Doorduin J, Sommer IE, de Vries EFJ (2021) The dual hit hypothesis of schizophrenia: evidence from animal models. Neurosci Biobehav Rev 131:1150–1168. https://doi.org/10.1016/j.neubiorev.2021.10.025
4. Trubetskoy V, Pardiñas AF, Qi T et al (2022) Mapping genomic loci implicates genes and synaptic biology in schizophrenia. Nature 604:502–508. https://doi.org/10.1038/s41586-022-04434-5
5. Jauhar S, Johnstone M, McKenna PJ (2022) Schizophrenia. Lancet 399:473–486. https://doi.org/10.1016/S0140-6736(21)01730-X
6. Owens DC, Johnstone EC (2018) The development of antipsychotic drugs. Brain Neurosci Adv 2:2398212818817498. https://doi.org/10.1177/2398212818817498
7. Carpenter WT Jr, Davis JM (2012) Another view of the history of antipsychotic drug discovery and development. Mol Psychiatry 17:1168–1173. https://doi.org/10.1038/mp.2012.121
8. Aringhieri S, Carli M, Kolachalam S, Verdesca V, Cini E, Rossi M, McCormick PJ, Corsini GU, Maggio R, Scarselli M (2018) Molecular targets of atypical antipsychotics: from mechanism of action to clinical differences. Pharmacol Ther 192:20–41. https://doi.org/10.1016/j.pharmthera.2018.06.012
9. Gomes FV, Grace AA (2021) Beyond dopamine receptor antagonism: new targets for schizophrenia treatment and prevention. Int J Mol Sci 22:4467. https://doi.org/10.3390/ijms22094467

References

10. Rognli EB, Bramness JG (2015) Understanding the relationship between amphetamines and psychosis. Curr Addict Rep 2:285–292. https://doi.org/10.1007/s40429-015-0077-4
11. Grayson DR (2010) Schizophrenia and the epigenetic hypothesis. Epigenomics 2:341–344. https://doi.org/10.2217/epi.10.22
12. De Gregorio D, Comai S, Posa L, Gobbi G (2016) D-lysergic acid diethylamide (LSD) as a model of psychosis: mechanism of action and pharmacology. Int J Mol Sci 17:1953. https://doi.org/10.3390/ijms17111953
13. Keshavan MS, Collin G, Guimond S, Kelly S, Prasad KM, Lizano P (2020) Neuroimaging in schizophrenia. Neuroimaging Clin N Am 30:73–83. https://doi.org/10.1016/j.nic.2019.09.007
14. Howes OD, Cummings C, Chapman GE, Shatalina E (2023) Neuroimaging in schizophrenia: an overview of findings and their implications for synaptic changes. Neuropsychopharmacology 48:151–167. https://doi.org/10.1038/s41386-022-01426-x
15. Fiorentini A, Cantù F, Crisanti C, Cereda G, Oldani L, Brambilla P (2021) Substance-induced psychoses: an updated literature review. Front Psychiatry 12:694863. https://doi.org/10.3389/fpsyt.2021.694863
16. Tansey EM (2006) Henry Dale and the discovery of acetylcholine. C R Biol 329:419–425. https://doi.org/10.1016/j.crvi.2006.03.012
17. Todman D (2008) John Eccles (1903-97) and the experiment that proved chemical synaptic transmission in the central nervous system. J Clin Neurosci 15:972–977. https://doi.org/10.1016/j.jocn.2008.01.001
18. Allen NC, Bagade S, McQueen MB, Ioannidis JPA, Kavvoura FK, Khoury MJ, Tanzi RE, Bertram L (2008) Systematic meta-analyses and field synopsis of genetic association studies in schizophrenia: the SzGene database. Nat Genet 40:827–834. https://doi.org/10.1038/ng.171
19. Nieratschker V, Nöthen MM, Rietschel M (2010) New genetic findings in schizophrenia: is there still room for the dopamine hypothesis of schizophrenia? Front Behav Neurosci 4:23. https://doi.org/10.3389/fnbeh.2010.00023
20. Brown AS (2011) The environment and susceptibility to schizophrenia. Prog Neurobiol 93:23–58. https://doi.org/10.1016/j.pneurobio.2010.09.003
21. Murray RM, Bhavsar V, Tripoli G, Howes O (2017) 30 years on: how the neurodevelopmental hypothesis of schizophrenia morphed into the developmental risk factor model of psychosis. Schizophr Bull 43:1190–1196. https://doi.org/10.1093/schbul/sbx121
22. Fišar Z (2023) Biological hypotheses, risk factors, and biomarkers of schizophrenia. Prog Neuro-Psychopharmacol Biol Psychiatry 120:110626. https://doi.org/10.1016/j.pnpbp.2022.110626
23. Miller CL (2022) The epigenetics of psychosis: a structured review with representative loci. Biomedicines 10:561. https://doi.org/10.3390/biomedicines10030561
24. Lee S-A, Huang K-C (2016) Epigenetic profiling of human brain differential DNA methylation networks in schizophrenia. BMC Med Genet 9:68. https://doi.org/10.1186/s12920-016-0229-y
25. Richetto J, Meyer U (2021) Epigenetic modifications in schizophrenia and related disorders: molecular scars of environmental exposures and source of phenotypic variability. Biol Psychiatry 89:215–226. https://doi.org/10.1016/j.biopsych.2020.03.008
26. Thome J, Foley P, Riederer P (1998) Neurotrophic factors and the maldevelopmental hypothesis of schizophrenic psychoses. J Neural Transm 105:85–100. https://doi.org/10.1007/s007020050040
27. Fernandes BS, Steiner J, Berk M, Molendijk ML, Gonzalez-Pinto A, Turck CW, Nardin P, Gonçalves C-A (2015) Peripheral brain-derived neurotrophic factor in schizophrenia and the role of antipsychotics: meta-analysis and implications. Mol Psychiatry 20:1108–1119. https://doi.org/10.1038/mp.2014.117
28. Rodrigues-Amorim D, Rivera-Baltanás T, Bessa J et al (2018) The neurobiological hypothesis of neurotrophins in the pathophysiology of schizophrenia: a meta-analysis. J Psychiatr Res 106:43–53. https://doi.org/10.1016/j.jpsychires.2018.09.007
29. Mehterov N, Minchev D, Gevezova M, Sarafian V, Maes M (2022) Interactions among brain-derived neurotrophic factor and neuroimmune pathways are key components of the major psychiatric disorders. Mol Neurobiol 59:4926–4952. https://doi.org/10.1007/s12035-022-02889-1

30. Seeman P (1987) Dopamine receptors and the dopamine hypothesis of schizophrenia. Synapse 1:133–152. https://doi.org/10.1002/syn.890010203
31. Brisch R, Saniotis A, Wolf R et al (2014) The role of dopamine in schizophrenia from a neurobiological and evolutionary perspective: old fashioned, but still in vogue. Front Psych 5:47. https://doi.org/10.3389/fpsyt.2014.00047
32. Millard SJ, Bearden CE, Karlsgodt KH, Sharpe MJ (2021) The prediction-error hypothesis of schizophrenia: new data point to circuit-specific changes in dopamine activity. Neuropsychopharmacology 47:628–640. https://doi.org/10.1038/s41386-021-01188-y
33. Perkovic MN, Erjavec GN, Strac DS, Uzun S, Kozumplik O, Pivac N (2017) Theranostic biomarkers for schizophrenia. Int J Mol Sci 18:733. https://doi.org/10.3390/ijms18040733
34. Stahl SM (2018) Beyond the dopamine hypothesis of schizophrenia to three neural networks of psychosis: dopamine, serotonin, and glutamate. CNS Spectr 23:187–191. https://doi.org/10.1017/S1092852918001013
35. Eggers AE (2013) A serotonin hypothesis of schizophrenia. Med Hypotheses 80:791–794. https://doi.org/10.1016/j.mehy.2013.03.013
36. Veerman SRT, Schulte PFJ, de Haan L (2014) The glutamate hypothesis: a pathogenic pathway from which pharmacological interventions have emerged. Pharmacopsychiatry 47:121–130. https://doi.org/10.1055/s-0034-1383657
37. Mei Y-Y, Wu DC, Zhou N (2018) Astrocytic regulation of glutamate transmission in schizophrenia. Front Psychiatry 9:544. https://doi.org/10.3389/fpsyt.2018.00544
38. Uno Y, Coyle JT (2019) Glutamate hypothesis in schizophrenia. Psychiatry Clin Neurosci 73:204–215. https://doi.org/10.1111/pcn.12823
39. Rajasekaran A, Venkatasubramanian G, Berk M, Debnath M (2015) Mitochondrial dysfunction in schizophrenia: pathways, mechanisms and implications. Neurosci Biobehav Rev 48:10–21. https://doi.org/10.1016/j.neubiorev.2014.11.005
40. Wegrzyn D, Juckel G, Faissner A (2022) Structural and functional deviations of the hippocampus in schizophrenia and schizophrenia animal models. Int J Mol Sci 23:5482. https://doi.org/10.3390/ijms23105482
41. Lieberman JA, Girgis RR, Brucato G et al (2018) Hippocampal dysfunction in the pathophysiology of schizophrenia: a selective review and hypothesis for early detection and intervention. Mol Psychiatry 23:1764–1772. https://doi.org/10.1038/mp.2017.249
42. Maletic V, Eramo A, Gwin K, Offord SJ, Duffy RA (2017) The role of norepinephrine and its α-adrenergic receptors in the pathophysiology and treatment of major depressive disorder and schizophrenia: a systematic review. Front Psychiatry 8:42. https://doi.org/10.3389/fpsyt.2017.00042
43. Snyder GL, Vanover KE, Davis RE, Li P, Fienberg A, Mates S (2021) A review of the pharmacology and clinical profile of lumateperone for the treatment of schizophrenia. Adv Pharmacol 90:253–276. https://doi.org/10.1016/bs.apha.2020.09.001
44. Mäki-Marttunen V, Andreassen OA, Espeseth T (2020) The role of norepinephrine in the pathophysiology of schizophrenia. Neurosci Biobehav Rev 118:298–314. https://doi.org/10.1016/j.neubiorev.2020.07.038
45. Müller-Vahl KR, Emrich HM (2008) Cannabis and schizophrenia: towards a cannabinoid hypothesis of schizophrenia. Expert Rev Neurother 8:1037–1048. https://doi.org/10.1586/14737175.8.7.1037
46. Mikulska J, Juszczyk G, Gawrońska-Grzywacz M, Herbet M (2021) HPA axis in the pathomechanism of depression and schizophrenia: new therapeutic strategies based on its participation. Brain Sci 11:1298. https://doi.org/10.3390/brainsci11101298
47. Generoso JS, Giridharan VV, Lee J, Macedo D, Barichello T (2021) The role of the microbiota-gut-brain axis in neuropsychiatric disorders. Braz J Psychiatry 43:293–305. https://doi.org/10.1590/1516-4446-2020-0987
48. Ruggiero RN, Rossignoli MT, De Ross JB, Hallak JEC, Leite JP, Bueno-Junior LS (2017) Cannabinoids and vanilloids in schizophrenia: neurophysiological evidence and directions for basic research. Front Pharmacol 8:399. https://doi.org/10.3389/fphar.2017.00399

References

49. Morris G, Sominsky L, Walder KR, Berk M, Marx W, Carvalho AF, Bortolasci CC, Maes M, Puri BK (2022) Inflammation and nitro-oxidative stress as drivers of endocannabinoid system aberrations in mood disorders and schizophrenia. Mol Neurobiol 59:3485–3503. https://doi.org/10.1007/s12035-022-02800-y
50. Navarrete F, García-Gutiérrez MS, Jurado-Barba R, Rubio G, Gasparyan A, Austrich-Olivares A, Manzanares J (2020) Endocannabinoid system components as potential biomarkers in psychiatry. Front Psychiatry 11:315. https://doi.org/10.3389/fpsyt.2020.00315
51. Stephan KE, Friston KJ, Frith CD (2009) Dysconnection in schizophrenia: from abnormal synaptic plasticity to failures of self-monitoring. Schizophr Bull 35:509–527. https://doi.org/10.1093/schbul/sbn176
52. Rolls ET, Cheng W, Gilson M et al (2020) Beyond the disconnectivity hypothesis of schizophrenia. Cereb Cortex 30:1213–1233. https://doi.org/10.1093/cercor/bhz161
53. Yue W, Huang H, Duan J (2022) Potential diagnostic biomarkers for schizophrenia. Medical Review 2:385–416. https://doi.org/10.1515/mr-2022-0009
54. Ganguli R, Brar JS, Rabin BS (1994) Immune abnormalities in schizophrenia: evidence for the autoimmune hypothesis. Harv Rev Psychiatry 2:70–83. https://doi.org/10.3109/10673229409017120
55. Jézéquel J, Johansson EM, Dupuis JP et al (2017) Dynamic disorganization of synaptic NMDA receptors triggered by autoantibodies from psychotic patients. Nat Commun 8:1791. https://doi.org/10.1038/s41467-017-01700-3
56. Prüss H (2021) Autoantibodies in neurological disease. Nat Rev Immunol 21:798–813. https://doi.org/10.1038/s41577-021-00543-w
57. Shiwaku H, Katayama S, Kondo K et al (2022) Autoantibodies against NCAM1 from patients with schizophrenia cause schizophrenia-related behavior and changes in synapses in mice. Cell Rep Med 3:100597. https://doi.org/10.1016/j.xcrm.2022.100597
58. Smith RS, Maes M (1995) The macrophage-T-lymphocyte theory of schizophrenia: additional evidence. Med Hypotheses 45:135–141. https://doi.org/10.1016/0306-9877(95)90062-4
59. Kroken RA, Sommer IE, Steen VM, Dieset I, Johnsen E (2018) Constructing the immune signature of schizophrenia for clinical use and research; an integrative review translating descriptives into diagnostics. Front Psychiatry 9:753. https://doi.org/10.3389/fpsyt.2018.00753
60. Parksepp M, Haring L, Kilk K, Taalberg E, Kangro R, Zilmer M, Vasar E (2022) A marked low-grade inflammation and a significant deterioration in metabolic status in first-episode schizophrenia: a five-year follow-up study. Metabolites 12:983. https://doi.org/10.3390/metabo12100983
61. Borges S, Gayer-Anderson C, Mondelli V (2013) A systematic review of the activity of the hypothalamic-pituitary-adrenal axis in first episode psychosis. Psychoneuroendocrinology 38:603–611. https://doi.org/10.1016/j.psyneuen.2012.12.025
62. Karanikas E, Ntouros E, Oikonomou D, Floros G, Griveas I, Garyfallos G (2017) Evidence for hypothalamus-pituitary-adrenal axis and immune alterations at prodrome of psychosis in males. Psychiatry Investig 14:703–707. https://doi.org/10.4306/pi.2017.14.5.703
63. Munawar N, Ahsan K, Muhammad K, Ahmad A, Anwar MA, Shah I, Al Ameri AK, Al Mughairbi F (2021) Hidden role of gut microbiome dysbiosis in schizophrenia: antipsychotics or psychobiotics as therapeutics? Int J Mol Sci 22:7671. https://doi.org/10.3390/ijms22147671
64. Tsamakis K, Galinaki S, Alevyzakis E, Hortis I, Tsiptsios D, Kollintza E, Kympouropoulos S, Triantafyllou K, Smyrnis N, Rizos E (2022) Gut microbiome: a brief review on its role in schizophrenia and first episode of psychosis. Microorganisms 10:1121. https://doi.org/10.3390/microorganisms10061121
65. McGuinness AJ, Davis JA, Dawson SL et al (2022) A systematic review of gut microbiota composition in observational studies of major depressive disorder, bipolar disorder and schizophrenia. Mol Psychiatry 27:1920–1935. https://doi.org/10.1038/s41380-022-01456-3
66. Nikolova VL, Smith MRB, Hall LJ, Cleare AJ, Stone JM, Young AH (2021) Perturbations in gut microbiota composition in psychiatric disorders: a review and meta-analysis. JAMA Psychiatry 78:1343–1354. https://doi.org/10.1001/jamapsychiatry.2021.2573

67. Fan Y, Gao Y, Ma Q et al (2022) Multi-omics analysis reveals aberrant gut-metabolome-immune network in schizophrenia. Front Immunol 13:812293. https://doi.org/10.3389/fimmu.2022.812293/
68. Chevalier G, Siopi E, Guenin-Macé L et al (2020) Effect of gut microbiota on depressive-like behaviors in mice is mediated by the endocannabinoid system. Nat Commun 11:6363. https://doi.org/10.1038/s41467-020-19931-2
69. Zhu F, Ju Y, Wang W et al (2020) Metagenome-wide association of gut microbiome features for schizophrenia. Nat Commun 11:1612. https://doi.org/10.1038/s41467-020-15457-9
70. Singh R, Stogios N, Smith E et al (2022) Gut microbiome in schizophrenia and antipsychotic-induced metabolic alterations: a scoping review. Ther Adv Psychopharmacol 12:20451253221096525. https://doi.org/10.1177/20451253221096525
71. Picard M, McEwen BS, Epel ES, Sandi C (2018) An energetic view of stress: focus on mitochondria. Front Neuroendocrinol 49:72–85. https://doi.org/10.1016/j.yfrne.2018.01.001
72. Roberts RC (2021) Mitochondrial dysfunction in schizophrenia: with a focus on postmortem studies. Mitochondrion 56:91–101. https://doi.org/10.1016/j.mito.2020.11.009
73. Brand BA, de Boer JN, Sommer IEC (2021) Estrogens in schizophrenia: progress, current challenges and opportunities. Curr Opin Psychiatry 34:228–237. https://doi.org/10.1097/YCO.0000000000000699

Chapter 2
Biomarkers

2.1 Introduction

A major challenge in psychiatry lies in distinguishing between certain mental disorders, as many of them exhibit overlapping characteristics. Discovering a biomarker that serves this purpose would represent a clinical revolution. The quest for molecular biomarkers for schizophrenia spectrum disorders (SSD) that could predict, diagnose, or provide insights into treatment responses has naturally progressed alongside the development of pathophysiological hypotheses, even contributing to their validation [1–6]. In the early 1940s, Hemphill and Reiss were cognizant of the tremendous heterogeneity of SSD. They were also disheartened by the many unproductive studies that had been carried out thus far, focusing solely on physiological aspects in isolation [7]. They aspired to contribute something more substantial and thus decided "to mirror the physiological conditions of patients" by conducting a battery of biochemical tests on urine and blood. This included measurements of 17-ketosteroids and cortin-like compounds, in an attempt to indirectly assess the abnormal anterior pituitary activity observed in previous biopsies. While their study did not yield conclusive results regarding pituitary function, it did uncover a correlation between steroid hormone metabolism and certain stages of SSD. Similarly, Koch et al. sought to establish a biochemical distinction between SSD patients and healthy individuals by measuring phosphorus metabolites in serum to predict treatment outcomes with corticosteroids, but their attempts were ultimately unsuccessful [8]. Dissatisfied with the traditional diagnostic system for mental illness, which primarily relies on artificial divisions of the patient's emotional and behavioral disturbances, McGorry and colleagues initially proposed a heuristic strategy. This approach involved the utilization of molecular markers from oxidative stress and HPA axis dysregulation, as well as considerations of folate status and biological changes like cognitive deficits, to assist in staging psychotic disorders [9]. The most challenging aspect has been the selection of biomarkers during the mapping of clinical stages. This sense of powerlessness and skepticism toward SSD biomarkers persists among many dedicated researchers, despite occasional moments of excitement or passing scientific trends. Nearly a century after extensive research, Martinez-Cao et al. underscored the paramount importance of establishing a robust protocol with validated biomarkers for staging SSD [10]. After conducting a comprehensive and detailed review on the subject, Yue and colleagues propose the integration of multiomics and multidimensional data to overcome the challenges of dealing with a complex disease. This approach allows for a more comprehensive exploration of the pathogenicity of SSD, enabling the development of more sophisticated research strategies tailored for both diagnosis and treatment [11].

2.2 Biomarker Classification

As science has progressed, the term "biomarker" has also evolved in its meaning, culminating in its most current definition provided by the FDA-NIH Biomarker Working Group. This definition encompasses its primary functions: serving as an indicator of normal or pathological biological processes, or as a marker of response to a particular exposure or intervention [3, 12]. Different approaches have resulted in a wide range of classifications, some of which come with lengthy and intricate definitions. As a result, they often feature more commonly used synonyms in the literature. For instance, the FDA-NIH's definition of a monitoring biomarker as "a repeated measurement to assess the status of a disease or medical condition or for evidence of exposure to (or effect of) a medical product or environmental agent" encompasses a broad concept [13]. Due to this, the term "theranostic," which refers to a biomarker used to monitor or predict treatment outcomes and/or adverse drug effects [1, 2, 5, 14–16], has been employed in place of "monitoring" to offer a more specific context and to distinguish it from the term "exposure monitoring," which is used for detecting exposure to and effects of chemicals in environmental monitoring programs [17–19]. The term "trait" is synonymous with a "prognostic" biomarker and reflects a predisposition to a disease. On the other hand, the term "state" refers to a biomarker capable of indicating changes during the treatment or course of a disease, fitting the definition of a "predictive" biomarker [20, 21]. The term "transdiagnostic," used for biomarkers capable of distinguishing a disease or condition from similar ones, such as SSD, bipolar disorder (BD), and major depressive disorder (MDD), represents the newest trend in the scientific community [22–25]. Depending on the sample source—whether biofluids or from the central nervous system (brain or cerebrospinal fluid)—biomarkers are categorized as either central or peripheral. Additionally, based on the analytical technology employed, they can be classified as "omic biomarkers," such as genomic, proteomic, metabolomic, epigenomic, and transcriptomic [3].

All these synonyms signify a continuous evolution in the definitions pertaining to various types of biomarkers, making it challenging to standardize the field. This can be confusing for those who are new to the subject.

In this book, our emphasis will center on molecular biomarkers associated with the pathophysiological hypotheses outlined in Chap. 1. These biomarkers are categorized into distinct groups based on these hypotheses: neuroendocrine, neurotransmitter, neurotrophic, inflammatory, metabolic (related to human metabolism), oral/gut microbiota-related (pertaining to microorganisms and their metabolites), stress-related, and autoimmune biomarkers. Furthermore, Table 2.1 includes illustrative examples.

2.3 Development and Validation of Machine Learning-Based Diagnostic Models

Exploratory analysis of biological systems often yields a wealth of data, capturing intricate interactions among numerous variables. This abundance of information serves as an ideal starting point for uncovering biomarkers, particularly for constructing diagnostic models. In this context, machine learning (ML) emerges as a versatile and powerful tool, adept at identifying objective biomarkers associated with diseases that might elude traditional diagnostic methods.

At its core, ML operates as a system that achieves artificial intelligence by leveraging algorithms to discern intricate patterns and relationships within extensive datasets. Through a process of iterative learning, ML models extract valuable insights from data, subsequently applying this acquired knowledge to make accurate predictions or informed decisions when presented with new information [67, 68]. This ability to discern patterns within complex datasets positions ML as an asset in the realm of biomarker discovery and diagnostic model construction. The development of an ML-based model comprises steps related to (i) data collection and preparation, (ii) model construction, and (iii) model testing (Fig. 2.1) [69, 70].

2.3.1 Data Collection and Preparation

Data collection involves, first and foremost, a well-planned experimental design (considering sample size, collection methods, analytical platform, etc.) to help acquire high-quality data, which will ultimately determine how intelligent the model can become. The data obtained is then cleaned, pre-processed, and formatted to be suitable for posterior analyses. This includes some tasks such as the removal of duplicates, alignment, deconvolution, transformation, normalization, and scaling of features [71, 72]. Subsequently, feature engineering provides the model with the most relevant and informative input data through creating new features, encoding of categorical variables, discretization of numeric variables, removal or censoring of outliers, imputation of missing values, extraction, and selection of features [73–77].

Table 2.1 Classification of biomarkers according to the pathophysiological hypotheses of schizophrenia spectrum disorders

Inflammatory [11, 14, 21, 26–31]	Peripheral markers (Biofluids and blood cells)	**Proinflammatory cytokines:** IL-17, IL-21, IL-22, IL-23, tumor necrosis factor (TNF-α), tumor necrosis factor receptor type I (TNF-RI), and type II (TNF-RII). **The IL-1 family with proinflammatory function:** IL-1α, IL-1β, IL-18, IL-33, IL-36a, IL-36b, IL-36 g. Three receptor antagonists: IL-1RA, IL-36RA, and IL-38. **Anti-inflammatory cytokine:** IL-37. **Cytokines with mixed effect:** IL-2 and soluble IL-2 receptor (sIL-2R), IL-4, IL-5, IL-6 (and IL-6R), IL-8 (CXCL8), IL-10, IL-12B, IL-13, IL-15, interferon gamma (INF-γ), apoptosis inhibitor (CD5L), transforming growth factor alpha and beta (TGF-α, TGF-β), cluster of differentiation 40 (CD40), macrophage-derived chemokine, and TNF receptor like 2 protein and matrix metalloproteinase 8 (MMP-8). **Acute phase proteins:** C-reactive protein (CRP), albumin, fibrinogen, serum amyloid A protein, haptoglobin, α-2 macroglobulin, and α-1 antitrypsin (A1T). **Blood cells (number & function):** T lymphocytes (CD3-positive), T helper cells (CD4-positive), T helper/T cytotoxic (CD4/CD8) ratios, natural killer cells (CD56-positive), and neutrophils (granulocytes). **Others:** docosahexaenoic acid (DHA), homocysteine, leptin, myo-inositol, choline, Intercellular Adhesion Molecule 1 (ICAM-1), C-C motif chemokine ligand 2 (CCL-2), CCL-11 (eotaxin-1), and CCL-3 (MIP-1α).
	CNS markers (Brain and cerebrospinal fluid)	**Microglia activation:** IL-1β, IL-12, 18 kDa translocator protein (TSPO), quinolinic acid, CD11B, and human leukocyte antigen-DR isotype (HLA-DR, microglia density marker). **Astrocytes activation:** S100B protein, IL-6, IL-10, TGF-β, and kynurenic acid (KYNA). **Others:** alpha-1-antichymotrypsin (SERPINA3), interferon-induced transmembrane proteins 1 and 2, cyclooxygenase-2 (COX-2), prostaglandin E2 (PGE2), and IL-8.
Blood brain barrier hyperpermeability [29, 32]	They are nonspecific for SSD and are linked to neuro inflammation	S100 calcium-binding protein beta (S100B), vascular endothelial growth factor (VEGF), integrin receptor, MMP-9, ubiquitin C-terminal hydrolase-L1 (UCH-L1), glial fibrillary acidic protein (GFAP), the CSF-to-blood albumin ratio, claudin-5, vascular cell adhesion molecule 1 (vCAM1), intercellular adhesion molecule 1 and (iCAM1).
Neurotrophic [11, 14, 33]		BDNF, nerve growth factor (NGF) and receptor (NGFR), neurotrophin-3 (NT-3), neurotrophin-4 (NT-4), neurotrophin-5 (NT-5).

2.3 Development and Validation of Machine Learning-Based Diagnostic Models

Neuroendocrine [11, 14, 34–36]	Neurosteroids and derivatives (synthesized in the brain)	Cortisol, pregnenolone, pregnenolone sulfate, progesterone, dehydroepiandrosterone (DHEA), dehydroepiandrosterone sulfate (DHEA-S), estradiol, testosterone, dihydrotestosterone (DHT), and allopregnanolone.
	Neuroactive steroids (synthesized by endocrine glands)	Cortisol, cortisol awakening response (CAR), estradiol, and testosterone.
	Hormones	Chromogranin A, adrenocorticotropic hormone (ACTH), insulin, leptin, prolactin, pro-opiomelanocortin, and growth hormone.
Stress-related [2, 14, 22, 30, 37–44]	Oxidative damage products	Malondialdehyde (MDA; byproduct of lipid peroxidation), 3-nitroTyr (marker of protein damage), lipid hydroperoxides (LOOH), 8-hydroxy-2′-deoxyguanosine (8-OHdG; marker of DNA oxidation), advanced glycation end products (AGEs), pentosidine (marker for AGEs), advanced oxidation protein products (AOPP), dityrosine (DITYR), N-formylkynurenine (NFK), KYN, total oxidant status (TOS), nitric oxide (NO), Trp, asymmetric dimethylarginine, hydroxyisocaproic acid, and 1-methylguanosine.
	Antioxidant defense	Catalase (CAT), superoxide dismutase (SOD), glutathione peroxidase (GPx), paraoxonase 1 (PON1), nuclear factor kappa B (NFκB), reduced and oxidized glutathione (GSH and GSSG, respectively) expressed by the total glutathione (GSH-t), glutathione reductase (GR), total antioxidant capacity (TAC), total antioxidant status (TAS), oxidative stress index (OSI), ferric reducing ability of plasma (FRAP), thiol-disulfide homeostasis (TDH), myeloperoxidase (MPO), taurine, thiobarbituric acid reactive substances (TBARS), thyroxine, triiodothyronine, thyroid-stimulating hormone (TSH), hydroxylamine, pyroglutamic acid, α-tocopherol (and its metabolite α-CEHC), γ-tocopherol, L-carnitine, Tyr, Trp, Leu, acetyl L-carnitine, guaiacol, indole acrylic acid, biliverdin, and phytanic acid.
	Mitochondrial dysfunction	Acylcarnitines, complexes I, III, and IV (found reduced in brain tissues but not specific for SSD), increased lactic acid and cellular acidosis (marker for mitochondrial disease), glutathione, phosphodiester, phosphocreatine, and decreased pyruvate dehydrogenase (PDH; a marker for increased anaerobic metabolism of glucose and mitochondrial dysfunction).

(continued)

Table 2.1 (continued)

Neurotransmitter [11, 14, 45–52]	Dopaminergic	Dopamine (DA), homovanillic acid (HVA; the main metabolite of DA), L-3,4-dihydroxyphenylalanine (L-DOPA; downstream metabolite of Tyr and precursor of DA), DA 3-O-sulfate, DA receptor type 2 (DR2), DA uptake by platelets, tyrosine hydroxylase (TH), and DA transporter (DAT).
	Serotonergic	Serotonin (5-HT), 5-hydroxy-indoleacetic acid (5-HIAA; the main metabolite of 5-HT), Trp, 2-aminophenol, 5-hydroxy indole acetaldehyde, serotonin receptor type 2 (5-HTR2), 5-HTR7, serotonin receptor type 1A (5-HTR1A), 5-HT transporter (5-HTT).
	Adrenergic	Noradrenaline (NE), 3-methoxy-4-hydroxyphenylglycol (MHPG; the major brain metabolite of NE), metanephrine and normetanephrine (O-methylated metabolites of NE and adrenaline), NE sulfate, vanillylmandelic acid (end product), Tyr, homogentisic acid (Tyr intermediate), acetoacetic acid and hydroquinone (degradation products of Tyr), Phe, α- and β-adrenergic receptors (ARs), 3,4-dihydroxyphenylalanine (DOPA), DA β-hydroxylase, Tyr hydroxylase, L-aromatic amino acid decarboxylase, catechol O-methyltransferase, monoamine oxidase (MAO), and cyclic adenosine monophosphate.
	Cholinergic	α-7 acetylcholine receptor, α-7 nicotinic receptor, and M1 muscarinic acetylcholine receptor.
	Cannabinoid	Cannabinoid receptor 1 and 2 (CB1R, CB2R), anandamide, 2-AG, FAAH, N-acyltransferase, diacylglycerol lipase (DAGL), and mono-acylglycerol lipase (MAGL).
	Glutamatergic	NMDAR, glutamate (Glu), glycine, D-serine, agmatine, glutamine (Gln), KYNA, kynurenine (KYN), vascular endothelial growth factor (VEGF), Glu/Gln ratio, glutamate dehydrogenase, glutaminase, glutamine synthetase, glutamic acid decarboxylase, GABA transaminase, aspartate, and alanine aminotransferases.
	GABAergic	γ-aminobutyric acid (GABA), Glu, glutamic acid decarboxylase, postsynaptic GABA$_A$ and GABA$_B$ receptors, GABA transporters (GAT-1, responsible for the synaptic reuptake of GABA), and parvalbumin.

2.3 Development and Validation of Machine Learning-Based Diagnostic Models

Gut microbiota-brain axis [53–59]	Beta-diversity	Between-sample diversity is estimated by statistical tests such as Bray-Curtis's dissimilarity and Binary Jaccard dissimilarity, which do not take phylogenetic relatedness into account, or similarity analysis by unweighted UniFrac distances, which does.
	Alpha-diversity	Within-sample diversity is estimated by Shannon/Simpson index, community richness indices (ACE, Chao1, and observed species (number of different OTUs [60])), Phylogenetic Diversity index, Microbial Dysbiosis index.
	Metabolites	Imidazole propionic acid, urocanic acid, D-alanine, KYNA, secondary (nonconjugated) bile acids, lipopolysaccharides (LPS), 5-hydroxytryptophan, levodopa, adrenaline and NE hydrochloric acid, histidine, short-chain fatty acids (SCFA) such as butyric, isovaleric, propanoic, and isohexanoic acids.
Autoimmune [27, 61–63]	Prognostic	Abs associated with a particular disorder trajectory: *Toxoplasma gondii*-Ab, HSV 1-Ab, CMV-Ab. Prognostic and theranostic: EBV-Ab, NCAM1-Ab.
	Diagnostic	Abs associated with the psychosis phase: NMDAR-Ab (present in 3–10% SSD patients, in FE and in some HC). Diagnostic and theranostic (response to immunotherapy/ adverse effects of antipsychotics): Antinuclear antibodies (ANA).
	Predictive or Theranostic	Abs associated with the response to treatment: TPO-Ab, TGB-Ab (they respond well to immunosuppression), anti-gliadin and anti-transglutaminase (the benefit from gluten-free diets), and folate receptor-Ab (improvement with folinic acid supplementation).
Metabolic [11, 14, 21, 27, 37, 42, 52, 64–66]		**Fatty acids:** Nervonic acid (MUFA), long-chain omega-3 and omega-6 (PUFA), resolvin D2, eicosadienoic acid, oleic acid, alpha-linolenic acid, arachidonic acid, eicosanoic acid, and eicosenoic acid. **Vitamins:** E, B6 (pyridoxal), D, and B2 (riboflavin). **Amino acids:** Phe, Ser, Glu, Trp, betaine, Thr, Asp, and carnitines. **Purine and pyrimidine:** Deoxyguanosine, 3-methylxanthine, orotidine. **Organic acids:** Pyrrolidone carboxylic acid (pyroglutamic acid), and perillic acid. **Phospholipids:** LPCs, sphingomyelins, PEs, PCs, and LPEs. **Energy metabolism (TCA cycle):** Octanoic acid, maltose, valine, inositol, sorbitol, creatinine, fumaric acid, L-carnitine, isocitrate, succinic acid, itaconic acid, *cis*-aconitic acid, and L-2-hydroxyglutarate. **Others:** 1-methylnicotinamide (MNA), allantoin (urea), and primary (conjugated) bile acids.

Ab antibody, *ACE* abundance-based coverage estimator index, *CMV* cytomegalovirus, *EBV* Epstein-Barr virus, *FAAH* fatty acid amide hydrolase, *FE* first-episode psychosis, *HC* healthy control, *HLA-DR* major histocompatibility complex (MHC) II cell surface receptor, *HVS1* herpes simplex, *MUFA* monounsaturated fatty acid, *NCAM1* neural cell adhesion molecule, *NMDAR* N-methyl-D-aspartate receptor, *OTU* operational taxonomic unit, *Phe* phenylalanine, *PUFA* polyunsaturated fatty acid, *Ser* serine, *TCA* tricarboxylic acid, *TGB* thyroglobulin, *TPO* thyroid peroxidase, *Thr* threonine, *Trp* tryptophan, and *Tyr* tyrosine

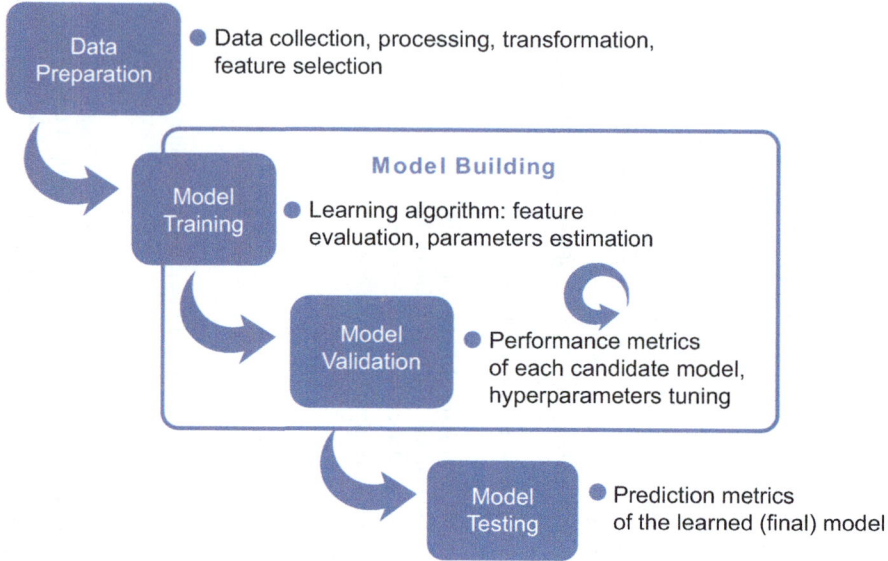

Fig. 2.1 Flowchart of development and validation of a machine learning-based diagnostic model

2.3.2 Model Construction

In the model-building step, the selected features (the input data) are divided into two subsets for training and validation: the proportion of each depends on the chosen learning algorithm, the number of samples, and the balance between bias and variance [78]. Most models can be used to predict quantitative response variables (e.g., tumor size), referred to as regression mode, or qualitative responses (e.g., health/disease), referred to as classification mode [79]. Some of the most commonly used examples are Linear Regression (LiR), Logistic Regression (LoR), Decision Tree (DT), Support Vector Machine (SVM), and k-Nearest Neighbors (kNN), which are very well described in [68, 74, 80].

The training set is tagged with the desired outcome (e.g., health/disease) and serves as a book where the algorithm learns how to identify labels by evaluating features (potential predictors) and estimating parameters (weights, coefficients, split points, leaf values, k values, etc.).

To identify the optimal parameters that will define the model with the most accurate predictions, the algorithm uses a measure of penalty for misclassifications—called loss function in some examples—that differs according to the algorithm used (see Table 2.2). An optimization logarithm, e.g., gradient descent, calculates the loss to adjust the parameters in the opposite direction of the gradient. This process is repeated several times until a convergence criterion is met. In each iteration (repetition), the algorithm updates its parameters to further minimize the loss function. The classification decision (stopping criterion) is obtained by defining, for example,

2.3 Development and Validation of Machine Learning-Based Diagnostic Models

Table 2.2 Building blocks of a learning algorithm

Learning algorithm	Loss function (LF)	Optimization logarithm	Goal
Linear regression	Mean squared error (MSE). It penalizes larger errors more heavily due to squaring operation	Gradient descent	To find the parameters (slope and intercept) that minimize the MSE
Logistic regression	Log loss (L), for binary classification. It penalizes models that predict with high confidence but are wrong	Gradient descent	To find the parameter values that maximize the likelihood of observing the given data
Hard-margin SVM (linearly separable case)	Hinge loss. It penalizes margin violations and points close to the decision boundary	Sequential minimal optimization (SMO), gradient descent	To find a hyperplane that perfectly separates the data
Soft-margin SVM (non-linearly separable case)	A combination of the hinge loss and a regularization term. It penalizes margin violations but tolerates some points on the wrong side of the decision boundary	Quadratic programming	To find a hyperplane that can separate the classes in a dataset with some margin of error, allowing for a certain number of classification errors
Decision tree	Gini impurity measures the probability of misclassifying a randomly chosen element if it was randomly labeled according to the class distribution in the node. Entropy measures the disorder or randomness in a set of data points	The feature space is partitioned based on impurity measurements without a global optimization process	To find the splits at each node that minimize the impurity measure, or equivalently, maximize the information gain (reduction in impurity) after the split
k-nearest neighbors	There is no LF involved. It stores the entire training dataset and makes predictions based on the "closeness" of data points	There is no optimization process	To make predictions for new data points by identifying the k closest data points from the training set based on a distance metric, e.g., Euclidean distance

a number of iterations, a threshold for the change in loss, or some other criterion indicating that further training is unnecessary. At this point, the model is constructed [68].

When the model predicts the labels of the training data well, it has a low bias. The model underfits when it presents high bias. In this case, the algorithm outputs erroneous predictions (overgeneralizes), indicating the model requires greater complexity, better parameter tuning, or larger training sets. The opposite scenario characterizes an overfitting model (high variance), which means that the model predicts the training data very well but makes inaccurate predictions (unable to generalize) on the hold-out sets—the validation or test sets—requiring regularization, which means building a less complex model [68]. Common types of regularization include L1 (LASSO) and L2 (Ridge) regularization, which will be discussed in detail in Chap. 10.

The validation set is used to address the above issues. In this step, the model architecture can be adjusted by manually tuning hyperparameters such as the number of features, number of iterations, proportion of splits, type of metric, and data regularization. These adjustments should be made incrementally, with intermediate checkpoints to track hyperparameter changes and associated performance metrics until reaching a stopping condition. This condition could be when the model's performance (e.g., in cross validation) ceases to improve or starts degrading, the loss reaches a certain threshold or stabilizes, or the desired level of accuracy is met [70].

Once validated, the final model is tested to predict classifications on a new set of data, called the test set, which is unknown and independent of the data used in the model-building stage. The resulting predictions are evaluated based on prior information (e.g., diagnosis based on clinical symptoms) and various performance metrics. The subsequent implementation of the model is a step that is not within our current focus [68].

2.4 Statistical Analysis of Biomarkers

A more detailed explanation of each statistical tool used in the process of building a diagnostic model will be provided in Chap. 10. However, some analyses merit prior discussion to facilitate the correct interpretation of the results presented in the upcoming chapters. These chapters will present the molecular biomarkers associated with the pathophysiological hypotheses of SSD that have been utilized in the past 12 years in constructing diagnostic models through Receiver Operating Characteristic (ROC) analysis.

Several univariate and multivariate statistical analyses are used during the development of an ML-based diagnostic model. In the process of selecting a feature, univariate analysis will look at one variable at a time—e.g., fold change is calculated to show the degree of variation in metabolite levels—to summarize the results through means and trends, scatter plots, frequency distribution tables and histograms [81, 82]. Multivariate analysis will verify the occurrence of patterns and correlations between several variables at the same time—e.g., the correlation within and between groups of the different analyzed phenotypes—to translate them into cause-and-effect relationships (Fig. 2.2) [80, 83, 84].

2.4.1 *Principal Component Analysis*

Principal component analysis (PCA) is an unsupervised learning algorithm, which means it analyzes large sets of data without prior knowledge of their labels. Its goal is to reduce the dimensionality of data in an interpretable way while preserving

2.4 Statistical Analysis of Biomarkers

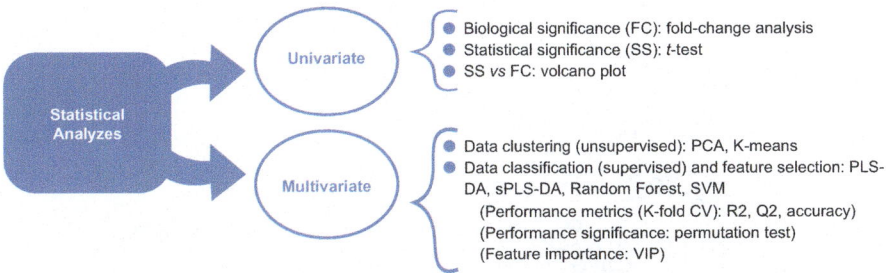

Fig. 2.2 Flowchart of statistical analysis used in the feature selection process for building a machine learning-based diagnostic model

most of the information [85]. It offers a general understanding of metabolite changes, the degree of variation between samples within a group, the presence of outliers, and the system suitability in relation to the quality control (QC) samples. Data from hydrogen nuclear magnetic resonance (1H-NMR), mass spectrometry (MS), or any other source of spectra are used as input in the data matrix obtained in the data processing step. The PCA algorithm performs a linear transformation of the data to reduce its dimensionality. The samples/features are reassembled into new combinations called principal components (PC), preserving the variance of the original data. The number of PCs formed is less than or at most equal to the number of original variables [86]. Each PC presents a percentage contribution to the grouping, which represents the ratio of the variance preserved during the transformation in relation to the original variance of the data. The goal is to obtain the smallest number of PCs (usually 2) that can explain the largest variance in X. The clustering performance is measured by the coefficient of determination (R2X), which represents the amount of variance preserved in X in each of the n PCs formed, or, in other words, the proportion of data (goodness-of-fit) that fits the linear transformation model. The closer the R2X value is to 1, the greater the percentage of data that fits the model, and the better the grouping [72, 87, 88].

In Fig. 2.3, the combined percentage contributions of PC1 and PC2 account for only 38.7% of the variance preserved in X. The 3D plot displays two groups, each containing representatives from three analyzed samples, indicating the model needs more informative features. Furthermore, the 2D plot shows heavily overlapping confidence interval ellipsoids. In a well-separable model, the data would cluster into non-overlapping ellipsoids. To enhance the model's performance, outliers —data points outside the ellipsoids—can be removed.

The effectiveness of PCA in clustering relies on the condition that the variation within groups is smaller than the variation between groups. However, it's important to acknowledge that other factors can lead to misclassifications in PCA. These may include sample preparation issues, experimental bias, or inadequate data preprocessing, or even that there is no difference between the groups analyzed [87]. As a linear model, PCA can fail to display informative patterns in the data if the variables of the system have non-linear relationships determining outcomes such as health/disease.

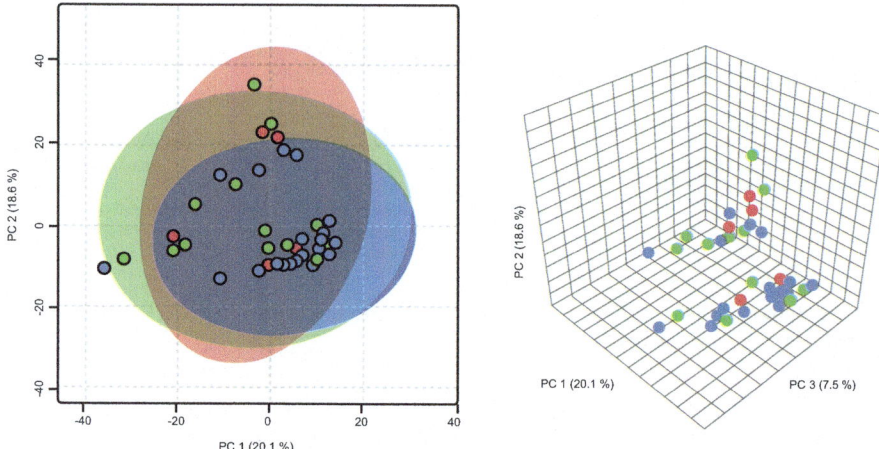

Fig. 2.3 PCA plots of three different samples: 2D graph containing ellipsoid confidence intervals (left) and 3D graph depicting the contribution percentage of each PC (right)

2.4.2 Discriminant Analysis

Another possibility for multivariate analysis is the use of classification algorithms such as partial least squares discriminant analysis (PLS-DA) or sparse PLS-DA (sPLS-DA). These techniques are considered supervised, which implies that the separation into classes is determined by maximizing the covariance between X (the data) and Y (the response) with the informed categories/labels of the samples, such as healthy control/disease [89]. PLS-DA reduces the dimensionality of the predictor variables by creating a set of new variables (latent variables or components) that are linear combinations of the original variables. This process can sometimes lead the model to overfit the data, which is a concern with PLS-DA. Similarly to PCA, where the first component explains the highest variance on X, the first component on PLS explains the most covariance between X and Y, the second component explains the second most, and so on. The number of components used is determined based on cross-validation or other criteria. The newly linear-transformed PLS components contain variables selected based on their importance in the projection. Variable Importance in the Projection (VIP) values greater than 1 indicate influential variables on the class separation, while VIP values less than 0.5 indicate irrelevant variables [90]. The most promising biomarker candidates for a discriminatory model have high VIP values.

To prevent the model from making inaccurate predictions (overfitting) when confronted with new data, it is necessary to evaluate how well the model "teaches" the algorithm to separate or predict the classes [87]. One of the most used techniques for this purpose is K-fold cross validation (K-fold CV). This method involves repeatedly subdividing the data into sets, using a portion for training the algorithm to identify features and parameters required for class prediction, and another portion

2.4 Statistical Analysis of Biomarkers

for validating the algorithm's "learning." In each iteration of training and validation, a set composed of K-1 subsets (e.g., 75–25%) is used for training, and the remaining subset is used for validation. This iterative process ensures that each subset of data is utilized for validation at some point in the model evaluation [91]. The performance measure includes accuracy (Acc), predictability estimate (Q2Y) and goodness of fit (R2Y). Good discriminatory models will result in both R2Y and Q2Y values greater than 0.7. When the difference between R2 and Q2 is too large (R2 >> > Q2), it indicates potential overfitting in supervised analyses [87, 92]. If Q2Y is negative, the model is not predictive at all or is overfitted. However, high values of these diagnostic statistics can be achieved purely by chance, due to luck in randomly choosing the samples in the training and validation sets. A new test is then necessary to quantify the statistical significance of the diagnostic statistics. The permutation test assesses the probability of obtaining a specific Q2Y or higher by randomly changing the data labels and fitting the model multiple times. The test assumes that observations are interchangeable since there is no difference between the conditions (the null hypothesis), so labels do not matter. Several permutations of the data are performed to estimate the population distribution. From there, the test assigns a probability value (p-value) of obtaining the observed data, assuming that the null hypothesis is true [93, 94]. In Fig. 2.4, the discriminatory model showed excellent results according to the values of R2Y and Q2Y but failed to pass the permutation test. The histogram originating from the permutation test represent the distribution of the swapped data (blue bars) and the observed original data (highlighted with a red arrow). The test statistic "separation distance" compares the variability within (W) the group and between (B) groups. If the original group B/W ratio is part of the permuted group distribution, the difference between the two groups is not considered significant [95]. A p-value greater than the predefined significance level (e.g., 0.05) may indicate that the model has no predictive ability or that the data are incompatible with the model (overfitting), which was the case as the sample size was too small (n of control/disease = 6/6) and did not contain enough data to teach the algorithm how to accurately predict the classes.

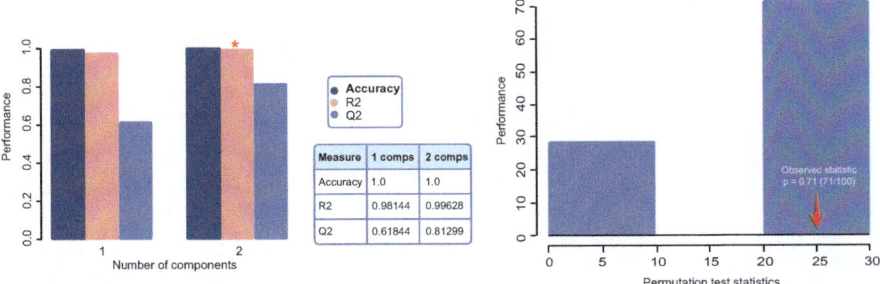

Fig. 2.4 PLS-DA performance metrics. Left: fivefold cross validation (the red asterisk denotes the highest Q2 value achieved with 2 principal components). Right: permutation test with 100 iterations (test statistic: separation distance)

2.4.3 ROC Curve Analysis

The Receiver Operating Characteristic (ROC) curve stands out as the "gold standard method" to describe and evaluate the overall performance of medical diagnostic models for binary outcomes (healthy people/SSD patients). The analysis follows the three main steps previously explained in Sect. 2.3, with the selection of biomarkers being the most critical step, since the features with maximum discriminatory power determine the success of a model. When the interest lies in using biomarkers separately (single biomarker diagnostic test), the biomarker selection will be carried out by univariate analysis (Fig. 2.5).

In the case of targeting two or more biomarkers (multi-biomarker diagnostic model), selection occurs through multivariate analysis (Fig. 2.6) [73].

During method testing (Fig. 2.7), if only the best scores for each feature (AUC, *t*-test, fold change, K-means (KM) clustering, least absolute shrinkage and selection operator (LASSO)) are considered for selecting biomarkers, there is a risk of overfitting. This is because biomarkers can be sample-specific and may not generalize well to unknown samples. This risk can be mitigated by considering prior knowledge, such as the relevance of the biomarker to a biological pathway or network, in conjunction with the overall biomarker score. In some cases, ratios between two

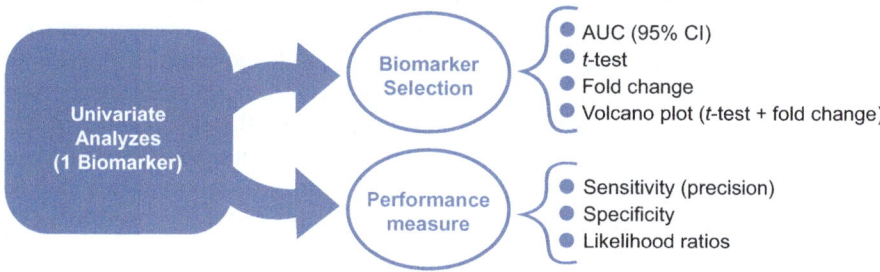

Fig. 2.5 Univariate approach to evaluate biomarkers during ROC curve analysis

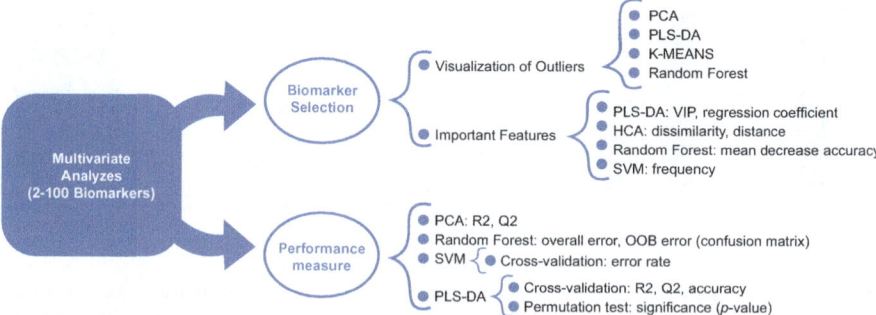

Fig. 2.6 Multivariate approach to evaluate biomarkers during ROC curve analysis

2.4 Statistical Analysis of Biomarkers

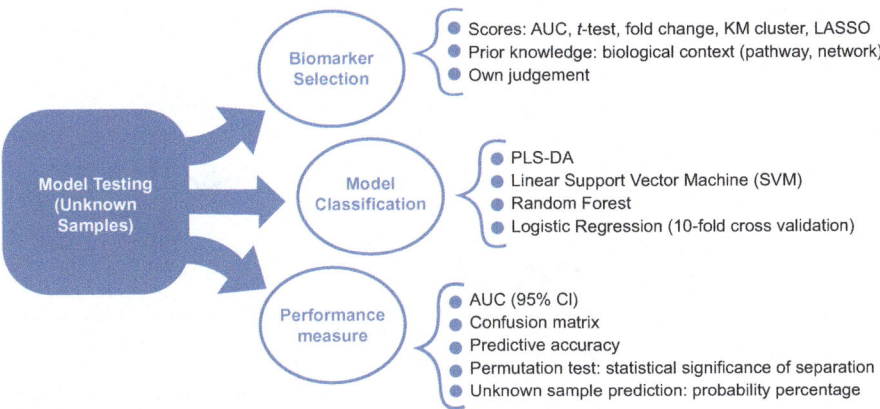

Fig. 2.7 ROC curve-based model testing

concentrations of metabolites can provide more informative results than using biomarker concentrations alone. Xia et al. recommend using a list of 1–10 biomarkers as predictors in a diagnostic model [73].

The performance of a single candidate biomarker (or a list of them, known as a biomarker signature) is determined by comparing the result predicted by the algorithm with the actual result established by the set of clinical signs and physiological measurements. The evaluation of this performance can be conducted in several ways. Visualization of the number of true and false classifications related to the original data classes, at a given threshold, is demonstrated in a confusion matrix. In this matrix, the original class labels of a binary model (e.g., disease-positive/control-negative) are represented by 0 and 1, respectively. In Fig. 2.8, the confusion matrix shows that of the 41 samples labeled disease, 30 were correctly classified as positive (outcome positive and test positive = true positive (TP)) and 11 incorrectly classified as negative (outcome positive and test negative = false negative (FN)). Of the 24 samples labeled control, 17 were correctly classified as negative (outcome negative, test negative = true negative (TN)) and 7 were incorrectly classified as positive (outcome negative and test positive = false positive (FP)). The predicted class probabilities plot shows that, at threshold 0.5 (50%), 7 healthy people who had a test result <0.5 were misclassified as sick (FP). Based on these results, one may decide to add more samples or features to help the learning algorithm better distinguish healthy people from sick people [68].

Additional inferences can be drawn from the confusion matrix. Accuracy represents the proportion of all correctly classified predictions (Acc = [(TP + TN)/(TP + TN + FN + FP)]*100) and implies that all misclassified outcomes are equally undesirable. The proportion of correct positive predictions to the overall number of positive predictions is expressed by predictive accuracy, also known as precision (P = [TP/(TP + FP)]*100). This estimate is particularly useful when the number of expected negatives is greater than the positives (e.g., a rare disease), since it does

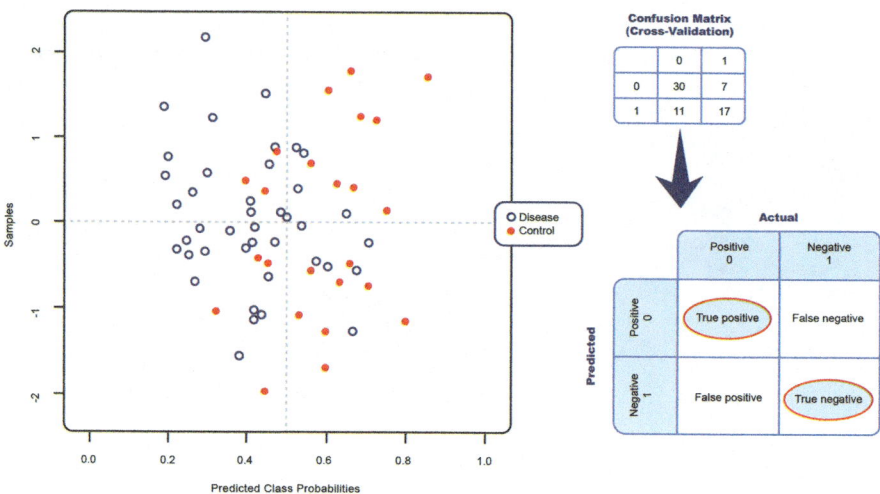

Fig. 2.8 Visualization of model performance. Left: predicted class probabilities plot (sort threshold at 0.5 on the X-axis, which means that all samples located to the left of the dashed vertical line on this cutoff are classified as positive, and to the right as negative). Right: confusion matrix

not consider the negative results. A precision score toward 1 means that the model did not miss any true positives, but it is not able to measure the FN. A low precision score (<0.5) indicates that the model has a high number of FPs, which could be due to imbalanced classes or untuned hyperparameters. In contrast, to account for FN, we measure recall (also called sensitivity), which is the proportion of true positive predictions to the overall number of actual positive instances in the dataset (Rec = [TP/(TP + FN)]*100) [68]. The positive likelihood ratio (LR(+) = TP/FP) corresponds to the probability that a person with a given disease will test positive, and the negative likelihood ratio (LR(−) = FN/TN) refers to the probability that a person with a given disease will test negative. In the example of Fig. 2.9, an LR(+) of 5.4 means that a person with a given disease is 5.4 times more likely to test positive than a person without the disease, while an LR(−) of 0.2 means that a person with a given disease is 0.2 times more likely to test negative than a person without the disease when using creatine as the predictor in the diagnostic model [96, 97].

The ROC curve (Fig. 2.9) summarizes all the confusion matrices produced by each tested threshold. The y-axis corresponds to the sensitivity of the model, expressed by the TP rate (TPR = TP/(TP + FN)), which indicates the proportion of subjects with an actual positive outcome (TP + FN) correctly assigned a positive outcome (TP only). The x-axis corresponds to 1-specificity, where specificity represents the proportion of actual negatives correctly classified as negative. This is expressed by the FP Rate (FPR = FP/(FP + TN)), informing about the proportion of misclassified samples. Any point on the diagonal line represents a threshold where the TPR equals the FPR, indicating the model's inability to discriminate between classes, predicting either a random or constant class in all cases. Any blue corner above the diagonal line represents the various thresholds tested to produce the ROC

2.4 Statistical Analysis of Biomarkers

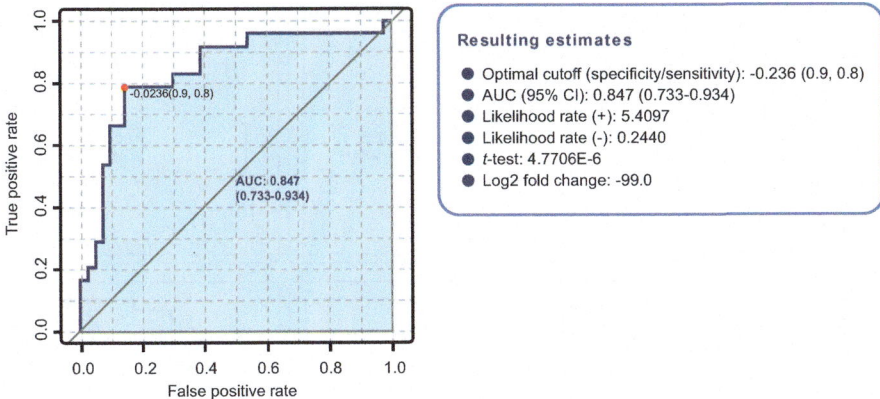

Fig. 2.9 ROC curve of a single candidate biomarker (creatine). The red point represents the optimal cutoff

curve. The red point corresponds to the optimal cutoff, i.e., the threshold (e.g., biomarker concentration) at which sensitivity and specificity consecutively have their highest values with the smallest difference between them. Thus, the closer the cutoff is to the upper left corner, the higher the AUC (blue shading on the ROC curve), indicating better predictor/classifier performance. One can assume that values above the optimal cutoff will be classified as positive (disease) and below the optimal cutoff will be classified as negative (health) [96].

The higher the AUC value (maximum = 1, corresponding to sensitivity and specificity = 1), the better the model is at predicting whether zero is truly zero (FPR) and 1 is truly 1 (TPR = sensitivity). This indicates a higher probability that the diagnostic method, with a given predictor, will classify a randomly chosen positive sample as positive than a randomly chosen negative sample [98, 99].

The AUC should be reported with its respective confidence interval (CI), which can be obtained by the bootstrapping technique, to inform the certainty range of the estimate. Unlike predictive accuracy, a ROC curve is not conditional on the prevalence of a given outcome. AUC values from different ROC curves can be directly compared to indicate the best model/predictor, following this order: 0.9–1.0 = excellent; 0.8–0.9 = good; 0.7–0.8 = fair; 0.6–0.7 = poor; 0.5–0.6 = fail [73]. Models exhibit similar AUC values, a partial AUC can be constructed at a specific threshold to differentiate the sensitivity or specificity of the models [96]. In Fig. 2.10, it is observed that the AUCs of creatine and glucose have very similar values. Constructing partial AUCs by constraining sensitivity to 0.8 (Y-axis: the minimum expected TPR), one of the points where the ROC curves intersect, reveals that the model created with creatine has a 10% higher correct negative rate than the model using glucose, as seen in the difference between the new values obtained for specificity (0.8 and 0.7, respectively). Conversely, by constraining specificity to 0.9 (X-axis: 1-specificity = 0.1, the maximal expected FPR), glucose generates a model that is 10% more sensitive (precise) than creatine. In short, the selection of a biomarker is linked to the level of sensitivity or specificity expected from the model

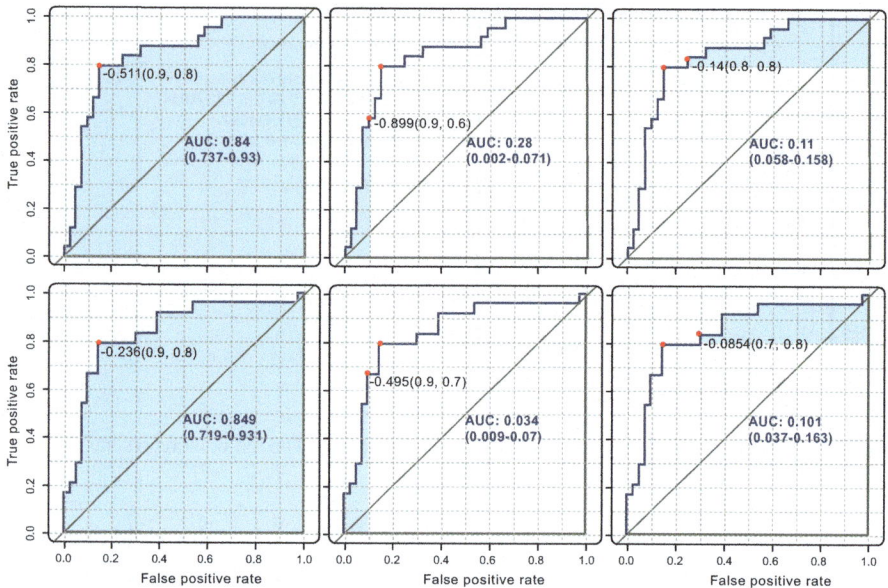

Fig. 2.10 Comparison of univariate AUCs and partial AUCs for two candidate biomarkers. Left: AUC (95% confidence interval); the red point indicates the optimal cutoff (specificity, sensitivity). Middle: partial AUCs with specificity constrained to 0.9. Right: partial AUCs with sensitivity constrained to 0.8

and the impact (cost) produced by misclassifications. Sometimes a deeper analysis is needed, depending on the case [73, 100, 101].

In multivariate analysis, the Monte Carlo cross-validation (MCCV) algorithm assesses feature importance using balanced subsampling. Sorting is done by predicting probability under the threshold 0.5, where the chance of being classified between the two labels is 50%, and consequent probability of the class is any number between 0 and 1. MCCV reserves 2/3 of the samples for model training, outputting 2–100 top features. Then the remaining 1/3 of the samples are set aside to validate the model with the selected feature signature and calculate predictive accuracy and confidence interval. This procedure is repeated multiple times (100, 500, or 1000 iterations) and results in a set of ROC curves created using different subsets of the selected biomarkers. Figure 2.11 exhibits on the left an ROC curve showing acceptable AUC values for a set of 3, 5, 10, 20, 41, or 82 tested sets of features. Observing the AUC values as more features are added to the model, the signature containing 5 features represents the best cost-benefit for model building. On the other hand, the ROC curve on the right shows that, whatever the signature used, the model will not be efficient enough to discriminate the analyzed phenotypes. The more metabolites added, the worse the model performance, which could be the result of insufficient data, useless variables, or even too noisy variables (e.g., outliers or mislabeled examples) [100, 101].

Now we are ready to go through the next chapters.

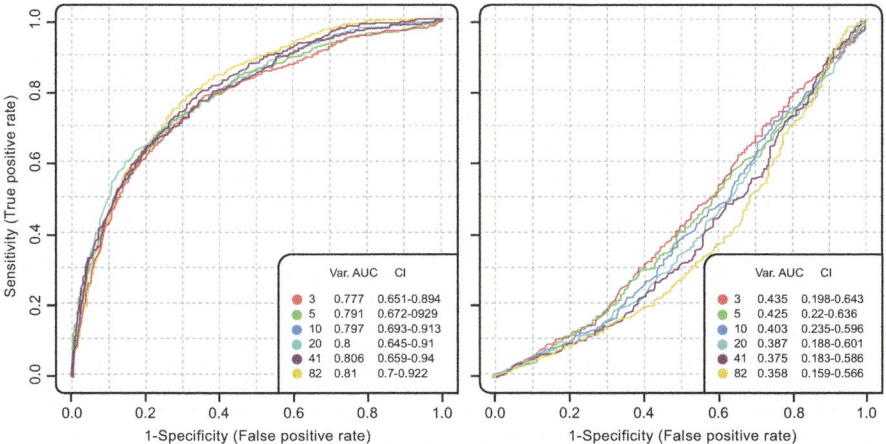

Fig. 2.11 ROC curves of multi-biomarkers. Left: fair results (AUC 0.7–0.8). Right: fail completely (AUC <0.5)

References

1. Weickert CS, Weickert TW, Pillai A, Buckley PF (2013) Biomarkers in schizophrenia: a brief conceptual consideration. Dis Markers 35:3–9. https://doi.org/10.1155/2013/510402
2. Lai C-Y, Scarr E, Udawela M, Everall I, Chen WJ, Dean B (2016) Biomarkers in schizophrenia: a focus on blood-based diagnostics and theranostics. World J Psychiatry 6:102–117. https://doi.org/10.5498/wjp.v6.i1.102
3. García-Gutiérrez MS, Navarrete F, Sala F, Gasparyan A, Austrich-Olivares A, Manzanares J (2020) Biomarkers in psychiatry: concept, definition, types and relevance to the clinical reality. Front Psych 11:432. https://doi.org/10.3389/fpsyt.2020.00432
4. Molina JD, Avila S, Rubio G, López-Muñoz F (2021) Metabolomic connections between schizophrenia, antipsychotic drugs and metabolic syndrome: a variety of players. Curr Pharm Des 27:4049–4061. https://doi.org/10.2174/1381612827666210804110139
5. Couttas TA, Jieu B, Rohleder C, Leweke FM (2022) Current state of fluid lipid biomarkers for personalized diagnostics and therapeutics in schizophrenia spectrum disorders and related psychoses: a narrative review. Front Psych 13:885904. https://doi.org/10.3389/fpsyt.2022.885904
6. Lin P, Sun J, Lou X et al (2022) Consensus on potential biomarkers developed for use in clinical tests for schizophrenia. Gen Psychiatry 35:e100685. https://doi.org/10.1136/gpsych-2021-100685
7. Hemphill RE, Reiss M (1948) Experimental investigations in the endocrinology of schizophrenia. Proc R Soc Med 41:533–540. https://www.ncbi.nlm.nih.gov/pubmed/18877130
8. Koch P, Laurin C, Lemieux R (1961) The influence of cortisone acetate on some serum phosphorus metabolites in young male schizophrenics. Am J Psychiatry 117:926–928. https://doi.org/10.1176/ajp.117.10.926
9. McGorry PD, Nelson B, Goldstone S, Yung AR (2010) Clinical staging: a heuristic and practical strategy for new research and better health and social outcomes for psychotic and related mood disorders. Can J Psychiatr 55:486–497. https://doi.org/10.1177/070674371005500803
10. Martínez-Cao C, de la Fuente-Tomás L, García-Fernández A, González-Blanco L, Sáiz PA, Garcia-Portilla MP, Bobes J (2022) Is it possible to stage schizophrenia? A systematic review. Transl Psychiatry 12:197. https://doi.org/10.1038/s41398-022-01889-y

11. Yue W, Huang H, Duan J (2022) Potential diagnostic biomarkers for schizophrenia. Med Rev 2:385–416. https://doi.org/10.1515/mr-2022-0009
12. FDA-NIH Biomarker Working Group. (2016). BEST (Biomarkers, EndpointS, and other Tools) Resource. 2016; https://www.ncbi.nlm.nih.gov/books/NBK326791/. Accessed 20 July 2023
13. FDA-NIH Biomarker Working Group. (2021). Monitoring Biomarker. BEST (Biomarkers, EndpointS, and other Tools) Resource. 2021; https://www.ncbi.nlm.nih.gov/books/NBK402282/. Accessed 20 July 2023
14. Perkovic MN, Erjavec GN, Strac DS, Uzun S, Kozumplik O, Pivac N (2017) Theranostic biomarkers for schizophrenia. Int J Mol Sci 18:733. https://doi.org/10.3390/ijms18040733
15. Yesilkaya UH, Gica S, Ilnem MC, Sen M, Ipekcioglu D (2021) Evaluation of IGF-1 as a novel theranostic biomarker for schizophrenia. J Psychiatr Res 140:172–179. https://doi.org/10.1016/j.jpsychires.2021.05.078
16. Dmitriy S, Regina N, Anastasia T, Evgeny E, Kristina S, Anna Z, Maria G, Maria P, Evgeny K (2017) Biomarkers in schizophrenia: a focus on blood based theranostics. Eur Neuropsychopharmacol 27:S398–S399. https://doi.org/10.1016/j.euroneuro.2016.09.437
17. Lam PKS, Gray JS (2003) The use of biomarkers in environmental monitoring programmes. Mar Pollut Bull 46:182–186. https://doi.org/10.1016/s0025-326x(02)00449-6
18. Silva MRF da, Souza KS, Assis CRD de, Sá RA de QC de, Santos MDV, Oliveira MBM de (2020) Biomarkers as a tool to monitor environmental impact on aquatic ecosystems. Braz J Dev 6:75702–75720. https://doi.org/10.34117/bjdv6n10-120
19. Lionetto MG, Caricato R, Giordano ME (2019) Pollution biomarkers in environmental and human biomonitoring. Open Biomark J 9:1–9. https://doi.org/10.2174/1875318301909010001
20. Lema YY, Gamo NJ, Yang K, Ishizuka K (2018) Trait and state biomarkers for psychiatric disorders: importance of infrastructure to bridge the gap between basic and clinical research and industry. Psychiatry Clin Neurosci 72:482–489. https://doi.org/10.1111/pcn.12669
21. Khoury R, Nasrallah HA (2018) Inflammatory biomarkers in individuals at clinical high risk for psychosis (CHR-P): state or trait? Schizophr Res 199:31–38. https://doi.org/10.1016/j.schres.2018.04.017
22. Pinto JV, Moulin TC, Amaral OB (2017) On the transdiagnostic nature of peripheral biomarkers in major psychiatric disorders: a systematic review. Neurosci Biobehav Rev 83:97–108. https://doi.org/10.1016/j.neubiorev.2017.10.001
23. Lalousis PA, Schmaal L, Wood SJ et al (2022) Neurobiologically based stratification of recent-onset depression and psychosis: identification of two distinct transdiagnostic phenotypes. Biol Psychiatry 92:552–562. https://doi.org/10.1016/j.biopsych.2022.03.021
24. Pelin H, Ising M, Stein F et al (2021) Identification of transdiagnostic psychiatric disorder subtypes using unsupervised learning. Neuropsychopharmacology 46:1895–1905. https://doi.org/10.1038/s41386-021-01051-0
25. Jauhar S, Nour MM, Veronese M, Rogdaki M, Bonoldi I, Azis M, Turkheimer F, McGuire P, Young AH, Howes OD (2017) A test of the transdiagnostic dopamine hypothesis of psychosis using positron emission tomographic imaging in bipolar affective disorder and schizophrenia. JAMA Psychiatry 74:1206–1213. https://doi.org/10.1001/jamapsychiatry.2017.2943
26. Tomasik J, Rahmoune H, Guest PC, Bahn S (2016) Neuroimmune biomarkers in schizophrenia. Schizophr Res 176:3–13. https://doi.org/10.1016/j.schres.2014.07.025
27. Carvalho AF, Solmi M, Sanches M et al (2020) Evidence-based umbrella review of 162 peripheral biomarkers for major mental disorders. Transl Psychiatry 10:152. https://doi.org/10.1038/s41398-020-0835-5
28. Fišar Z (2023) Biological hypotheses, risk factors, and biomarkers of schizophrenia. Prog Neuropsychopharmacol Biol Psychiatry 120:110626. https://doi.org/10.1016/j.pnpbp.2022.110626
29. Kroken RA, Sommer IE, Steen VM, Dieset I, Johnsen E (2018) Constructing the immune signature of schizophrenia for clinical use and research; an integrative review translating descriptives into diagnostics. Front Psych 9:753. https://doi.org/10.3389/fpsyt.2018.00753

References

30. Boll KM, Noto C, Bonifácio KL, Bortolasci CC, Gadelha A, Bressan RA, Barbosa DS, Maes M, Moreira EG (2017) Oxidative and nitrosative stress biomarkers in chronic schizophrenia. Psychiatry Res 253:43–48. https://doi.org/10.1016/j.psychres.2017.03.038
31. Khalfallah O, Barbosa S, Martinuzzi E, Davidovic L, Yolken R, Glaichenhaus N (2022) Monitoring inflammation in psychiatry: caveats and advice. Eur Neuropsychopharmacol 54:126–135. https://doi.org/10.1016/j.euroneuro.2021.09.003
32. Schrenk DA (2022) Faulty fences: blood-brain barrier dysfunction in schizophrenia. Curr Psychiatr Ther 21:28–32. https://doi.org/10.12788/cp.0278
33. Rodrigues-Amorim D, Rivera-Baltanás T, Bessa J et al (2018) The neurobiological hypothesis of neurotrophins in the pathophysiology of schizophrenia: a meta-analysis. J Psychiatr Res 106:43–53. https://doi.org/10.1016/j.jpsychires.2018.09.007
34. Misiak B, Frydecka D, Loska O, Moustafa AA, Samochowiec J, Kasznia J, Stańczykiewicz B (2018) Testosterone, DHEA and DHEA-S in patients with schizophrenia: a systematic review and meta-analysis. Psychoneuroendocrinology 89:92–102. https://doi.org/10.1016/j.psyneuen.2018.01.007
35. Zorn JV, Schür RR, Boks MP, Kahn RS, Joëls M, Vinkers CH (2017) Cortisol stress reactivity across psychiatric disorders: a systematic review and meta-analysis. Psychoneuroendocrinology 77:25–36. https://doi.org/10.1016/j.psyneuen.2016.11.036
36. Cai H, Cao T, Zhou X, Yao JK (2018) Neurosteroids in schizophrenia: pathogenic and therapeutic implications. Front Psych 9:73. https://doi.org/10.3389/fpsyt.2018.00073
37. Fan Y, Gao Y, Ma Q et al (2022) Multi-omics analysis reveals aberrant gut-metabolome-immune network in schizophrenia. Front Immunol 13:812293. https://doi.org/10.3389/fimmu.2022.812293
38. Juchnowicz D, Dzikowski M, Rog J, Waszkiewicz N, Zalewska A, Maciejczyk M, Karakuła-Juchnowicz H (2021) Oxidative stress biomarkers as a predictor of stage illness and clinical course of schizophrenia. Front Psych 12:728986. https://doi.org/10.3389/fpsyt.2021.728986
39. Samuelsson M, Skogh E, Lundberg K, Vrethem M, Öllinger K (2013) Taurine and glutathione in plasma and cerebrospinal fluid in olanzapine treated patients with schizophrenia. Psychiatry Res 210:819–824. https://doi.org/10.1016/j.psychres.2013.09.014
40. Guler EM, Kurtulmus A, Gul AZ, Kocyigit A, Kirpinar I (2021) Oxidative stress and schizophrenia: a comparative cross-sectional study of multiple oxidative markers in patients and their first-degree relatives. Int J Clin Pract 75:e14711. https://doi.org/10.1111/ijcp.14711
41. Buosi P, Borghi FA, Lopes AM et al (2021) Oxidative stress biomarkers in treatment-responsive and treatment-resistant schizophrenia patients. Trends Psychiatry Psychother 43:278–285. https://doi.org/10.47626/2237-6089-2020-0078
42. Liu M-L, Zheng P, Liu Z, Xu Y, Mu J, Guo J, Huang T, Meng H-Q, Xie P (2014) GC-MS based metabolomics identification of possible novel biomarkers for schizophrenia in peripheral blood mononuclear cells. Mol BioSyst 10:2398–2406. https://doi.org/10.1039/c4mb00157e
43. Rajasekaran A, Venkatasubramanian G, Berk M, Debnath M (2015) Mitochondrial dysfunction in schizophrenia: pathways, mechanisms and implications. Neurosci Biobehav Rev 48:10–21. https://doi.org/10.1016/j.neubiorev.2014.11.005
44. Mednova IA, Chernonosov AA, Kornetova EG, Semke AV, Bokhan NA, Koval VV, Ivanova SA (2022) Levels of acylcarnitines and branched-chain amino acids in antipsychotic-treated patients with paranoid schizophrenia with metabolic syndrome. Metabolites 12. https://doi.org/10.3390/metabo12090850
45. Maletic V, Eramo A, Gwin K, Offord SJ, Duffy RA (2017) The role of norepinephrine and its α-adrenergic receptors in the pathophysiology and treatment of major depressive disorder and schizophrenia: a systematic review. Front Psych 8:42. https://doi.org/10.3389/fpsyt.2017.00042
46. Caton M, Ochoa ELM, Barrantes FJ (2020) The role of nicotinic cholinergic neurotransmission in delusional thinking. NPJ Schizophr 6:16. https://doi.org/10.1038/s41537-020-0105-9
47. Navarrete F, García-Gutiérrez MS, Jurado-Barba R, Rubio G, Gasparyan A, Austrich-Olivares A, Manzanares J (2020) Endocannabinoid system components as potential biomarkers in psychiatry. Front Psych 11:315. https://doi.org/10.3389/fpsyt.2020.00315

48. Mei Y-Y, Wu DC, Zhou N (2018) Astrocytic regulation of glutamate transmission in schizophrenia. Front Psych 9:544. https://doi.org/10.3389/fpsyt.2018.00544
49. Lieberman JA, Girgis RR, Brucato G et al (2018) Hippocampal dysfunction in the pathophysiology of schizophrenia: a selective review and hypothesis for early detection and intervention. Mol Psychiatry 23:1764–1772. https://doi.org/10.1038/mp.2017.249
50. de Jonge JC, Vinkers CH, Hulshoff Pol HE, Marsman A (2017) GABAergic mechanisms in schizophrenia: linking postmortem and *in vivo* studies. Front Psych 8:118. https://doi.org/10.3389/fpsyt.2017.00118
51. Morris G, Sominsky L, Walder KR, Berk M, Marx W, Carvalho AF, Bortolasci CC, Maes M, Puri BK (2022) Inflammation and nitro-oxidative stress as drivers of endocannabinoid system aberrations in mood disorders and schizophrenia. Mol Neurobiol 59:3485–3503. https://doi.org/10.1007/s12035-022-02800-y
52. Cui G, Qing Y, Li M, Sun L, Zhang J, Feng L, Li J, Chen T, Wang J, Wan C (2021) Salivary metabolomics reveals that metabolic alterations precede the onset of schizophrenia. J Proteome Res 20:5010–5023. https://doi.org/10.1021/acs.jproteome.1c00504
53. Shen Y, Xu J, Li Z, Huang Y, Yuan Y, Wang J, Zhang M, Hu S, Liang Y (2018) Analysis of gut microbiota diversity and auxiliary diagnosis as a biomarker in patients with schizophrenia: a cross-sectional study. Schizophr Res 197:470–477. https://doi.org/10.1016/j.schres.2018.01.002
54. Li Z, Zhou J, Liang H et al (2022) Differences in alpha diversity of gut microbiota in neurological diseases. Front Neurosci 16:879318. https://doi.org/10.3389/fnins.2022.879318
55. Nguyen TT, Kosciolek T, Daly RE, Vázquez-Baeza Y, Swafford A, Knight R, Jeste DV (2021) Gut microbiome in schizophrenia: altered functional pathways related to immune modulation and atherosclerotic risk. Brain Behav Immun 91:245–256. https://doi.org/10.1016/j.bbi.2020.10.003
56. Jiang Y, Sun X, Hu M, Zhang L, Zhao N, Shen Y, Yu S, Huang J, Li H, Yu W (2022) Plasma metabolomics of schizophrenia with cognitive impairment: a pilot study. Front Psych 13:950602. https://doi.org/10.3389/fpsyt.2022.950602
57. Kim S, Okazaki S, Otsuka I et al (2022) Searching for biomarkers in schizophrenia and psychosis: case-control study using capillary electrophoresis and liquid chromatography time-of-flight mass spectrometry and systematic review for biofluid metabolites. Neuropsychopharmacol Rep 42:42–51. https://doi.org/10.1002/npr2.12223
58. Qing Y, Wang P, Cui G, Zhang J, Liang K, Xia Z, Wang P, He L, Jia W (2022) Targeted metabolomics reveals aberrant profiles of serum bile acids in patients with schizophrenia. Schizophrenia (Heidelb) 8:65. https://doi.org/10.1038/s41537-022-00273-5
59. Deng H, He L, Wang C, Zhang T, Guo H, Zhang H, Song Y, Chen B (2022) Altered gut microbiota and its metabolites correlate with plasma cytokines in schizophrenia inpatients with aggression. BMC Psychiatry 22:629. https://doi.org/10.1186/s12888-022-04255-w
60. Hughes JB, Hellmann JJ, Ricketts TH, Bohannan BJ (2001) Counting the uncountable: statistical approaches to estimating microbial diversity. Appl Environ Microbiol 67:4399–4406. https://doi.org/10.1128/AEM.67.10.4399-4406.2001
61. Prüss H (2021) Autoantibodies in neurological disease. Nature Rev Immunol 21:798–813. https://doi.org/10.1038/s41577-021-00543-w
62. Shiwaku H, Katayama S, Kondo K et al (2022) Autoantibodies against NCAM1 from patients with schizophrenia cause schizophrenia-related behavior and changes in synapses in mice. Cell Rep Med 3:100597. https://doi.org/10.1016/j.xcrm.2022.100597
63. Pollak TA, Rogers JP, Nagele RG, Peakman M, Stone JM, David AS, McGuire P (2019) Antibodies in the diagnosis, prognosis, and prediction of psychotic disorders. Schizophr Bull 45:233–246. https://doi.org/10.1093/schbul/sby021
64. Davison J, O'Gorman A, Brennan L, Cotter DR (2018) A systematic review of metabolite biomarkers of schizophrenia. Schizophr Res 195:32–50. https://doi.org/10.1016/j.schres.2017.09.021

65. De Luca V, Viggiano E, Messina G, Viggiano A, Borlido C, Viggiano A, Monda M (2008) Peripheral amino acid levels in schizophrenia and antipsychotic treatment. Psychiatry Investig 5:203–208. https://doi.org/10.4306/pi.2008.5.4.203
66. Zhu J-L, Luo W-W, Cheng X, Li Y, Zhang Q-Z, Peng W-X (2020) Vitamin D deficiency and schizophrenia in adults: a systematic review and meta-analysis of observational studies. Psychiatry Res 288:112959. https://doi.org/10.1016/j.psychres.2020.112959
67. Hulten G (2018) Building intelligent systems: a guide to machine learning engeneering. Apress, Washington. https://doi.org/10.1007/978-1-4842-3432-7
68. Burkov A (2019) The hundred-page machine learning book. 1999579518, 9781999579517 Published by Andriy Burkov
69. Jain AK, Duin PW, Mao J (2000) Statistical pattern recognition: a review. IEEE Trans Pattern Anal Mach Intell 22:4–37. https://doi.org/10.1109/34.824819
70. Harrington P (2012) Machine learning in action. Manning, Shelter Island
71. Walach J, Filzmoser P, Hron K (2018) Data normalization and scaling: consequences for the analysis in omics sciences. In: Jaumot J, Bedia C, Tauler R (eds) Comprehensive analytical chemistry, vol 82. Elsevier, pp 165–196
72. Zhou B, Xiao JF, Tuli L, Ressom HW (2012) LC-MS-based metabolomics. Mol BioSyst 8:470–481. https://doi.org/10.1039/c1mb05350g
73. Xia J, Broadhurst DI, Wilson M, Wishart DS (2013) Translational biomarker discovery in clinical metabolomics: an introductory tutorial. Metabolomics 9:280–299. https://doi.org/10.1007/s11306-012-0482-9
74. Webb AR, Copsey KD, Cawley G (2011) Statistical pattern recognition, 3rd edn. John Wiley & Sons, UK. https://doi.org/10.1002/9781119952954
75. Do KT, Wahl S, Raffler J et al (2018) Characterization of missing values in untargeted MS-based metabolomics data and evaluation of missing data handling strategies. Metabolomics 14:1–18. https://doi.org/10.1007/s11306-018-1420-2
76. Antonelli J, Claggett BL, Henglin M et al (2019) Statistical workflow for feature selection in human metabolomics data. Meta 9:143. https://doi.org/10.3390/metabo9070143
77. Sun J, Xia Y (2023) Pretreating and normalizing metabolomics data for statistical analysis. Genes & Diseases 11:100979. https://doi.org/10.1016/j.gendis.2023.04.018
78. Greener JG, Kandathil SM, Moffat L, Jones DT (2021) A guide to machine learning for biologists. Nat Rev Mol Cell Biol 23:40–55. https://doi.org/10.1038/s41580-021-00407-0
79. Sarker IH (2021) Machine learning: algorithms, real-world applications and research directions. SN Computer Science 2:1–21. https://doi.org/10.1007/s42979-021-00592-x
80. Varmuza K, Filzmoser P (2009) Introduction to multivariate statistical analysis in Chemometrics, 1st edn. CRC Press, London. https://doi.org/10.1201/9781420059496
81. Vinaixa M, Samino S, Saez I, Duran J, Guinovart JJ, Yanes O (2012) A guideline to univariate statistical analysis for LC/MS-based untargeted metabolomics-derived data. Meta 2:775–795. https://doi.org/10.3390/metabo2040775
82. Saccenti E, Hoefsloot HCJ, Smilde AK, Westerhuis JA, Hendriks MMWB (2014) Reflections on univariate and multivariate analysis of metabolomics data. Metabolomics 10:361–374. https://doi.org/10.1007/s11306-013-0598-6
83. Liland KH (2011) Multivariate methods in metabolomics—from pre-processing to dimension reduction and statistical analysis. Trends Analyt Chem 30:827–841. https://doi.org/10.1016/j.trac.2011.02.007
84. Ren S, Hinzman AA, Kang EL, Szczesniak RD, Lu LJ (2015) Computational and statistical analysis of metabolomics data. Metabolomics 11:1492–1513. https://doi.org/10.1007/s11306-015-0823-6
85. Jolliffe IT, Cadima J (2016) Principal component analysis: a review and recent developments. Philos Trans A Math Phys Eng Sci 374:20150202. https://royalsocietypublishing.org/doi/10.1098/rsta.2015.0202
86. Abdi H, Williams LJ (2010) Principal component analysis. Wires Comput Stat 2:433–459. https://doi.org/10.1002/wics.101

87. Worley B, Powers R (2013) Multivariate analysis in metabolomics. Curr. Metabolomics 1:92–107. https://doi.org/10.2174/2213235X11301010092
88. Li B, Martin E, Morris J (2001) Latent variable selection in partial least squares modelling. IFAC Proc 34:463–468. https://doi.org/10.1016/S1474-6670(17)33867-3
89. Gromski PS, Muhamadali H, Ellis DI, Xu Y, Correa E, Turner ML, Goodacre R (2015) A tutorial review: metabolomics and partial least squares-discriminant analysis—a marriage of convenience or a shotgun wedding. Anal Chim Acta 879:10–23. https://doi.org/10.1016/j.aca.2015.02.012
90. Galindo-Prieto B, Eriksson L, Trygg J (2014) Variable influence on projection (VIP) for orthogonal projections to latent structures (OPLS). J Chemom 28:623–632. https://doi.org/10.1002/cem.2627
91. James G, Witten D, Hastie T, Tibshirani R, Taylor J (2023) An introduction to statistical learning: with applications in R, online version. Springer Science, New York
92. Bevilacqua M, Bro R (2020) Can we trust score plots? Meta 10:278. https://doi.org/10.3390/metabo10070278
93. Westerhuis JA, Hoefsloot HCJ, Smit S, Vis DJ, Smilde AK, van Velzen EJJ, van Duijnhoven JPM, van Dorsten FA (2008) Assessment of PLSDA cross validation. Metabolomics 4:81–89. https://doi.org/10.1007/s11306-007-0099-6
94. Lindgren F, Hansen B, Karcher W, Sjöström M, Eriksson L (1996) Model validation by permutation tests: applications to variable selection. J Chemom 10:521–532. https://doi.org/10.1002/(SICI)1099-128X(199609)10:5/6<521::AID-CEM448>3.0.CO;2-J
95. Ojala M, Garriga GC (2010) Permutation tests for studying classifier performance. J Mach Learn Res 11:1833–1863. https://www.jmlr.org/papers/volume11/ojala10a/ojala10a.pdf. Accessed 8 Dec 2023
96. Nahm FS (2022) Receiver operating characteristic curve: overview and practical use for clinicians. Korean J Anesthesiol 75:25–36. https://doi.org/10.4097/kja.21209
97. Ranganathan P, Aggarwal R (2018) Understanding the properties of diagnostic tests—part 2: likelihood ratios. Perspect Clin Res 9:99–102. https://doi.org/10.4103/picr.PICR_41_18
98. Mandrekar JN (2010) Receiver operating characteristic curve in diagnostic test assessment. J Thorac Oncol 5:1315–1316. https://doi.org/10.1097/JTO.0b013e3181ec173d
99. Chan MK, Krebs M-O, Cox D et al (2015) Development of a blood-based molecular biomarker test for identification of schizophrenia before disease onset. Transl Psychiatry 5:e601. https://doi.org/10.1038/tp.2015.91
100. Walter SD (2005) The partial area under the summary ROC curve. Stat Med 24:2025–2040. https://doi.org/10.1002/sim.2103
101. Zou KH, James O'Malley A, Mauri L (2007) Receiver-operating characteristic analysis for evaluating diagnostic tests and predictive models. Circulation 115:654–657. https://doi.org/10.1161/CIRCULATIONAHA.105.594929

Chapter 3
Biomarkers Related to the Hypotheses of the Pathophysiology of Schizophrenia Spectrum Disorders

3.1 Genetic, Epigenetic, and Proteomic Biomarkers

As discussed in Chap. 1, one of the determining causes of schizophrenia spectrum disorders (SSD) is heritability. However, genetic markers per se and copy number variations are prone to the influence of various factors, including the environment, which can induce epigenetic modifications. These markers also indicate pathways shared by converging diseases such as SSD, major depression disorder (MDD), autism, and bipolar disorder (BD) [1]. Despite their significance, there is currently no consensus on their clinical utility, and the same goes for epigenetic markers [2–4]. In fact, research combining several techniques, including genome-wide association studies (GWAS), brain structural endophenotypes, protein-wide association studies (PWAS), and human single-cell RNA sequencing data, has provided evidence of a shared genetic and molecular basis between major psychiatric and neurodegenerative disorders, revealing common mechanisms among them [5]. Leveraging the diverse findings from peripheral blood proteomic studies on SSD, BD, and MDD, Fernandes et al. aimed to concatenate these findings into biological pathways to achieve a deeper understanding of pathophysiology and to potentially repurpose existing drugs [6]. In both studies, the most altered pathways were found to belong to the immunoinflammatory system and synaptic transmissions. Half of the differentially expressed proteins were common to at least two disorders, and some uniquely expressed proteins belonged to shared pathways among SSD, BD, and MDD. This observation makes a blood protein-based test unfeasible due to its likely low diagnostic accuracy. The authors emphasized the importance of considering the characteristic stages of each disease and the necessity for the redefinition of many nosological categories.

In the upcoming years, the crucial factor for achieving an effective and accurate panel of protein markers will be to concentrate on the differential diagnosis of phenotypically similar disorders, utilizing integrative techniques and multi-modal complementary phenotyping [6].

3.2 Metabolic Biomarkers for SSD Assessed Through ROC Analysis

A systematic review spanning from January 1970 to June 2016 revealed the concurrent presence of multiple metabolites and dysregulated pathways in both drug-free/naïve and drug-treated SSD patients. This suggests that metabolic disturbances are already present at the onset of the disease, with some persisting or emerging during antipsychotic treatment. Notably, certain metabolites returned to normal levels after antipsychotic treatment [7]. Subsequently, there was a glimmer of hope that these altered metabolites could serve as a foundation for developing methods to monitor the progress of patients undergoing treatment. Pillinger and colleagues conducted an analysis of 18 antipsychotics, aiming to rank them based on their metabolic side effects and identify predictors of antipsychotic-induced metabolic dysregulation [8]. Their network meta-analysis reaffirmed the known correlation between medication efficacy and a broad metabolic disturbance. They associated improvements in psychopathology with concurrent increases in body weight, total cholesterol, and low-density lipoprotein (LDL), along with reductions in high-density lipoprotein (HDL) level. However, they noted low certainty of evidence across all parameters, high inconsistencies in data for triglycerides, fasting-glucose, and body weight, and a large proportion of studies (84%) indicating evidence of bias. These findings prevented them from designating these metabolites as the most reliable predictors. Following an extensive review of the subject, Quintero et al. put forth a list of more than 120 potential biomarkers for SSD and its associated pathways [9].

The quest for efficient metabolic biomarkers with diverse functions has unfolded throughout history, as outlined in Table 3.1. In 2011, Xuan et al. conducted a profile of serum metabolites in SSD patients undergoing risperidone treatment. They successfully established a model with 20 metabolites, revealing pathways such as impaired glycolysis, a reduced tricarboxylic acid (TCA) cycle, and deficits in mitochondrial respiration [10].

In the same vein, Yang and colleagues developed a diagnostic model for distinguishing SSD patients undergoing 4-week treatment from healthy controls based on 22 significantly expressed metabolites. Only four of these metabolites were needed to differentiate the cohorts, with an AUC of 94.5% [11]. Although the model has not transitioned into clinical practice, the altered metabolites unveiled a compromised antioxidant system, indicating low serum cystine levels and subsequent reduction of the antioxidant glutathione. This observation confirmed a shift in the patients' energy metabolism toward fatty acid catabolism and ketosis—a core feature of SSD. This revelation may have inspired further preclinical and clinical trials, leading to the exploration of ketogenic therapy. This therapeutic approach involves

3.2 Metabolic Biomarkers for SSD Assessed Through ROC Analysis 39

Table 3.1 Potential metabolic biomarkers for schizophrenia spectrum disorders (SSD) assessed by ROC curve analysis

Cohort (effect in concentration): biomarker	Sample analytical platform biomarker class	Cohort (n = male/female) / age[a]	AUC (95% CI)/sensitivity/specificity (%) [predictor] learning algorithm	Pathway/symptom association/ correlations
bSSD vs HC (↓): Citrate (FC = −2.66, p = 0.009), palmitic acid (FC = −1.77, p = 0.004). bSSD vs HC (↑): Myo-inositol (FC = 1.46, p = 0.02), urea (allantoin; FC = 3.5, p = 0.03). Risp-treat vs bSSD (↓): Myo-inositol (FC = −1.49, p = 0.025). Risp-treat vs bSSD (↑): Uric acid (FC = 1.81, p = 0.039), Trp (FC = 1.87, p = 0.007). Responder vs nonresponder (↑): Uric acid (FC = 2.33), Trp (FC = 2.40), p = 0.016 for both. Xuan et al., 2011 [10]	Serum. GC/MS. Diagnostic and Theranostic.	Baseline SSD (10/8) / 41.3 (16.1). HC (10/8) / 41.0 (15.0). 8-week risp-treat SSD (14) / 45.6 (15.7). Responder (7) / 36.0 (17.2). Nonresponder (7) / 55.1 (5.0). Note: Age was a confounding factor for discrimination between responder and nonresponder to treatment. Cohort nationality: China (Han ethnicity).	**bSSD vs HC:** 95.8 [citrate, palmitic acid, myo-inositol, and urea]. **Risp-treat SSD vs bSSD:** 94.6 [myo-inositol, uric acid, and tryptophan]. Note: CI, sensitivity, and specificity were not reported. **PCA (4PCs) and PLS-DA:** bSSD vs HC: R2X = 0.305, R2Y = 0.995, Q2Y = 0.923. **Risp-treat SSD vs bSSD:** R2X = 0.330, R2Y = 0.997, Q2Y = 0.988. **Responder vs nonresponder (3PCs):** R2X = 0.377, R2Y = 0.997, Q2Y = 0.955. Five-fold cross validation, 200 permutation tests, VIP > 1, p < 0.05. **Support vector machine** for predictor selection with double cross-validation.	Citrate: TCA cycle. Palmitic acid: Fatty acid metabolism. Myo-inositol: Inositol phosphate metabolism. Urea: Uric acid metabolism. Uric acid: Purine metabolism. Tryptophan: Trp metabolism.

(continued)

Table 3.1 (continued)

Cohort (effect in concentration): biomarker	Sample analytical platform/ biomarker class	Cohort (n = male/female) / age[a]	AUC (95% CI)/sensitivity/ specificity (%) [predictor] learning algorithm	Pathway/symptom association/ correlations
TC (↑): Glycerate (FC = 2.57), pyruvate (FC = 1.88), β-hydroxybutyrate (FC = 2.45), eicosenoic acid (FC = 1.96). TC (↓): Cystine (FC = −1.36). Yang et al., 2013 [11]	Serum. GC-TOF/MS for untargeted and targeted metabolomics. Diagnostic.	**Training cohort (TC):** 4-week-treat SSD (26/36) / 36.9 (11.9). HC (26/36) / 36.9 (9.3). **Testing cohort (TeC):** 4-week-treat SSD (21/29) / 36.9 (12.6) vs HC (0/48) / 28.3 (8.6). Cohort nationality: China (Anhui Province).	TC: 94.5 (90.0–99.1)/ 95.2/98.4 TeC: 89.5 (82.9–96.1)/ 82.0/89.6 [glycerate, pyruvate, β-hydroxybutyrate, eicosenoic acid, cystine]. **PCA and OPLS-DA:** TC: R2Y = 0.72, Q2 = 0.47. Seven-fold cross validation, 999 permutation tests, p and FDR < 0.05, VIP > 1.5.	β-Hydroxybutyrate, eicosenoic acid, and glycerate indicate increased catabolism of fatty acid, serving as an alternative source of energy. Pyruvate suggests increased energy production, likely a compensatory response to inefficient brain circuits. β-Hydroxybutyrate, cystine: Reduced synthesis of the antioxidant glutathione in cells with immune function, compromising the brain under oxidative stress.

3.2 Metabolic Biomarkers for SSD Assessed Through ROC Analysis

TC (↓): Pyroglutamic acid (FC = 0.13), sorbitol (FC = 0.31), alpha-tocopherol (FC = 0.19). Liu M-L. et al., 2014 [15]	PBMC. GC/MS for untargeted metabolomics, to assess the disturbed metabolisms and for ROC analysis; for targeted metabolomics, to quantitate the 3 most significant metabolites used in the ROC analysis. Diagnostic.	**Training cohort (TC), untargeted metabolomics:** Drug-treat + drug-naïve SSD (18/27) / 33.22 (12.93). HC (22/28) / 37.26 (8.67). **Testing cohort (TeC) 1, untargeted metabolomics:** Drug-treat SSD (5/19) / 24.46 (10.52). HC (19/16) / 38.86 (15.61). **Testing cohort 2, targeted metabolomics:** SSD (10/10) / 31.95 (10.22). HC (10/10) / 32.60 (3.32). Cohort nationality: China (Chongqing).	**TC:** 82 (74–91) **TeC 1:** 71 (56–87) [pyroglutamic acid, sorbitol, tocopherol-α]. Note: Sensitivity and specificity were not reported. **OPLS-DA:** **TC:** R2X = 0.55, R2Y = 0.54, Q2Y = 0.29. 300 permutation tests, $p < 0.001$, FDR < 0.05, VIP > 1. Linear regression with AIC for classifier evaluation.	The diminished levels of pyroglutamic acid, a by-product of glutathione metabolism linked to oxidative stress, and tocopherol-α, a peroxyl radical scavenger, indicate a reduction in antioxidant defense in SSD patients. Additionally, the presence of sorbitol suggests a potential impact on energy metabolism, specifically the tricarboxylic acid (TCA) cycle.

(continued)

Table 3.1 (continued)

Cohort (effect in concentration): biomarker	Sample analytical platform/biomarker class	Cohort (n = male/female)/age[a]	AUC (95% CI)/sensitivity/specificity (%) [predictor] learning algorithm	Pathway/symptom association/correlations
Drug-treat SSD vs HC (↓) μM: 13.1 D-lactate, 101.0 L-Trp. **Drug-treat SSD vs HC (↑)** μM: 159.0 linoleic acid, 2.1 γ-GluCys. Fukushima et al., 2014 [16]	Serum. HPLC/UV detector for targeted metabolites. HPLC/fluorescence detector for quantitation of AA. HPLC/MS for KYN and Trp. HPLC/Electrochemical detector for 5-HT. Diagnostic.	Drug-treat SSD (11/14) / 28.2 (4.4). HC (12/15) / 26.5 (5.6). Cohort nationality: Japan (Koufu-shi and Shirakawa-shi cities).	**Drug-treat SSD vs HC:** 88.7/88.0/81.5 [γ-GluCys]. 84.9/80.0/77.8 [linoleic acid]. 84.2/88.0/77.8 [D-lactate]. Note: CI was not reported. **OPLS-DA:** R2Y = 0.861, Q2 = 0.720. Seven-fold cross validation, $p < 0.001$, VIP > 1.	γ-GluCys serves as a precursor to the endogenous antioxidant glutathione (GSH). D-lactate, a metabolite of methylglyoxal, which is a breakdown product of the triose phosphate intermediates of glycolysis not only plays a role in modulating neuronal excitability but also contributes to maintaining homeostasis and consolidating memory [39]. Linoleic acid: Fatty acid metabolism. Trp: Tryptophan metabolism. D-lactate and γ-GluCys showed a significant negative correlation.
Drug-treat SSD vs HC (↓): 6-DMF, galactose oxime, eicosanoic acid, cholesterol. **Drug-treat SSD vs HC (↑):** 2-PCA. Note: $p < 0.001$, FDR < 0.05 for each biomarker. Al Awam et al., 2015 [40]	Serum. GC/MS. Diagnostic.	Drug-treat SSD (20/6) / 37.27 (12.4). HC (20/6) / 37.04 (10.7). Cohort nationality: Germany (Bochum city).	**Drug-treat SSD vs HC:** 87/83/92 [6-DMF], 93/82/92 [2-PCA], 86/88/92 [galactose oxime], 82/75/81 [eicosanoic acid], 76/71/69 [cholesterol]. **PCA:** CI, R2X and Q2 were not reported.	2-PCA and galactose oxime: GABA release/uptake and GABA precursor, respectively. Eicosanoic acid: Monoaminergic neurotransmitter. 6-DMF: Neurotransmitter metabolism. Cholesterol: Myelin maintenance and cognition.

3.2 Metabolic Biomarkers for SSD Assessed Through ROC Analysis

bSSD vs HC (↑): C4-OH(C3DC) (FC = 1.39), C16:1 (FC = 1.28) ($p < $ 1E-06). **bSSD vs HC (↓):** C8 (FC = 0.59), C10:1 (FC = 0.56), C12 (FC = 0.69), C14:1-OH (FC = 0.8), C14:2 (FC = 0.84) ($p < $ 1E-05). **bSSD vs treat-SSD (↓):** All predictors ($p < $ 1E-04) but C4 (FC = 1.47, $p < $ 1E-11). **Treat-SSD vs HC (↓):** All predictors ($p < $ 1E-12). Note: p and FDR < 0.05, VIP > 1 for each biomarker. Cao B. et al., 2019 [18]	Serum. RSLC-quadrupole-Orbitrap/MS. Diagnostic and theranostic. Note: Data were adjusted for age, sex, BMI, smoking, and drinking.	Baseline SSD (90/135) / 37.31. HC (54/121) / 39.44. 8-week-treat-SSD (156). Cohort nationality: China (Shandong Province).	**bSSD vs HC:** 92.6 (90.1–95.2) [C4-OH (C3DC), C8, C10:1, C12, C14:1-OH, C14:2, C16:1]. R2Y = 0.585, Q2 = 0.535. **bSSD vs treat-SSD:** 85.1 (81.0–89.2) [C2, C4, C4-OH (C3DC), C10:1, C14:1, C14:1-OH, C14:2, C16:1-OH, C16:2, C18:1]. R2Y = 0.352, Q2 = 0.297. **Treat-SSD vs HC:** 95.0 (92.8–97.2) [C2, C8, C10:1, C12, C12:1, C14:1-OH, C14:2-OH, C16:2-OH, C18]. R2Y = 0.583, Q2 = 0.532. **OPLS-DA:** Seven-fold cross validation, 300 permutation tests. **Logistic regression** (backward method) for selection of predictors using Akaike information criterion. Note: Sensitivity and specificity were not reported.	Acyl-carnitines (C) facilitate the transport of long-chain FAs from the cytosol across the mitochondrial membrane, directing them toward the mitochondrial matrix where β-oxidation occurs. Elevated rates of β-oxidation in acyl-carnitines may confer increased antioxidant activity, regulation of immune functions, stimulation of certain proteins and enzymes, and enhancement of cholinergic neurotransmission. Baseline C levels did not show a correlation with changes in PANSS total scores after treatment. Alterations in BMI and waist circumference were inversely associated with several short-chain and medium-chain C at baseline. Differences in C levels between pre- and post-treated patients point to an imbalance in bioenergetic pathways.

(continued)

Table 3.1 (continued)

Cohort (effect in concentration): biomarker	Sample analytical platform biomarker class	Cohort (n = male/female) / age[a]	AUC (95% CI)/sensitivity/ specificity (%) [predictor] learning algorithm	Pathway/symptom association/ correlations
TC (↑): Total LPE and SM, LPE (20:4). TC (↓): Total LPC and PE, PC(18:2/18:2), PC(O-16:0/18:2), LPC(18:0), LPC(20:0), PE(P-18:0/18:2). Note: FDR and $p < 0.05$, VIP > 1, correlation coefficient (r) > 0.7 for each metabolite. Wang D. et al., 2019 [41]	Serum. UPLC-quadrupole-Orbitrap/MS Diagnostic. Note: Data were adjusted for age, gender, BMI, smoking, drinking and psychiatric family history.	Training cohort (TC): FE drug naïve SSD (10/10) / 28.00 (7.31) + recurrent SSD (29/35) / 29.48 (5.62). HC (25/52) / 30.61 (4.87). Testing cohort (TeC): FE drug naïve SSD (3/5) / 25.13 (7.38) + recurrent SSD (10/17) / 29.15 (6.05). HC (25/52) / 30.61 (4.87). Cohort nationality: China (Shandong Province).	TC: 99.1 (98.0–100.0)/97.6/92.5 [LPC(18:0), LPC(20:0), PC(18:2/18:2), PC(O-16:0/18:2), LPE(20:4), PE(P-18:0/18:2)]. TeC: 98.0 (95.1–100.0)/97.1/96.9. PCA: R2X = 0.831, Q2 = 0.632. PLS-DA: R2Y = 0.923, Q2 = 0.857. Seven-fold cross validation, 300 permutation tests.	The high inverse correlation between SMs and LPCs implies a distinct mechanism in the systemic regulation of lipid balance across various physiological compartments. The reduced PE to increased LPE ratio indicates hyperactivation of the enzyme phospholipase A2, a phenomenon associated with heightened inflammation.
SSD vs HC (↑): Oleoylcarnitine (FC = 1.58). SSD vs HC (↓): LPC(15:0) (FC = 0.76), LPC(14:0) and 9-decenoylcarnitine (FC = 0.61 for both). Note: $p < 0.0001$, VIP > 1, FDR < 0.05 for each metabolite. Cao T. et al., 2020 [17]	Serum. HILIC in RSLC and UPLC-quadrupole-Orbitrap/MS. Diagnostic. Note: Data were adjusted for age, gender, BMI, smoking, and current drinking.	FE/recurrent drug-naïve SSD (43/70) / 29.36 (6.19). HC (39/72) / 30.66 (4.61). Cohort nationality: China (Shandong Province).	SSD vs HC: 99.3 (98.7–100) [oleoyl carnitine, 9-decenoylcarnitine, LPC(15:0), LPC(14:0)]. PLS-DA: R2Y = 0.962, Q2 = 0.938. Note: Sensitivity and specificity were not reported.	No correlation was observed between the metabolites and PANSS total scores. The 11 most significant predictors were primarily associated with cellular bioenergetics and oxidative stress, with a particular emphasis on fatty acid and amino acid metabolism.

3.2 Metabolic Biomarkers for SSD Assessed Through ROC Analysis

SSD vs HC (↑): Arachidonic acid, 9-HODE, α-CEHC, leukotriene A4, thromboxane A2, 5-HETE, phytanic acid. **SSD vs HC (↓):** Trp, KYN, Leu, indoleacrylic acid, Phe, Tyr, phenylpyruvate, O-Tyr, carnitine, acetyl carnitine, guaiacol, Gln, biliverdin. Note: p and FDR < 0.05, VIP > 1, FC > 1.5 for each metabolite. Cui G. et al., 2020 [14]	Serum. FT-ICR/MS for untargeted metabolomics. Diagnostic. Note: Carnitine was identified against its standard using LC/MS. Data were adjusted for age, sex, BMI, and smoking.	SSD (18/36) / 38.44 (11.63). HC (13/41) / 33.59 (9.87). Cohort nationality: China (Anhui Province).	**SSD vs HC:** 99.7 (99.2–100)/100/96.3 [carnitine]. **PLS-DA:** R2Y = 0.976, Q2 = 0.948. 999 permutation tests.	The significantly altered metabolites were found to correlate with the inflammatory response, as evidenced by the upregulated arachidonic acid pathway and the downregulated aromatic amino acid pathway. Carnitine was associated with an imbalance in redox homeostasis, while 8-OH-deoxyguanosine, a marker of oxidative stress, exhibited a negative correlation with the antioxidant biliverdin and a positive correlation with oxidation products 9-HODE and O-tyrosine. These findings led the authors to propose an enhanced oxidative stress in SSD, supporting the related pathophysiology hypothesis.

(continued)

Table 3.1 (continued)

Cohort (effect in concentration): biomarker	Sample analytical platform biomarker class	Cohort (n = male/female) / age[a]	AUC (95% CI)/sensitivity/ specificity (%) [predictor] learning algorithm	Pathway/symptom association/ correlations
Predictors: (−)-*trans*-caryophyllene, 3-OH-3-methyl butyric acid, nonadecylic acid, *N*-acetylmethionine, 13-OxoODE, urocanic acid, glyceroPC, L-Arg, hexyl acetate, 15-deoxy-d-12,14-PGJ2, LPC (18:1), glycerol 3-phosphate, 9-OxoODE, 2,2-dimethyl succinic acid, 3-methylglutaric acid, taurine, 7-methylguanine, 2-oxovaleric acid, adipic acid, L-Phe, hexyl butyrate, methanesulfonic acid, 9,10-DHOME, acetaminophen glucuronide, naphthylacetic acid Note: $p < 0.05$, VIP > 1.5 for each metabolite. Du et al., 2021 [27]	Exosome from serum. UPLC-Qtrap/MS. Diagnostic. Note: Data were adjusted for age and disease severity.	**Training cohort (TC)** SSD (67/11) / 31.7 (7.73). HC (28/38) / 24.46 (5.06). Huangshan second People's hospital (Han ethnicity) **Testing cohort 1 (TeC1)** SSD (93/14) / 31.45 (8.85). HC (19/43) / 30.66 (6.34). Urumqi fourth People's hospital (Han/Uyghur ethnicity) **Testing cohort 2 (TeC2)** SSD (54/50) / 26.69 (8.2). HC (104/38) / 28.59 (5.44). Third People's Hospital of Foshan (Han ethnicity) **Subgroup TeC2 (STeC2)** **Testing cohort 3 (TeC3)** SSD (50/46) / 29.98 (5.76). HC (28/34) / 29.31 (5.23).	TC: 95.7 (92.6–98.9)/85.9/95.5. TeC1: 91 (85.7–96.3)/86.9/87.1. TeC2: 82.7 (77.6–87.9)/73.1/80.3. STeC2: 86.4 (80.3–92.6)/77.5/85.9. TeC3: 99 (97.7–100)/97.9/95.2. STeC3: 100 (100–100)/100/100. The above models were constructed using 25 predictors. **FE drug-free STeC2 vs chronically treated SSD from TC:** 73.8 (63.9–83.6)/73.8 [tauro-UDCA, uracil, 6-β-OH-testosterone, methyl stearate, 4-*tert*-octylphenol, 2,4-di-*tert*-butylphenol]. **FE drug-free STeC3 vs chronically treated SSD from TC:** 100 (100–100)/100 [6 predictors].	The most significantly altered analytes tested (with 10 up- and 15 down-regulated) exhibited enrichment (rich factor > 0.25) for pathways related to Phe, Trp, and Tyr biosynthesis, taurine and hypotaurine metabolism, as well as Phe metabolism and glycerophospholipid metabolism. While the 25 metabolites did not show correlation with PANSS scores in TC and TeC1, they demonstrated significant correlations with PANSS negative scores in TeC2 and TeC3. These results lend support to the membrane phospholipids hypothesis as the biochemical foundation for the neurodevelopmental concept of SSD, as proposed by Horrobin in 1998 [42]. The performance of the model containing 25 predictors showed no influence from gender.

3.2 Metabolic Biomarkers for SSD Assessed Through ROC Analysis

		Third People's Hospital of Foshan (Han ethnicity) **Subgroup TeC3 (STeC3)** FE drug-free SSD (49) vs HC (62). Cohort nationality: China	**36 Uyghur SSD from TeC1 vs HC:** 93.7 (88.9–98.6)/92.9/88.7 [23 predictors]. **PCA and OPLS-DA:** No performance measures were reported.	The authors hypothesized hexyl butyrate as a potential metabolite of *Cannabis sativa*, while urocanic acid may be a metabolite of *Escherichia coli*.
TC (↓): α-CEHC (FC = 0.44), neuraminic acid (FC = 0.31), L-Asn (1H NMR). **TC (↑):** Glyceraldehyde (FC = 8.62). Note: FDR ≤ 0.004, VIP > 1.5, FC < 0.5 or > 2 for each metabolite. Liu L. et al., 2020 [43]	Serum, brain tissue. Diagnostic. UPLC-QTOF/MS. 1H NMR (500 MHz). Microarray for transcriptome ($p < 0.05$; FC > 1.2).	**Training cohort (TC):** SSD (20/35) / 33.39 (12.76). HC (21/36) / 44.42 (13.49). **Testing cohort (TeC):** 60 (no other data was reported) **Transcriptomic cohort:** *Post-mortem* brain tissue of SSD (20/8) / 73.32 (15.20). HC (11/12) / 69.04 (21.55). Cohort nationality: China (Harbin, Heilongjiang)	**TeC:** 99.2/96.6/93.6 [α-CEHC, neuraminic acid, glyceraldehyde, L-Asn]. Note: CI was not reported. **PCA/PLS-DA (UPLC/QTOF-MS):** 200 permutation tests. R2X = 0.59, Q2 = 0.55. R2Y = 0.85, Q2 = 0.78. **PCA/PLS-DA (1H NMR):** R2X = 0.32, Q2 = 0.26. R2Y = 0.86, Q2 = 0.78. **PCA/PLS-DA (transcriptome):** R2X = 0.24, Q2 = 0.14. R2Y = 0.93, Q2 = 0.79.	Enriched pathways, combining metabolomic data from UPLC/QTOF-MS (44 significant metabolites) and 1H NMR (28 metabolites), along with transcriptomic data (705 differentially expressed genes), included amino acid metabolism (I = 0.3), aminoacyl-tRNA biosynthesis (I = 0.2), glyoxylate and dicarboxylate metabolism (I = 0.35), tricarboxylic acid cycle (I = 0.6), Arg, Val, Leu, Ile biosynthesis (I = 0.4), glucose metabolism (I = 0.38), and Phe, Tyr, and Trp biosynthesis (I = 1.2). The enriched pathways glycerolipid metabolism and Phe, Tyr, and Trp biosynthesis exhibited *p*-values of 0.01 and 0.045, respectively.

(continued)

Table 3.1 (continued)

Cohort (effect in concentration): biomarker	Sample analytical platform biomarker class	Cohort (n = male/female) / age[a]	AUC (95% CI)/sensitivity/specificity (%) [predictor] learning algorithm	Pathway/symptom association/correlations
TC: Vanillylmandelic acid (FC = 2.6; p = 0.02), L-asparagine/L-aspartic acid ratio (FC = 0.77; p = 0.003), glutaric acid (FC = 0.65; p = 0.001). Chen et al., 2020 [44]	Plasma. Predictive of violence in SSD patients. GC/MS.	**Training cohort (TC):** Violent SSD: (38/15) / 32.6 (4.7). Non-violent SSD: (13/11) 30.9 (5.2). Cohort nationality: China (Sichuan, Han ethnicity)	**TC:** 80.8 [L-asparagine/L-aspartic acid ratio, glutaric acid, vanillylmandelic acid]. Note: Sensitivity, specificity and CI were not reported. **PCA and OPLS-DA:** R2Y = 0.67, Q2 = 0.11. Seven-fold cross validation, 100 permutation tests.	
TC (FC > 1, p < 0.001): Cer(d18:1/16:0), Cer(d18:1/18:0), Cer(d18:1/24:0). **TeC SSD vs HC** (FC = 1.76, p < 0.001): Cer(d18:1/18:0). **TeC FE SSD vs HC** (FC > 1, p < 0.05): All 3 Cer. Tkachev et al., 2021 [45]	Plasma. Predictive of cardiovascular disease in SSD patients. UPLC-quadrupole-Orbitrap/MS	**Training cohort (TC):** SSD (63/19) / 31.2 (8.4). HC (108/30) / 29.5 (8.3). **Testing cohort (TeC):** FE SSD (9/10) / 27 (7). SSD (9/15) / 35 (12). HC (36/19) / 32 (8). Cohort nationality: Russia (Moscow).	**TC:** 81.0 [Cer(d18:1/18:0)]. **TeC SSD vs HC:** 73.0 [Cer(d18:1/18:0)]. **TeC FE SSD vs HC:** 81.0 [Cer(d18:1/18:0)]. Other Cer showed AUC ≤ 68%. Note: Sensitivity, specificity and CI were not informed. **Logistic regression.**	Ceramides (Cer) actively participate in fundamental processes such as apoptosis and inflammation, playing a crucial role in the functioning of cells within brain tissue.

Predictors tested: 22 lipid clusters comprising the following classes: CE, Cer, LPC, PC, PE, SM, and TGs. Note: $p < 0.05$ for each metabolite. Lamichhane et al., 2021 [25]	Serum. UHPLC-QTOF/MS for lipidomic. Theranostic. Note: Lipid internal standards (IS) had coefficient of variation <11%. Lipids were quantified against IS calibration curves. Data was adjusted for sex.	Baseline FE SSD (30 Turku, 14 UK (26/18)) / 26.9 (6.3). HC (31 Turku, 17 UK (31/17)) / 27.5 (5.9). Baseline CHR SSD (22 Turku (11/11)) / 26.2 (5.0). 1-year follow-up of BMI in chlorpromazine-treat: HC (21), FE (23 Turku, 2 UK), CHR (9 Turku). Cohort nationality: Finland (Turku) and UK (London).	**bFE SSD vs 1-year chlorp-treat FE:** 74 (60–85) [TG(45:0), TG(48:0)]. **bCHR vs 1-year chlorp-treat CHR:** 73 (61–83) [TG(45:0), TG(48:0)]. **Logistic ridge regression.** Note: Sensitivity and specificity were not informed.	Nine triglycerides (TGs), specifically (48:0), (47:0), (48:1), (47:1), (14:0/16:0/18:1), (16:0/16:0/16:0), and (49:0), were found to be significantly associated with weight gain, with regression coefficients ≥4. The results suggest that among FE SSD patients, those who are likely to gain weight in the future exhibit high levels of liver fat, rather than being influenced by dietary lipids.

(continued)

Table 3.1 (continued)

Cohort (effect in concentration): biomarker	Sample analytical platform biomarker class	Cohort (n = male/female) / age[a]	AUC (95% CI)/sensitivity/ specificity (%) [predictor] learning algorithm	Pathway/symptom association/ correlations
Drug-treat SSD vs HC (↑): Hydroxyindole, piperidine, tetrahydro uridine, isatin, phenylacetylglutamine, Glu-Glu, melamine, N-acetyl-2-aminoadipate, lipoamide. **Drug-treat SSD vs HC (↓):** γ-Glu-Trp, γ-Glu-his, γ-Glu-Val, γ-Glu-Phe, γ-Glu-Ile, γ-Glu-Leu, urea, N-acetylAla, isethionic acid, creatinine. Note: FDR < 0.05 for each metabolite. Okamoto et al., 2021 [46]	Serum. CE-FT/MS. Diagnostic.	Chronic (26 years sick) drug-treat SSD (17/13) / 48 (9.0). HC (5/5) 48 (8.7). Cohort nationality: Japan (Kitakyushu).	**Drug-treat SSD vs HC:** 95/93/100 [20 metabolites]. Factor loading[b] of upregulated metabolites: 0.22 to 0.38. Factor loading of downregulated metabolites: −0.77 to −0.83. **PCA and logistic regression.** Note: CI, R2X and Q2 values were not informed.	Gamma-glutamyl-valine (γ-Glu-Val) exhibited a significant negative correlation with PANSS total scores ($r = -0.45$, $p = 0.012$) and overall scores ($r = -0.49$, $p = 0.0055$). Similarly, gamma-glutamyl-phenylalanine (γ-Glu-Phe) displayed a significant negative correlation with PANSS total scores ($r = -0.40$, $p = 0.031$) and overall scores ($r = -0.41$, $p = 0.025$). Tetrahydro uridine showed a significant positive correlation with PANSS negative scores ($r = 0.53$, $p = 0.0061$). The most affected pathways included Glu, Phe, and Tyr metabolism, the urea cycle, energy conversion through the TCA cycle, and choline metabolism.

3.2 Metabolic Biomarkers for SSD Assessed Through ROC Analysis

Drug naïve SSD vs HC (↑): LPE(16:1sn1), LPE(16:1sn2), LPE(22:5sn2), LPC(16:0 sn2), PE(34:2), PC(32:1). **Drug naïve SSD vs HC (↓):** Carnitines (C8:1, C10:1, C10:2), free FA (16:2, 18:2, 18:3, 20:4, 22:6), PE(O-34:3), PE(O-36:6), PC(O-34:2), aspartic acid, lithocholic acid. Note: FDR < 0.05 for each Metabolite. Liu Y. et al., 2021 [30]	Plasma. UHPLC-TQ/MS. Diagnostic. Note: Targeted metabolomics used internal standard calibration. The relative standard deviation of metabolites was <30%.	Drug naïve SSD (16/22)/ 34.82 (13.33). HC (15/10) 33.68 (12.48). Drug-treat SSD (20/18) 30.53 (12.34). HC (11/14) 32.84 (9.73). Cohort nationality: China (Chongqing).	**Drug naïve SSD vs HC:** 93.6 (87.5–99.7/97.3/84 [PC(32:1), PE(34:2), PE(O-34:3), aspartic acid]. **Drug-treat SSD vs HC:** 96.3 (92.6–100)/89.2/92 [PC(32:1), PE(34:2), PE(O-34:3), aspartic acid]. **Drug-treat vs drug naïve SSD:** The differences were not significant. **PCA and binary logistic regression** (enter model, forward elimination, and backward elimination method).	Carnitines transport long-chain FAs into the mitochondria, where they undergo oxidation to produce energy. They also participate in lipid metabolism. Aspartate not only plays a regulatory role in the synthesis and secretion of steroid hormones but also exhibits a high affinity for NMDA glutamate receptors, thereby influencing the neuroendocrine system in the brain. Elevated levels of LPE can be indicative of chronic stress. Phospholipids, such as PC and PE, are essential constituents of the cell membrane and serve as precursors to neuroprotective molecules like endocannabinoids.

(continued)

Table 3.1 (continued)

Cohort (effect in concentration): biomarker	Sample analytical platform biomarker class	Cohort (n = male/female) / age[a]	AUC (95% CI)/sensitivity/ specificity (%) [predictor] learning algorithm	Pathway/symptom association/ correlations
bCHR male vs female (↑): TG, LPC. **bCHR male vs female (↓):** SM. **CHRt vs CHRr or CHRp (↓):** PC(O-22:2/22:3), PC(O-32:0). **CHRr vs HC (↑):** SM(d34:2), SM(d18:1/24:0). Note: FDR < 0.01, FC > 1.2, VIP > 1 for each biomarker. Dickens et al., 2021 [24]	Plasma. UHPLC-QTOF/MS for lipidomics. Diagnostic and predictive. Note: Data were adjusted for age, sex, BMI, site, subject status, transition status, ethnicity, medication.	Baseline CHR (141/122) / 22.47 (4.84). HC (29/22) / 22.76 (4.13). **5-year follow-up:** bCHR: Patient at baseline. CHRt (transition period): 57 bCHR developed psychosis within 2 years, 8 developed after 2 years, Some had remission (CHRr), Some persisted as CHR (CHRp). Cohort nationality: Netherlands (Amsterdam), Switzerland (Basel), Spain (Barcelona), Germany (Cologne), Denmark (Copenhagen), UK (London), France (Paris), Australia (Melbourne), Brazil (São Paulo), and Austria (Vienna).	**bCHR vs HC:** 83 (71–95)/69/83 [PC(38:4), Cer(d18:1/24:0), LPC(22:5), PC(40:5), PC(O-32:0), SM(d18:1/24:0), SM(d36:0), SM(d36:1)]. **CHRt vs CHRp + r:** 81 (69–93)/52/88 [Cer(d18:1/24:0), LPC(22:5), PC(38:4), PC(40:5), PC(O-32:0)]. **CHRr vs CHRp + t:** 83 (72–95)/64/90 [Cer(d18:1/24:0), LPC(22:5), PC(40:5), PC(38:4), PC(O-32:0), SM(d18:1/24:0)]. **Sparse PLS-DA:** Seven-fold cross-validation 70–30% for training/validation **Logistic ridge regression:** 1000 bootstrapping Ten-fold cross-validation 70–30% for training/validation	TGs with shorter and more saturated fatty acyl chains, as well as SM levels, exhibited distinct variations based on gender. In bCHR individuals, blood levels of numerous lipids were higher. Both TG and SM were found to be associated with obesity, insulin resistance, and type 2 diabetes. CHRt were distinguished from CHRp+r by a reduction in ether phospholipid free radical scavengers. This observation suggests that CHRt individuals may be more vulnerable to oxidative stress.

3.2 Metabolic Biomarkers for SSD Assessed Through ROC Analysis

FE SSD vs HC (↓): 5-HIAA. **FE SSD vs HC** (↑): Lysyl serine. **CHR vs HC** (↓): 4-OH-retinal, IL-8, IL-1β, IFNγ, PG(18:2(9Z,12Z)/16:0), PGD2, S100B. **CHR vs HC** (↑): Lysyl serine, prenyl glucoside. Note: FDR < 0.05, VIP > 1 for each biomarker. Cui G. et al., 2021 [23]	Saliva: UPLC-QTOF/MS for metabolomics. Serum: Cytokine antibody array for cytokines, S100B, and CRP. ELISA for thioredoxin. Diagnostic and predictive. Note: Data were adjusted for age, gender, and education.	FE SSD (48/35) / 25.60 (7.42). HC (39/39) / 23.03 (6.45). CHR (30/12) / 17.45 (2.34). Cohort nationality: China (Shanghai).	**FE SSD vs HC:** 80.4 (73.5–87.2)/91.6/61.5 [5-HIAA, lysyl serine]. 87.1 (80.9–93.3)/93.1/70.7 [5-HIAA, lysyl serine +9 H$_2$S-producing bacteria]. **CHR vs HC:** 94 (89.8–96.0) /90.5/84.6 [4-OH-retinal, PGD2, (PG(18:2(9Z,12Z)/16:0), prenyl glucoside]. 95.8 (92.6–98.9) /92.9/85.5 [PG(18:2(9Z,12Z)/16:0), 4-OH-retinal, prenyl glucoside, PGD2 + 6 H$_2$S-producing bacteria]. **OPLS-DA** **Logistic regression**	**FE SSD vs CHR**: L-carnitine biosynthesis (energy metabolism & stress adaptation), ketosis. **FE SSD vs HC**: Glu, 5-HT and DA receptors signaling; NE, 5-HT, Gln and adrenaline degradation. **CHR vs HC**: Tyrosine degradation and nucleotide metabolism. Strong correlations were observed between saliva metabolites and the composition of the oral microbiota.

(continued)

Table 3.1 (continued)

Cohort (effect in concentration): biomarker	Sample analytical platform biomarker class	Cohort (n = male/female) / age[a]	AUC (95% CI)/sensitivity/specificity (%) [predictor] learning algorithm	Pathway/symptom association/correlations
Chronic SSD vs HC (↑): Alanine (D-ala), glutamate (Glu), lactic acid, ornithine, serine (Ser). **Chronic SSD vs HC (↓):** Urea. **Acute SSD vs remission SSD (↑):** Fatty acid (15:0), (17:0), and (19:1), cis-11-eicosenoic acid, thyroxine. Note: $p < 0.001$ for each metabolite. Kim et al., 2022 [28]	Serum. CE-TOF/MS and LC-TOF/MS. Diagnostic and state markers. Note: Data were adjusted for age, sex, and medication.	Acute psychotic SSD (11/9) / 42.5 (12.7), with remission after treatment. HC (10/10) / 41.8 (12.2). Chronic SSD (10/10) / 43.4 (8.8). Cohort nationality: Japan (Hyogo prefecture, Kansai region).	**Chronic SSD vs HC:** 97.2 (93.4–100) [alanine]. 95.8 (90.2–100) [glutamate]. 96 (90.1–100) [lactic acid]. 90 (80.7–99.3) [ornithine]. 95 (89.2–100) [serine]. 90.5 (81.5–99.5) [urea]. **PCA (all groups), OPLS-DA (chronic vs HC) and OPLS-effect projection** (acute vs remission SSD): **Chronic SSD vs HC:** R2Y = 0.79, Q2 = 0.76. **Acute SSD vs remission SSD:** R2Y = 0.66, Q2 = 0.40,	The elevation of Glu implies deregulation of the glutamatergic system, yet comprehensive large-scale studies are necessary to discern the potential impact of medication. Ser and ala function as co-agonists of NMDA receptors, contributing to the glutamatergic activation of these receptors. D-alanine, an extrinsic D-amino acid, originates from the gut microbiome. Ornithine, a component of the urea cycle, undergoes breakdown by the sentinel enzyme ornithine decarboxylase 1 (ODC1) into putrescine, a compound associated with various diseases and neurodevelopmental disorders [47]. The increase in lactic acid suggests mitochondrial dysfunction, while FAs with high odd-numbers indicate dysbiosis of the gut microbiota.

3.2 Metabolic Biomarkers for SSD Assessed Through ROC Analysis

TC (↓): Cholic acid (CA), CDCA, DCA, βDCA, UDCA, βCDCA, 7-ketoLCA, 3-DHCA, TBA, unconjugated bile acids (BA). TC (↑): apoCA, norCA, CA/CDCA ratio. Note: FDR < 0.05 and VIP > 1 for each biomarker. Qing et al., 2022 [31]	Serum. UPLC-TQ/MS for target metabolomics. Diagnostic. Note: 40 bile acids were detected and quantified using calibration curves.	**Training cohort (TC):** Drug-free SSD (23/36) / 37.48 (11.80). HC (25/35) / 36.80 (9.74). **Testing cohort (TeC):** Drug-free SSD (20/29) / 36.13 (12.68). HC (0/48) / 28.29 (8.60). Cohort nationality: China (Anhui Province).	**TC:** 75.8/69.5/76.7 [total CDCA, βCDCA, 6-ketoLCA, apoCA, norCA, GHCA, THCA, total primary BAs]. **TeC:** 73.2/67.3/70.8 [the same above]. **PCA (all groups), OPLS-DA:** **TC:** R2Y = 0.59, Q2 = 0.45. **TeC:** R2Y = 0.64, Q2 = 0.52. Permutations tests with 1000 iterations ($p < 0.001$).	Chenodeoxycholic acid (CDCA) functions as an antagonist for NMDA and GABA receptors. Decreased levels of CDCA in SSD patients suggest disruptions in glutamatergic and GABAergic neurotransmission. Furthermore, reduced levels of total bile acids (TBA) signify impaired bile acid (BA) synthesis, a linkage observed in type 2 diabetic peripheral neuropathy. The diminished levels of unconjugated BA point to the impairment of species-specific bacterial deconjugation of CDCA.

(continued)

Table 3.1 (continued)

Cohort (effect in concentration): biomarker	Sample analytical platform biomarker class	Cohort (n = male/female) / age[a]	AUC (95% CI)/sensitivity/ specificity (%) [predictor] learning algorithm	Pathway/symptom association/ correlations
Cimp SSD vs CN SSD (↑): Imidazole propionic acid, homoserine, aspartic acid. **Cimp SSD vs CN SSD (↓)**: Erythronic acid. Note: $p < 0.05$ and VIP > 1.0 for each metabolite. Jiang et al., 2022 [26]	Serum. UPLC/MS. Predictive for the development of cognitive impairment (Cimp).	Cimp SSD (12/5) / 54.18 (13.97). HC (13/7) / 32.05 (4.62). CN SSD (11/6) / 49.53 (12.17). Cohort nationality: Germany (Magdeburg).	**Cimp SSD vs HC**: 97.4. **Cimp SSD vs CN SSD**: 84.1. **CN SSD vs HC**: 79.1 [imidazole propionic acid, erythronic acid, homoserine, aspartic acid]. **PCA (all groups) and OPLS-DA**: **HC vs Cimp SSD**: R2Y = 0.863, Q2 = 0.644. **HC vs CN SSD**: R2Y = 0.818, Q2 = 0.438. **CN SSD vs CI SSD**: R2Y = 0.771, Q2 = −0.128.	Ala, asp and Glu metabolism ($p < 1e-4, I = 0.57$). D-Gln and D-Glu metabolism ($p = 4.67e-4, I = 0.33$). Citrate cycle (TCA cycle) ($p = 1.72e-4, I = 0.25$).

3.2 Metabolic Biomarkers for SSD Assessed Through ROC Analysis

AH/non-AH SSD vs HC (↓): BDNF, IL-2, NGF-β, TNF-α, cystine. **AH/non-AH SSD vs HC** (↑): Phe, glycerate, serine, dihydrosphingosine, Glu. **AH vs non-AH SSD** (↑): Phe. **AH vs non-AH SSD** (↓): Pyruvate. Note: $p < 0.01$ and VIP > 1.5 for each metabolite. Li X. et al., 2022 [48]	Serum. ELISA for proteins. UPLC-QTOF/MS for metabolites. Diagnostic and state biomarker between patients with (AH) or without (non-AH) auditory hallucination.	AH/non-AH SSD (56), disease duration >10 years. AH SSD (8/10) / 38.06 (5.91). Non-AH SSD (15/13) / 37.25 (7.52). HC (20/23) / 39.51 (7.68). Cohort nationality: China (Changsha, Hunan Province).	**Non-AH SSD vs HC:** 81.2 [cystine]. 88.0 [phenylalanine]. **AH SSD vs HC:** 88.8 [cystine]. 98.8 [phenylalanine]. **AH vs non-AH SSD:** 83.9 [pyruvate]. **PCA (all groups) and PLS-DA:** CI, sensitivity, specificity, goodness-of-fit (R2) and predictability (Q2) of the model were not informed.	Pathways enriched in AH/non-AH SSD: Phe, Tyr, and Trp biosynthesis, Phe metabolism, D-Gln and D-Glu metabolism. AH SSD patients showed IL-2 negatively correlated with Glu ($p = 0.03$) and cystine ($p = 0.002$). Myelin basic protein levels did not differ among the 3 groups.
SSD vs HC (↑): LPC, LPE, LPI, PIP. **SSD vs HC** (↓): AcCa, PE. **SSD vs MDD** (↑): Cer, LPC, LPE, LPI. Note: $p < 0.01$, FDR < 0.05, FC > 2 or < 0.5, and VIP > 1.5 for each metabolite. Wang F. et al., 2022 [35]	Plasma. UHPLC-quadrupole-Orbitrap/MS. Diagnostic and transdiagnostic between SSD and MDD.	SSD (12/19) / [22.0–45.0]. MDD (12/23) / [25.0–57.0]. HC (10/22) / [22.0–53.0]. Cohort nationality: China (Xian, Shaanxi Province)	**SSD vs HC:** 95.3 (89.7–100) [103 lipid species] **SSD vs MDD:** 92.0 (84.2–99.7) [111 lipid species] **PLS-DA (all groups) and OPLS-DA** (200 permutation tests)**:** **SSD vs HC:** R2Y = 0.458 and Q2 = −0.684 Note: Sensitivity and specificity were not reported.	**SSD vs HC:** Reduced levels of PE and AcCa exhibited a negative correlation with both positive and negative PANSS scores ($r \leq -0.3, p < 0.01$), while increased LPC, LPE, and PIP correlated positively ($r \geq 0.3, p < 0.01$). Enriched pathways observed in SSD patients included linoleic acid metabolism, glycerophospholipid metabolism, GPI-anchor biosynthesis, and α-linolenic acid metabolism.

(continued)

Table 3.1 (continued)

Cohort (effect in concentration): biomarker	Sample analytical platform biomarker class	Cohort (n = male/female) / age[a]	AUC (95% CI)/sensitivity/specificity (%) [predictor] learning algorithm	Pathway/symptom association/correlations
SSD vs HC (↑): Methylamine **SSD vs HC (↓):** Oleamide, dimethylamine, N-(1-deoxy-1-fructosyl) isoleucine, LPA (18:1(9Z)/0:0), phenylalanyl phenylalanine. Note: FDR < 0.05, FC > 2, and VIP > 2 for each metabolite. Wang T. et al., 2022 [49]	Plasma. HPLC-QTOF/MS, ^{1}H-NMR. GEO database for transcriptomic analysis. Diagnostic.	SSD (64) / 44.56 (9.53). HC (40) / 43.76 (13.87). Transcriptome: SSD (18), HC (12). Cohort nationality: China (Harpin, Heilongjiang Province). Note: Gender was not specified.	**SSD vs HC:** 100 [6 metabolites]. 100 [CCDC184, ERAS]. **OPLS-DA** (200 permutation tests): HPLC/QTOF-MS positive mode: R2X = 0.734, R2Y = 0.863, Q2 = 0.838. Negative mode: R2X = 0.522, R2Y = 0.881, Q2 = 0.903. ^{1}H-NMR: R2X = 0.618, R2Y = 0.804, Q2 = 0.690. Transcriptome: R2X = 0.799, R2Y = 0.953, Q2 = 0.732. Note: CI, sensitivity, and specificity were not reported. **LASSO regression analysis.**	Methylamine: Amine catabolism and neuronal potassium channels. Dimethylamine: Asymmetric dimethylarginine metabolism. LPA: Glycerophospholipid metabolism. N-(1-deoxy-1-fructosyl) Ile: Amino acid metabolism. Phenylalanyl phenylalanine: Catecholamines biosynthesis. Oleamide (structurally related to the endogenous cannabinoid anandamide): Cannabinoid type 1 receptor full agonist.

3.2 Metabolic Biomarkers for SSD Assessed Through ROC Analysis

SSD vs HC (↑): Lignoceric & quinolinic acid, GLA, ICA. **SSD vs HC (↓)**: Capric, caprylic, and nervonic acid, indole, tritridecanoin, NFK. **SSD vs MDD (↑)**: Erucic acid, ILA, ICA. **SSD vs MDD (↓)**: Nervonic, docosadienoic, palmitoleic, elaidic, behenic, stearic and eicosatrienoic acid. Note: $p < 0.05$, VIP > 1 for each metabolite. Liu JC et al., 2023 [50]	Plasma. UHPLC-QTRAP/MS Diagnostic and transdiagnostic between SSD and MDD.	SSD (7/15) / 22.85 (3.65). MDD (8/16) / 21.01 (2.51). HC (7/16) / 20.91 (2.78). Cohort nationality: China (Xian, Shaanxi Province).	**SSD vs HC**: 99.6 (96.4–100) [10 metabolites]. **SSD vs MDD**: 98.1 (87.2–100) [10 metabolites]. **Random Forest.** Note: Specificity and sensitivity were not informed.	Trp metabolism: Indole-3-carboxaldehyde (ICA), N-formyl-kynurenine (NFK), indole, indole-3-lactic acid (ILA), quinolinic acid. Fatty acid metabolism: Gamma-linolenic acid (GLA), capric acid and all the other acids mentioned in column 1. **SSD vs HC**: The level of omega-3 PUFA exhibited a negative correlation with negative and total PANSS scores ($r = -0.5$, $p < 0.05$). A decreased omega-3/omega-6 ratio was negatively correlated with positive and total PANSS scores ($r = -0.25$, $p < 0.05$).

(continued)

Table 3.1 (continued)

Cohort (effect in concentration): biomarker	Sample analytical platform biomarker class	Cohort (n = male/female) / age[a]	AUC (95% CI)/sensitivity/ specificity (%) [predictor] learning algorithm	Pathway/symptom association/ correlations
HSD-M vs LSD-M (↑): 3-methylhistamine, indole-3-lactate. **HSD-M vs LSD-M** (↓): Creatine, O-phosphocholine. **HOH-M vs LOH-M** (↑): Gluconate. **HOH-M vs LOH-M** (↓): 3-phenyllactate. **HSD-F vs LSD-F** (↓): 4-OH-proline, galactose, cysteine, 5-aminolevulinic acid, lactose, chlorogenate. **HSD-F vs LSD-F** (↑): 5-methoxytryptamine, 3-nitrotyrosine, tyramine. Note: $p < 0.05$ for each metabolite or pathway. Heynen et al., 2023 [37]	Urine. ^{1}H-NMR (700 MHz). Predictive, considering maternal stress caused by an environmental disaster (flood). Note: Males and females were analyzed separately due to significant sex differences in the urinary metabolome.	4-years old offspring (50/39). High maternal objective hardship/ male offspring HOH-M (26) vs low LOH-M (24). HOH/female offspring HOH-F (20) vs LOH-F (19). High maternal subjective distress/ male offspring HSD-M (25) vs LSD-M. HSD-F (23) vs LSD-F (16). Cohort nationality: Australia (Brisbane, Queensland).	**HSD-M vs LSD-M:** 96.2 (86.3–100) [4 metabolites] R2Y = 0.757, Q2 = 0.594. **HOH-M vs LOH-M:** 95.2 (82.8–100) [2 metabolites] R2Y = 0.837, Q2 = 0.635. **HSD-F vs LSD-F:** 99.5 (96.7–100) [9 metabolites] R2Y = 0.912, Q2 = 0.84. **HOH-F vs LOH-F:** 98.5 (93.4–100) [5 metabolites] R2Y = 0.935, Q2 = 0.791. **OPLS-DA:** Double ten-fold cross validation. 2000 permutation tests. Note: Sensitivity and specificity were not informed.	Enriched pathways in **HOH-M vs LOH-M:** Lys degradation (I = 0.14), taurine/ Hypotaurine metabolism (I = 0.43), pentose phosphate (I = 0.05). **HSD-F vs LSD-F:** Cys/met (I = 0.30), Gly/Ser/Thr (I = 0.22), Tyr (I = 0.04) and gal (I = 0.20) metabolism, CoA/ pantothenate (I = 0.01) and aminoacyl-tRNA biosynthesis (I = 0.17). **HOH-F vs LOH-F:** Aminoacyl-tRNA (I = 0.17) and Arg biosynthesis (I = 0.12), Gly/Ser/Thr (I = 0.56), starch/ sucrose (I = 0.13), glyoxylate/ dicarboxylate (I = 0.15), D-Gln/D-Glu (I = 0.5), gal (I = 0.15), ascorbate/aldarate (I = 0.5), GSH (I = 0.11); porphyrin/chlorophyll (I = 0.03) and Arg/pro metabolism (I = 0.22); pentose/ Glucuronate interconversions (I = 0.20).

3.2 Metabolic Biomarkers for SSD Assessed Through ROC Analysis

Treat-SSD vs HC (↓): m/z 60 (trimethylamine—TMA), 69 (isoprene), 74 (N-butyl amine), 88 (not identified), 90 (butyric acid). **Treat-MDD vs HC (↓):** m/z 88, isoprene, butyric acid. **Treat-SSD vs treat-MDD (↓):** TMA. Note: $p < 0.001$ and FDR < 0.05 for each metabolite. Henning et al., 2023 [38] Held et al., 2023 [51] Jiang et al., 2022 [51, 52]	Exhaled breath. PTR/MS. Diagnostic and transdiagnostic between SSD and MDD. Note: Data showed correlation with BMI (m/z 59 and 60) and with gender (m/z 69 and 70).	Treated-SSD (18/16) / 38.6 (10.1). Treated-MDD (14/22) / 39.11 (13.20). HC (17/17) / 32.9 (7.70). Cohort nationality: Germany (Magdeburg).	**Treat-SSD vs HC:** 83.6 [m/z 60, 69, 74, 88, 90]. **Treat-SSD vs HC:** 87.9 [m/z 60, 90]. **MDD vs HC:** 76.8 [5 metabolites] **SSD vs MDD:** 80.9 [5 metabolites] Note: CI, sensitivity, and specificity were not reported. **Multinomial logistic regression.** Bootstrap analysis with 1000 samples.	Trimethylamine (TMA) is synthesized by the gut microbiota from choline and transformed in the liver into TMAO, which serves as a neuronal protein stabilizer. Reduced levels of TMA have been associated with obesity and hyperglycemia. TMA is the primary agonist interacting with trace amine-associated receptors. Butyric acid, a product of the gut microbiota, influences histone acetylation, providing protection to neurons from cell death. Isoprene is involved in cholesterol synthesis, maintaining cell membrane integrity, and contributing to neuroprotection. The compound m/z 63, corresponding to dimethyl sulfide, was found to be associated with illness duration and antipsychotic therapy and compound m/z 74 was positively associated with PANSS total score.

(continued)

Table 3.1 (continued)

Cohort (effect in concentration): biomarker	Sample analytical platform / biomarker class	Cohort (n = male/female) / age[a]	AUC (95% CI)/sensitivity/ specificity (%) [predictor] learning algorithm	Pathway/symptom association/ correlations
bSSD vs HC (↓): α-dimorphecolic, sulfate, PC(16:0/18:1), L-Trp, 1- methylnicotinamide (1-MNA), PE(20:2/18:2). **Treat-SSD vs bSSD (↑):** PC(16:0/18:1(11Z)), sulfate, PE(20:2(11Z,14Z)/18:2(9Z,12Z)), C16 sphinganine. **Treat-SSD vs bSSD (↓):** γ-linolenic acid, linoleic acid. Note: $p < 0.05$, VIP > 1 and FC = 0.5–0.9 or 1.1–1.5 for each metabolite. Song et al., 2023 [36]	Serum. UPLC-QTOF/MS. Diagnostic and theranostic.	Baseline SSD (25/28) / 27.3 (6.0). Treated SSD (4–6 weeks) (19/18) / 27.4 (6.4). HC (13/19) / 27.1 (4.2). Cohort nationality: China (Xinxiang, Henan Province).	**bSSD vs HC:** 97.2 (93.4–100) [1- MNA, α-dimorphecolic, sulfate, L-Trp, PC(16:0/18:1), PE(20:2/18:2)]. **Treat-SSD vs bSSD:** 90.5 (81.3–96.7) [linoleic acid, C16 sphinganine, sulfate, gamma-linolenic acid, PC(16:0/18:1), PE(20:2/18:2)]. Note: Sensitivity and specificity were not informed. **PCA (all groups) and OPLS-DA** (200 permutation tests): R2X = 0.67, Q2 = 0.48. **bSSD vs HC:** R2Y = 0.91, Q2 = 0.75. **Treat-SSD vs bSSD:** R2Y = 0.62, Q2 = 0.34.	**bSSD vs HC:** L-Trp metabolism, PPAR signaling pathway, sulfur metabolism, bile secretion, lipid metabolism. **Treat-SSD vs bSSD:** Monotherapy with either olanzapine or risperidone had an impact on sulfur, lipid, and fatty acid metabolism. DHEA-sulfate, stearic acid, eicosatrienoic acid, and coenzyme Q9 demonstrated a correlation with the total PANSS score ($r > 0.4$, FDR < 0.05).

3.2 Metabolic Biomarkers for SSD Assessed Through ROC Analysis

BD vs HC or vs SSD (↑): SM(41:1;O2), SM(42:1;O2), vit D3 der, TG(53:7). **BD vs HC or vs SSD (↓):** Cer(42:2;O), Cer(44:2;O), cerebroside 3-sulfate. **SSD vs HC (↑):** LPE(18:0), LPS(21:0), LPC(18:1), LPC(16:1), FA(20:4;O4), Cer(44:2;O). **SSD vs HC (↓):** CE(16:1), SM(32:0;O2), SM(32:1;O2), SM(40:1;O2), SM(41:1;O2), SM(42:1;O2), SM(43:1;O2), PA(O-40:4), PC(46:7), PS(43:2), TG(53:7). Note: $p < 0.05$, VIP > 1, FC > ± 1 for each metabolite. Costa et al., 2023 [34]	Plasma. UHPLC-Orbitrap/MS. Diagnostic and transdiagnostic between SSD and BD.	Drug-naïve SSD (16/14) / 26.5 (6.8). Drug-naïve BD (11/19) / 26.6 (4.4). HC (15/15) / 26.5 (2.2). Cohort nationality: Brazil (São Paulo city).	**SSD vs HC:** 86.5 (72.5–98.8)/ 83.3/73.3 [44 lipid species]. **BD vs HC:** 73.5 (55.2–89.5)/76.7/66.7 [90 lipid species]. **BD vs SSD:** 80.4 (61.5–98.1)/70.0/73.3 [23 lipid species]. **PCA (all groups) and OPLS-DA:** **SSD vs HC:** R2Y = 0.299, Q2 = 0.169. **BD vs HC:** R2Y = 0.310, Q2 = 0.045. **BD vs SSD:** R2Y = 0.183, Q2 = −0.054.	Shared enriched pathways in SSD and BD vs HC: Androgen/estrogen biosynthesis and metabolism, bile acid biosynthesis, glycosphingolipid (GSP), linoleate, PIP and glycerophospholipid metabolism. **BD vs HC (exclusive):** Prostaglandin formation from arachidonate, vitamin D3 metabolism. **SSD vs HC (exclusive):** GSP biosynthesis—Ganglio series. **BD vs SSD:** Urea cycle and metabolism, lysine, arginine, proline, glutamate, aspartate, and asparagine metabolism.

(continued)

Table 3.1 (continued)

Cohort (effect in concentration): biomarker	Sample analytical platform biomarker class	Cohort (n = male/female) / age[a]	AUC (95% CI)/sensitivity/specificity (%) [predictor] learning algorithm	Pathway/symptom association/correlations
SSD vs HC (↑): Iron (FC = 1.29, t = 2.88, p = 0.005), iron/ferritin (t = 3.03, p = 0.003). **SSD vs HC (↓):** Ferritin (t = −2.41, p = 0.017) Copper and zinc (no significant difference). Lotan et al., 2023 [53]	*Postmortem* PFC (prefrontal cortex). ICP/MS for Fe, cu and Zn. Western blot for ferritin. Diagnostic. Note: Data were adjusted for all confounding covariables.	SSD (59/27) / 52.6 (16.2). HC (61/24) / 54.4 (14.9). Cohort nationality: Australia (Sydney, Melbourne), USA (Columbia/Virginia).	**SSD vs HC:** 83.7 (63.3–96.4)/71/89 [iron levels in individuals who died younger than 35], cutoff: 0.99 (nearest to method). 76.6 (63.9–86.6)/79/62 [ferritin in individuals with below-mean iron younger than 35], cutoff: 1.17 (Youden method). 67.4 (58.6–75.4)/66/68 [iron/ferritin], cutoff: 0.15 (Youden method). 71.0 (63.0–77.0)/69/72 [iron, ferritin, iron/ferritin], cutoff: 0.21 (Youden method). **Logistic regression.**	The disparity in iron levels between groups was most pronounced in young adults who succumbed to death before the age of 35, with elevated iron levels correlating with a higher likelihood of being diagnosed with SSD. Even among young adults with lower iron levels, a decreased ferritin level still posed a potential risk for an SSD diagnosis. The accumulation of iron is linked to the aging process, potentially contributing to the altered molecular structure and lipidome characteristics observed in SSD. There was no observed association between variants in the ferritin genes and SSD. The decline in ferritin levels was attributed to autophagy activated by lipid peroxidation products.

3.2 Metabolic Biomarkers for SSD Assessed Through ROC Analysis

SSD vs HC (↑): Sarcosine, CE (15:0), CE (16:1), cortisol, proline betaine, cholic acid. **SSD vs HC (↓):** 20 metabolites such as lysoPC, PC, spermidine, SM, choline, Cer, AcCa, GABA. Note: $p < 0.05$, VIP > 1 for all metabolites. Su et al. 2023 [54]	Serum. LC-QTRAP/MS for target metabolomics (630 metabolites). Diagnostic. Note: Data were adjusted for sex, age, and BMI.	SSD (17/43) / 38.08 (10.48). HC: (16/20) / 38.03 (9.66). Cohort nationality: China (Tianjin).	**SSD vs HC:** 73.6 [cortisol] 71.9 [Cer (d18:1/22:0)] 73.6 [AcCa (C2)] 77.3 [GABA] 86.7 [4 metabolites] **PCA and OPLS-DA:** 200 permutation tests. R2Y = 0.849, Q2 = 0.57 Note: CI, specificity, and sensitivity were not reported.	Cholic acid was positively correlated with the PANSS negative score ($r = 0.292$, $p = 0.029$). Positive correlations were observed between PCs levels and PANSS positive score ($r = 0.296$, $p = 0.027$).
Treated SSD vs HC: GABA (FC = 0.71), CE (20:3) (FC = 1.97), cholic acid (FC = 2.33), glycocholic acid (FC = 1.96). **Naïve SSD vs HC:** GABA (FC = 0.61), CE (20:3) (FC = 1.57), cholic acid (FC = 3.08), glycocholic acid (FC = 2.60). **Treated SSD vs naïve SSD:** GABA (FC = 1.17), CE (20:3) (FC = 1.26), cholic acid (FC = 0.76), glycocholic acid (FC = 0.75). Note: $p < 0.05$ for each metabolite. Wang X. et al. 2023 [55]	Serum. FIA-MS and LC-MS for target metabolomics (630 metabolites). Diagnostic. Note: Mass spectrometer and chromatographic conditions were not reported.	Naïve SSD (38 females) / 39.74 (9.83). HC (19 females) / 40.0 (9.1). 8 weeks treated SSD. Cohort nationality: China (Tianjin).	**Naïve SSD vs HC:** 89.1 [4 metabolites] **OPLS-DA:** 100 permutation tests ($p < 0.01$). R2Y = 0.799, Q2 = 0.557. Note: CI, specificity, and sensitivity were not reported.	Baseline CE (20:3) levels were correlated with PANSS total score ($r = 0.444$, $p = 0.005$), positive score ($r = 0.384$, $p = 0.017$), and general psychopathology score ($r = 0.390$, $p = 0.016$).

(continued)

Table 3.1 (continued)

Note: The distinction between plasma and serum lies in the use of an anticoagulant during blood collection. In instances where the original article does not specify the use of an anticoagulant, the sample is presumed to be serum, regardless of any contrary statements

AA amino acids, *AcCa* acylcarnitine, *AIC* Akaike's information criterion, *AP* antipsychotic, *ApoCA* apocholic acid, *Asn* asparagine, *AUC* area under the curve, *βCDCA* 3β-chenodeoxycholic acid, *BD* bipolar disorder, *BMI* body mass index, *CAARMS* Comprehensive Assessment of At Risk Mental States, *CE* cholesterol esters, *CE-FT/MS* Capillary Electrophoresis Fourier Transform Mass Spectrometry, *Cer* ceramides, *CHR* patients at clinical high risk for psychosis/prodromal period, *CI* confidence interval, *Cimp SSD* patients with cognitive impairment, *CN SSD* patients with normal cognition, *DA* dopamine, *DCA* bacterial 7α-dehydroxylated cholic acid, *βDCA* beta epimer of DCA, *3-DHCA* 3-dehydrocholic acid, *6-DMF* 6-deoxy-mannofuranose, *ELISA* Enzyme Linked Immunosorbent Assay, *FA* fatty acid, *FC* fold change, *FDR* false discovery rate, *FE* first-episode, *FIA-MS/MS* Flow Injection Analysis Tandem Mass Spectrometry *FT-ICR/MS* Fourier transform-ion cyclotron resonance mass spectrometry, *GABA* gamma aminobutyric acid, *GC-TOF/MS* gas chromatography coupled to time-of-flight mass spectrometer, *GHCA* glycohyocholic acid, *Glu* glutamate, *Gln* glutamine, *GSH* glutathione, *γ-GluCys* γ-glutamylcysteine, *HC* healthy control, *5-HIAA* 5-hydroxyindoleacetic acid, *HILIC* Hydrophilic Interaction Liquid Chromatography, ^{1}*H-NMR* or ^{1}*H-MRS* hydrogen nuclear magnetic resonance, *5-HT* serotonin, *I* pathway impact, *ICP-MS* inductively coupled plasma-mass spectrometry, *7-keto LCA* 7-ketolithocholic acid, *KYN* kynurenine, *LPC* lysophosphatidylcholine, *LPE* lysophosphatidylethanolamine, *LPI* lysophosphatidylinositol, *LPS* lysophosphatidylserine, *NE* norepinephrine/noradrenaline, *NGF* nerve growth factor, *NI* not informed, *NMDA* N-methyl-D-aspartate, *norCA* norcholic acid, *OPLS-DA* orthogonal partial least square discriminant analysis, *PANSS* Positive and Negative Syndrome Scale, *PBMC* peripheral blood mononuclear cells, *PC* phosphatidylcholine, *2-PCA* 2-piperidine carboxylic acid, *PE* phosphatidylethanolamine, *PGD2* prostaglandin D2, *PG* phosphatidylglycerol, *PGJ2* prostaglandin J2, *PIP* phosphatidylinositol phosphate, *PS* phosphatidylserine, *PTR-MS* proton transfer–reaction mass spectrometry, *PUFA* polyunsaturated fatty acid, *risp-treat* risperidone-treated, *ROC* receiver operating characteristic curve, *RSLC* rapid separation liquid chromatography, *SIPS* Structured Interview of Prodromal Syndromes, *SM* sphingomyelin, *TCA* tricarboxylic acid cycle, *TG* triacylglycerol, *THCA* taurohyocholic acid, *Trp* tryptophan, *UDCA* ursodeoxycholic acid, *UHPLC-QTOF* ultra-high performance liquid chromatography coupled to quadrupole-time of flight mass spectrometer, *UPLC-QTrap* UPLC coupled to Triple Quadrupole/Linear Ion Trap mass spectrometer, *UPLC-TQ/MS* UPLC coupled to triple quadrupole mass spectrometer, *VIP* variable importance in the projection

[a]Age is given as mean (standard deviation) or median [interquartile].
[b]Factor loading is defined as the correlation coefficient between the principal component score and each metabolite level [56].

subjecting the patient to a high-fat, moderate-protein, and very low-carbohydrate diet, offering an alternative source of energy to glucose through high production of β-hydroxybutyrate. This is aimed at counteracting oxidative stress, mitochondrial dysfunction, and inflammation observed in SSD [12, 13]. Disturbances in antioxidant defense were also identified by other research groups in peripheral blood mononuclear cells (PBMC) and serum [14–18].

Among the top predictors for baseline SSD was the reduced level of *myo*-inositol, an abundant metabolite in astroglia and a precursor of membrane inositol phospholipid synthesis. A meta-analysis, encompassing 19 small-scale studies on the medial frontal cortex, established a correlation between *myo*-inositol reduction and astroglial dysfunction occurring in some early-stage patients, potentially those affected by depression, which is known to trigger glutamatergic abnormalities [19]. In a subsequent long-term study, similar findings were observed in early-stage disease, and it was revealed that after 6 months of treatment, the patient's *myo*-inositol level returned to normal. This suggests that magnetic resonance spectroscopy (MRS) *myo*-inositol could serve as a monitoring marker for improvement in early-stage psychosis [20].

The pursuit of prognostic biomarkers capable of anticipating the disease even before the onset of psychosis has been one of the most sought-after goals in the twentieth century [21, 22]. Cui G. et al. analyzed the saliva of individuals at clinical high risk (CHR) and first-episode (FE) patients, identifying metabolic and inflammatory disturbances in the prodromal stage of SSD. They constructed diagnostic models using different combinations of the most significant metabolites and subsequently augmented these models by incorporating hydrogen sulfide (H_2S)-producing bacteria from the oral microbiota, including *Leptotrichia, Megasphaera, Actinomyces, Veillonella, Fusobacterium, Atopobium, Desulfobulbus, Granulicatella, Campylobacter, Selenomonas,* and *Prevotella*. While this addition improved the model's performance in distinguishing FE patients by 8%, it exhibited only a 2% improvement for CHR patients [23].

Dickens et al. developed a model based on lipidomic data from CHR individuals in a long-term (5-year) multiethnic cohort, aiming to differentiate patients in the transition to full-blown psychosis [24]. They identified abnormal levels of several plasma lipids before the onset of psychosis. However, their model performed less effectively than that of Cui and colleagues [23], which incorporated more polar compounds involved in a broader range of pathways. In a separate study, Lamichhane et al. attempted to distinguish between CHR and FE patients prone to weight gain after a year of treatment using a model based on triglycerides, yielding moderate results [25].

Aiming to differentiate SSD patients with cognitive impairment from those without cognitive impairment and healthy control, Jiang et al. selected four out of 46 serum metabolites with significant expression to build a diagnostic model, achieving AUCs of 97.4% and 84.1%, respectively [26]. Notably, one of the pivotal predictors contributing to the model's high performance was imidazole propionic acid, a metabolite formed from histidine via a urocanate intermediate in the gut microbiota. Urocanic acid, a metabolite from *Escherichia coli*, was among the 25 predictors used by Du et al. to differentiate SSD patients from healthy individuals [27]. Kim et al. also incorporated a serum metabolite originating from the gut microbiota, D-alanine, to create the most accurate model [28]. Various intestinal microorganisms identified in the feces of SSD patients, including *Clostridia*, *Blautia*, and *Eggerthella* (refer to Table 5.1), act on bile acids (BAs), deconjugating the amino acids glycine and taurine to form secondary BAs. These secondary BAs can be further degraded into compounds linked to various diseases [29]. Liu et al. discovered reduced levels of lithocholic acid in drug-naïve SSD patients, but the model used to differentiate these patients from drug-treated individuals did not include the molecule [30]. This omission may be attributed to its non-specificity to SSD, functioning instead as an indicator of other diseases, such as biliary cirrhosis and liver diseases.

Qing et al. profiled BAs in SSD patients by quantifying 40 compounds and constructed a discriminatory model based on eight related predictors, including hyocholic acid derivatives, nonconjugated and primary BAs [31]. The results were satisfactory and later corroborated by a Japanese publication [32], indicating BAs as theranostic biomarkers with therapeutic potential—a trend toward precision psychiatry, aiming for "the right drug for the right patient at the right time" [33].

Costa et al. employed OPLS-DA to construct a model featuring 44 lipids, primarily glycerophospholipids and sphingolipids, to differentiate SSD patients from healthy individuals [34]. Despite achieving an AUC greater than 85%, the model exhibited low values of R2 and Q2, indicating potential overfitting. A similar scenario was observed in the model developed by Wang F. et al., utilizing PLS-DA and more than 100 lipids [35]. In contrast, Song et al. built a model using OPLS-DA and six metabolites from different classes, yielding promising results [36]. On the other hand, Liu et al. achieved greater success in their model by employing binary logistic regression and incorporating only three lipids and aspartic acid [30].

The predominant focus in existing research (87.5% of studies) has centered around the exploration of biomarkers within blood samples. However, a study conducted by a Canadian research group delved into understanding environmental impacts on fetal development. This investigation examined the urinary metabolome of 4-year-old offspring whose mothers had experienced stress due to a flood during pregnancy [37]. In a distinct approach, another research team analyzed exhaled breath samples to ascertain the transdiagnostic potential of compounds in distinguishing between individuals with SSD and MDD [38]. Additionally, a study involving PBMC sought to differentiate SSD patients from healthy individuals [15]. While satisfactory Area Under the Curve (AUC) values were obtained, it is noteworthy that essential metrics for a comprehensive model evaluation, such as sensitivity and specificity, were not reported.

3.2.1 Discussion

When planning this review, a preliminary search on the PubMed database using the keywords "metabolite schizophrenia" yielded nearly 2000 articles spanning 63 years (1960–2023). Subsequently, filtering these results with "metabolite biomarker

3.2 Metabolic Biomarkers for SSD Assessed Through ROC Analysis

schizophrenia" reduced the count to 10% of the initial articles, covering a period of 50 years (1973–2023). A more specific filter for this review, "metabolite biomarker AUC schizophrenia," narrowed down the search to 11 articles over 12 years (2011–2023). This specific filter aimed to identify studies evaluating the potential of biomarkers in constructing diagnostic models using the ROC curve, a gold standard tool for assessing binary medical models. Additionally, Google Scholar, Medline, and reference searches were conducted until a total of 32 articles were included in this review, spanning the period from 2011 to 2023. It is noteworthy that 62% of these studies originated in China, with 9% from Germany and Japan each, and 3% from Russia, Finland, Brazil, and Australia each. Only two studies involved multinational cohorts [24, 25], and just one considered different ethnicities [27].

Remarkably, 88% of the reviewed articles utilized up to 10 biomarkers, aligning with the recommendation of Xia et al. [57]. Only four articles ventured beyond this norm, employing a range of 20 to 110 biomarkers to construct diagnostic models. This observation underscores the prevailing trend in biomarker studies for SSD.

During the compilation of Table 3.1, it became evident that there is a lack of standardization in the nomenclature of certain terms, contributing to potential confusion in their usage. For instance, the term "classifier," denoting an algorithm categorizing data into classes, is sometimes used interchangeably with "predictor," representing a variable used to predict the target variable [58]. Additionally, there is often confusion in the steps of model development, with validation and testing erroneously treated as synonymous.

Further, a lack of standardization is also noted in the description of algorithms, parameters, and result presentations. Notably, 34% of studies did not report values for goodness of fit (R2) and goodness of prediction (Q2) statistics. Only a few articles mentioned p-values from permutation tests, indicating the significance of the obtained statistics. Comprehending model fit is crucial for identifying the underlying reasons for suboptimal model accuracy. This understanding serves as a guide for implementing corrective measures.

The reporting of sample preparation methodologies and equipment used is often oversimplified, hindering experiment reproducibility. The description of sample types is incomplete, often making assumptions about plasma use without specifying the anticoagulant, leaving room for potential serum use. Moreover, participant ages are inconsistently expressed as means with standard deviations or medians with interquartile ranges. Concentrations of metabolites are presented in micromolar units, fold change values, or simply described as increased or reduced. Fold changes are indicated as positive and negative or as greater/lesser than one, signifying upregulation or downregulation, respectively. In some instances, the significance of these changes is not reported.

The preference for certain learning algorithms is evident in the graph depicted in Fig. 3.1, emphasizing the concurrent use of principal component analysis (PCA) and orthogonal partial least square discriminant analysis (OPLS-DA).

The inconsistent naming of many metabolites in relation to the databases posed challenges in identifying them during enrichment (overrepresentation) analysis. This analysis aimed to explore whether the biomarkers deemed significant in

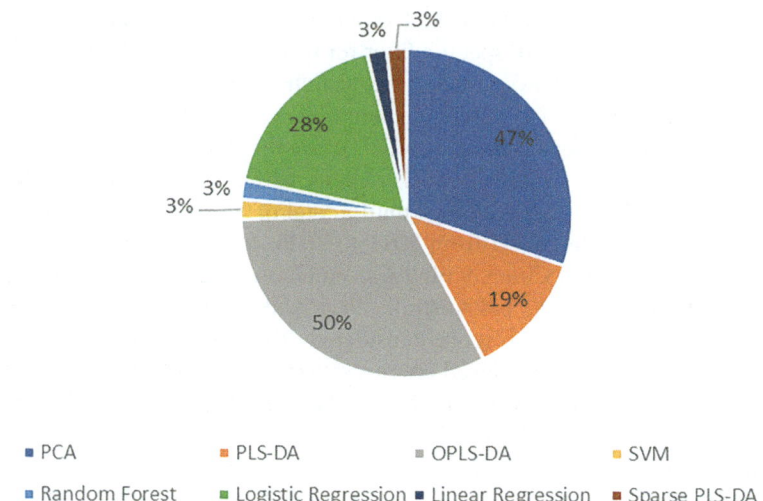

Fig. 3.1 Preferential use of learning algorithms observed in the analyzes of 32 articles

constructing diverse diagnostic models could provide insights into the pathways most affected by SSD. To accomplish this, we utilized 123 compounds identified in studies comparing SSD patients to healthy individuals (SSD vs HC; Table 3.1). However, 39 compounds were excluded from the analysis due to discrepancies in their names compared to the registered entries in the databases. Despite observing pathway enrichment in accordance with the findings from the considered studies, none retained statistical significance after adjusting for p-values (Fig. 3.2; FDR < 0.05).

During the enrichment analysis, when highlighting the disease signature, SSD ranked twelfth ($p = 0.01$, FDR = 0.20). The signature (metabolite-specific assay) provided by MetaboAnalyst 5.0 was characterized by only five amino acids and taurine (Fig. 3.3). The variability in this result may be attributed to the heterogeneity within the cohort, encompassing SSD patients at different disease stages, including FE patients, recurrent cases, those under various treatment regimens (both treated and untreated with drugs), and individuals of diverse nationalities and ethnicities. However, it may also stem from the potential inaccuracy of disease-based libraries, as highlighted in a study conducted by Marco-Ramell et al. [59]. Indeed, with the release of MetaboAnalyst version 6.0 in December 2023, the list of metabolites for the disease signature enrichment saw a significant expansion from 32 (referenced from PubMed) to 102 compounds (referenced from HMDB). Upon reanalysis of the same set of 84 metabolites, SSD ascended to the second position (p-value = 1.08E-32, FDR = 1.51E-30), with 19 of the input metabolites being included in the updated list.

3.2 Metabolic Biomarkers for SSD Assessed Through ROC Analysis

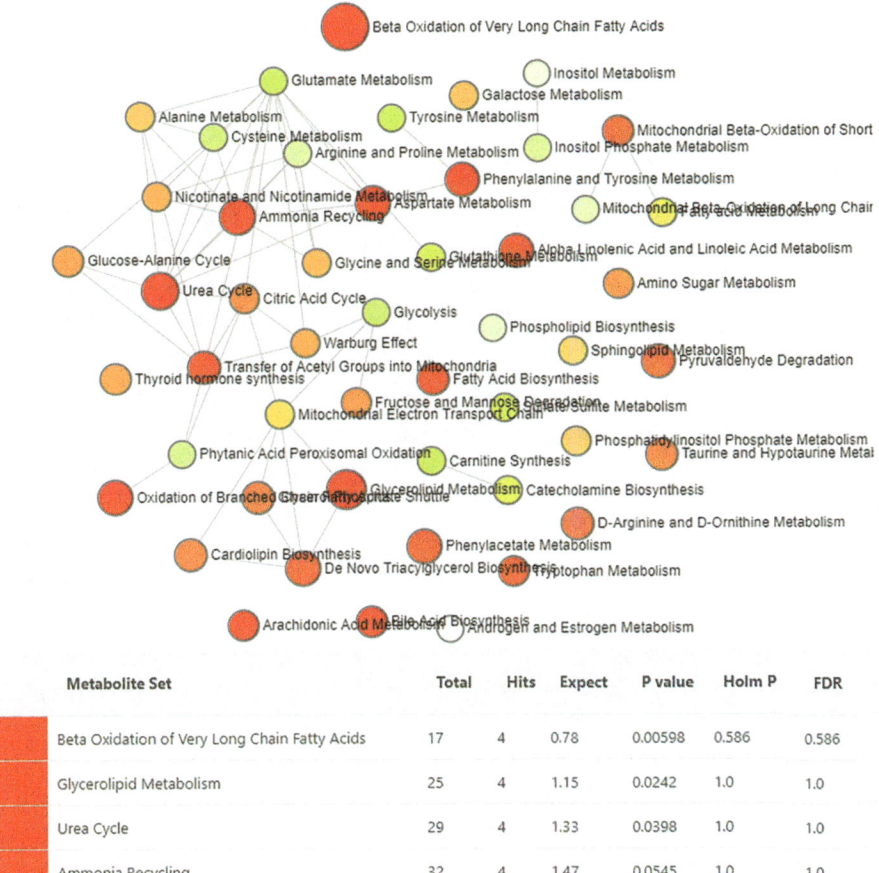

Metabolite Set	Total	Hits	Expect	P value	Holm P	FDR
Beta Oxidation of Very Long Chain Fatty Acids	17	4	0.78	0.00598	0.586	0.586
Glycerolipid Metabolism	25	4	1.15	0.0242	1.0	1.0
Urea Cycle	29	4	1.33	0.0398	1.0	1.0
Ammonia Recycling	32	4	1.47	0.0545	1.0	1.0

Fig. 3.2 Network view of overrepresentation analysis using MetaboAnalyst 5.0, the small molecule pathway database and 84 serum metabolites extracted from cohorts of SSD patients vs healthy controls, as detailed in Table 3.1

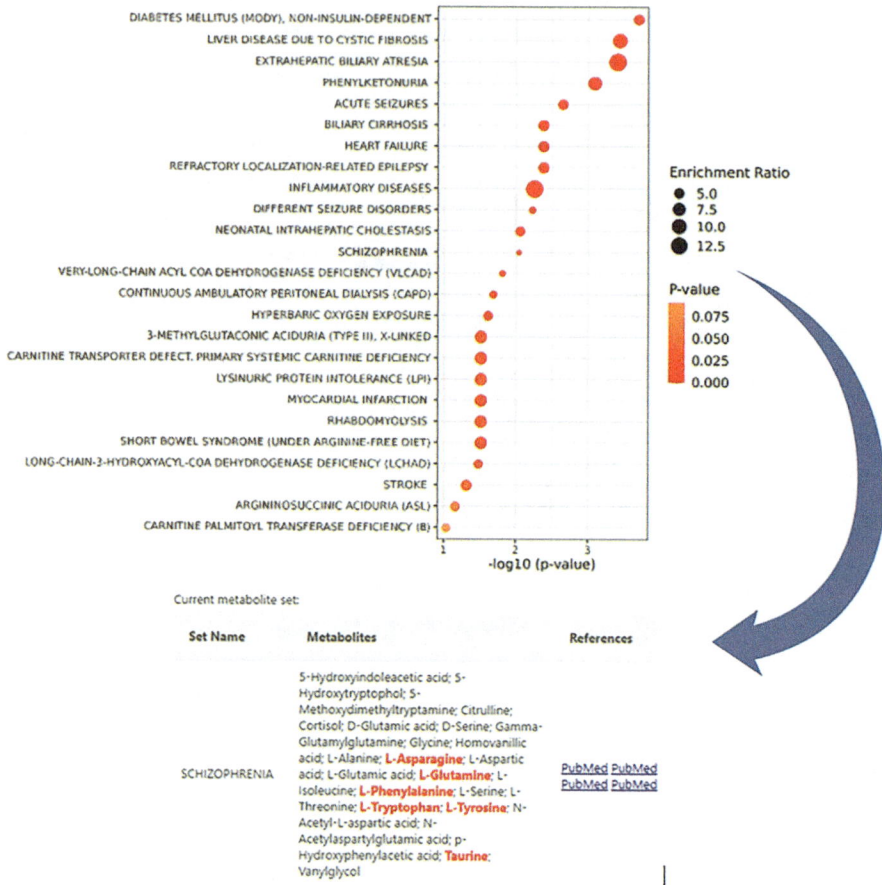

Fig. 3.3 Overview of enriched disease signatures (Top 25) using MetaboAnalyst 5.0 and 84 serum metabolites extracted from cohorts of SSD patients vs healthy controls, as detailed in Table 3.1. Enrichment is highlighted in red

References

1. Jutla A, Foss-Feig J, Veenstra-VanderWeele J (2022) Autism spectrum disorder and schizophrenia: an updated conceptual review. Autism Res 15:384–412. https://doi.org/10.1002/aur.2659
2. Martínez-Cao C, de la Fuente-Tomás L, García-Fernández A, González-Blanco L, Sáiz PA, Garcia-Portilla MP, Bobes J (2022) Is it possible to stage schizophrenia? A systematic review. Transl Psychiatry 12:197. https://doi.org/10.1038/s41398-022-01889-y
3. Lin P, Sun J, Lou X et al (2022) Consensus on potential biomarkers developed for use in clinical tests for schizophrenia. Gen Psychiatry 35:e100685. https://doi.org/10.1136/gpsych-2021-100685
4. Perkovic MN, Erjavec GN, Strac DS, Uzun S, Kozumplik O, Pivac N (2017) Theranostic biomarkers for schizophrenia. Int J Mol Sci 18:733. https://doi.org/10.3390/ijms18040733

5. Wingo TS, Liu Y, Gerasimov ES et al (2022) Shared mechanisms across the major psychiatric and neurodegenerative diseases. Nat Commun 13:4314. https://doi.org/10.1038/s41467-022-31873-5
6. Fernandes BS, Dai Y, Jia P, Zhao Z (2022) Charting the proteome landscape in major psychiatric disorders: from biomarkers to biological pathways towards drug discovery. Eur Neuropsychopharmacol 61:43–59. https://doi.org/10.1016/j.euroneuro.2022.06.001
7. Davison J, O'Gorman A, Brennan L, Cotter DR (2018) A systematic review of metabolite biomarkers of schizophrenia. Schizophr Res 195:32–50. https://doi.org/10.1016/j.schres.2017.09.021
8. Pillinger T, McCutcheon RA, Vano L et al (2020) Comparative effects of 18 antipsychotics on metabolic function in patients with schizophrenia, predictors of metabolic dysregulation, and association with psychopathology: a systematic review and network meta-analysis. Lancet Psychiatry 7:64–77. https://doi.org/10.1016/S2215-0366(19)30416-X
9. Quintero M, Stanisic D, Cruz G, Pontes JGM, Costa TBBC, Tasic L (2019) Metabolomic biomarkers in mental disorders: bipolar disorder and schizophrenia. Adv Exp Med Biol 1118:271–293. https://doi.org/10.1007/978-3-030-05542-4_14
10. Xuan J, Pan G, Qiu Y et al (2011) Metabolomic profiling to identify potential serum biomarkers for schizophrenia and risperidone action. J Proteome Res 10:5433–5443. https://doi.org/10.1021/pr2006796
11. Yang J, Chen T, Sun L et al (2013) Potential metabolite markers of schizophrenia. Mol Psychiatry 18:67–78. https://doi.org/10.1038/mp.2011.131
12. Morris G, Puri BK, Carvalho A, Maes M, Berk M, Ruusunen A, Olive L (2020) Induced ketosis as a treatment for neuroprogressive disorders: food for thought? Int J Neuropsychopharmacol 23:366–384. https://doi.org/10.1093/ijnp/pyaa008
13. Sarnyai Z, Palmer CM (2020) Ketogenic therapy in serious mental illness: emerging evidence. Int J Neuropsychopharmacol 23:434–439. https://doi.org/10.1093/ijnp/pyaa036
14. Cui G, Qing Y, Hu X et al (2020) Serum metabolomic profiling based onFourier transform-ion cyclotron resonance-mass spectrometry: do the dysfunctions of metabolic pathways reveal a universal risk of oxidative stress in schizophrenia? Antioxid Redox Signal 33:679–688. https://doi.org/10.1089/ars.2020.8141
15. Liu M-L, Zheng P, Liu Z, Xu Y, Mu J, Guo J, Huang T, Meng H-Q, Xie P (2014) GC-MS based metabolomics identification of possible novel biomarkers for schizophrenia in peripheral blood mononuclear cells. Mol BioSyst 10:2398–2406. https://doi.org/10.1039/c4mb00157e
16. Fukushima T, Iizuka H, Yokota A et al (2014) Quantitative analyses of schizophrenia-associated metabolites in serum: serum D-lactate levels are negatively correlated with gamma-glutamylcysteine in medicated schizophrenia patients. PLoS One 9:e101652. https://doi.org/10.1371/journal.pone.0101652
17. Cao B, Wang D, Pan Z, McIntyre RS, Brietzke E, Subramanieapillai M, Nozari Y, Wang J (2020) Metabolic profiling for water-soluble metabolites in patients with schizophrenia and healthy controls in a Chinese population: a case-control study. World J Biol Psychiatry 21:357–367. https://doi.org/10.1080/15622975.2019.1615639
18. Cao B, Wang D, Pan Z et al (2019) Characterizing acyl-carnitine biosignatures for schizophrenia: a longitudinal pre- and post-treatment study. Transl Psychiatry 9:19. https://doi.org/10.1038/s41398-018-0353-x
19. Das TK, Dey A, Sabesan P, Javadzadeh A, Théberge J, Radua J, Palaniyappan L (2018) Putative astroglial dysfunction in schizophrenia: a meta-analysis of H-MRS studies of medial prefrontal myo-inositol. Front Psych 9:438. https://doi.org/10.3389/fpsyt.2018.00438
20. Jeon P, Mackinley M, Théberge J, Palaniyappan L (2021) The trajectory of putative astroglial dysfunction in first episode schizophrenia: a longitudinal 7-Tesla MRS study. Sci Rep 11:22333. https://doi.org/10.1038/s41598-021-01773-7
21. Ferrarelli F, Mathalon D (2020) The prodromal phase: time to broaden the scope beyond transition to psychosis? Schizophr Res 216:5–6. https://doi.org/10.1016/j.schres.2019.12.035

22. Lai C-Y, Scarr E, Udawela M, Everall I, Chen WJ, Dean B (2016) Biomarkers in schizophrenia: a focus on blood-based diagnostics and theranostics. World J Psychiatry 6:102–117. https://doi.org/10.5498/wjp.v6.i1.102
23. Cui G, Qing Y, Li M, Sun L, Zhang J, Feng L, Li J, Chen T, Wang J, Wan C (2021) Salivary metabolomics reveals that metabolic alterations precede the onset of schizophrenia. J Proteome Res 20:5010–5023. https://doi.org/10.1021/acs.jproteome.1c00504
24. Dickens AM, Sen P, Kempton MJ et al (2021) Dysregulated lipid metabolism precedes onset of psychosis. Biol Psychiatry 89:288–297. https://doi.org/10.1016/j.biopsych.2020.07.012
25. Lamichhane S, Dickens AM, Sen P, Laurikainen H, Borgan F, Suvisaari J, Hyötyläinen T, Howes O, Hietala J, Orešič M (2020) Association between circulating lipids and future weight gain in individuals with an at-risk mental state and in first-episode psychosis. Schizophr Bull 47:160–169. https://doi.org/10.1093/schbul/sbaa087
26. Jiang Y, Sun X, Hu M, Zhang L, Zhao N, Shen Y, Yu S, Huang J, Li H, Yu W (2022) Plasma metabolomics of schizophrenia with cognitive impairment: a pilot study. Front Psych 13:950602. https://doi.org/10.3389/fpsyt.2022.950602
27. Du Y, Chen L, Li X-S et al (2021) Metabolomic identification of exosome-derived biomarkers for schizophrenia: a large multicenter study. Schizophr Bull 47:615–623. https://doi.org/10.1093/schbul/sbaa166
28. Kim S, Okazaki S, Otsuka I et al (2022) Searching for biomarkers in schizophrenia and psychosis: case-control study using capillary electrophoresis and liquid chromatography time-of-flight mass spectrometry and systematic review for biofluid metabolites. Neuropsychopharmacol Rep 42:42–51. https://doi.org/10.1002/npr2.12223
29. Guzior DV, Quinn RA (2021) Review: microbial transformations of human bile acids. Microbiome 9:140. https://doi.org/10.1186/s40168-021-01101-1
30. Liu Y, Song X, Liu X et al (2021) Alteration of lipids and amino acids in plasma distinguish schizophrenia patients from controls: a targeted metabolomics study. Psychiatry Clin Neurosci 75:138–144. https://doi.org/10.1111/pcn.13194
31. Qing Y, Wang P, Cui G, Zhang J, Liang K, Xia Z, Wang P, He L, Jia W (2022) Targeted metabolomics reveals aberrant profiles of serum bile acids in patients with schizophrenia. Schizophrenia (Heidelb) 8:65. https://doi.org/10.1038/s41537-022-00273-5
32. Koike S, Miyaji Y, Suzuki K, Miyashita M, Itokawa M, Arai M, Ogasawara Y (2022) Plasma unconjugated bile acids as novel biomarker for schizophrenia. Biochem Biophys Res Commun 634:70–74. https://doi.org/10.1016/j.bbrc.2022.09.110
33. Fernandes BS, Quevedo J, Zhao Z (2022) Fostering precision psychiatry through bioinformatics. Braz J Psychiatry 44:119–120. https://doi.org/10.1590/1516-4446-2021-2083
34. Costa AC, Riça LB, van de Bilt M, Zandonadi FS, Gattaz WF, Talib LL, Sussulini A (2023) Application of lipidomics in psychiatry: plasma-based potential biomarkers in schizophrenia and bipolar disorder. Meta 13:600. https://doi.org/10.3390/metabo13050600
35. Wang F, Guo L, Zhang T et al (2022) Alterations in plasma lipidomic profiles in adult patients with schizophrenia and major depressive disorder. Medicina 58:1509. https://doi.org/10.3390/medicina58111509
36. Song M, Liu Y, Zhou J et al (2023) Potential plasma biomarker panels identification for the diagnosis of first-episode schizophrenia and monitoring antipsychotic monotherapy with the use of metabolomics analyses. Psychiatry Res 321:115070. https://doi.org/10.1016/j.psychres.2023.115070
37. Magistretti PJ, Allaman I (2018) Lactate in the brain: from metabolic end-product to signalling molecule. Nat Rev Neurosci 19:235–249. https://doi.org/10.1038/nrn.2018.19
38. Al Awam K, Haußleiter IS, Dudley E, Donev R, Brüne M, Juckel G, Thome J (2015) Multiplatform metabolome and proteome profiling identifies serum metabolite and protein signatures as prospective biomarkers for schizophrenia. J Neural Transm 122(Suppl 1):S111–S122. https://doi.org/10.1007/s00702-014-1224-0
39. Wang D, Cheng SL, Fei Q, Gu H, Raftery D, Cao B, Sun X, Yan J, Zhang C, Wang J (2019) Metabolic profiling identifies phospholipids as potential serum biomarkers for schizophrenia. Psychiatry Res 272:18–29. https://doi.org/10.1016/j.psychres.2018.12.008

40. Horrobin DF (1998) The membrane phospholipid hypothesis as a biochemical basis for the neurodevelopmental concept of schizophrenia. Schizophr Res 30:193–208. https://doi.org/10.1007/978-3-642-47076-9_19
41. Liu L, Zhao J, Chen Y, Feng R (2020) Metabolomics strategy assisted by transcriptomics analysis to identify biomarkers associated with schizophrenia. Anal Chim Acta 1140:18–29. https://doi.org/10.1016/j.aca.2020.09.054
42. Chen X, Xu J, Tang J, Dai X, Huang H, Cao R, Hu J (2020) Dysregulation of amino acids and lipids metabolism in schizophrenia with violence. BMC Psychiatry 20:97. https://doi.org/10.1186/s12888-020-02499-y
43. Tkachev AI, Stekolshchikova EA, Morozova AY et al (2021) Ceramides: shared lipid biomarkers of cardiovascular disease and schizophrenia. Consortium Psychiatricum 2:35–43. https://doi.org/10.17816/CP101
44. Okamoto N, Ikenouchi A, Watanabe K, Igata R, Fujii R, Yoshimura R (2021) A metabolomics study of serum in hospitalized patients with chronic schizophrenia. Front Psych 12:763547. https://doi.org/10.3389/fpsyt.2021.763547
45. Prokop JW, Bupp CP, Frisch A et al (2021) Emerging role of in neurodevelopmental disorders and brain development. Genes 12:470. https://doi.org/10.3390/genes12040470
46. Li X, Yang C, Liang X, Li D, Zhou Z, Xiao H, Liu X, Li J, Yang D, Li M (2022) Metabolomics and cytokine analysis for identification of schizophrenia with auditory hallucination. Clin Invest Med 45:E39–E48. https://doi.org/10.25011/cim.v45i2.38096
47. Wang T, Li P, Meng X et al (2022) An integrated pathological research for precise diagnosis of schizophrenia combining LC-MS/H NMR metabolomics and transcriptomics. Clin Chim Acta 524:84–95. https://doi.org/10.1016/j.cca.2021.11.028
48. Liu J-C, Yu H, Li R, Zhou C-H, Shi Q-Q, Guo L, He H (2023) A preliminary comparison of plasma tryptophan metabolites and medium- and long-chain fatty acids in adult patients with major depressive disorder and schizophrenia. Medicina 59:413. https://doi.org/10.3390/medicina59020413
49. Heynen JP, McHugh RR, Boora NS, Simcock G, Kildea S, Austin M-P, Laplante DP, King S, Montina T, Metz GAS (2023) Urinary H NMR Metabolomic analysis of prenatal maternal stress due to a natural disaster reveals metabolic risk factors for non-communicable diseases: the QF2011 Queensland flood study. Meta 13. https://doi.org/10.3390/metabo13040579
50. Henning D, Lüno M, Jiang C, Meyer-Lotz G, Hoeschen C, Frodl T (2023) Gut-brain axis volatile organic compounds derived from breath distinguish between schizophrenia and major depressive disorder. J Psychiatry Neurosci 48:E117–E125. https://doi.org/10.1503/jpn.220139
51. Held A, Henning D, Jiang C, Hoeschen C, Frodl T (2023) Dynamic stability of volatile organic compounds in respiratory air in schizophrenic patients and its potential predicting efficacy of TAAR agonists. Molecules 28:4385. https://doi.org/10.3390/molecules28114385
52. Jiang C, Dobrowolny H, Gescher DM, Meyer-Lotz G, Steiner J, Hoeschen C, Frodl T (2022) Volatile organic compounds from exhaled breath in schizophrenia. World J Biol Psychiatry 23:773–784. https://doi.org/10.1080/15622975.2022.2040052
53. Lotan A, Luza S, Opazo CM et al (2023) Perturbed iron biology in the prefrontal cortex of people with schizophrenia. Mol Psychiatry 28:2058–2070. https://doi.org/10.1038/s41380-023-01979-3
54. Su Q, Bi F, Yang S, Yan H, Sun X, Wang J, Qiu Y, Li M, Li S, Li J (2023) Identification of plasma biomarkers in drug-naïve schizophrenia using targeted metabolomics. Psychiatry Investig 20:818–825. https://doi.org/10.30773/pi.2023.0121
55. Wang X, Xie J, Ma H et al (2023) The relationship between alterations in plasma metabolites and treatment responses in antipsychotic-naïve female patients with schizophrenia. World J Biol Psychiatry 1–10:106. https://doi.org/10.1080/15622975.2023.2271965
56. Yamamoto H, Fujimori T, Sato H, Ishikawa G, Kami K, Ohashi Y (2014) Statistical hypothesis testing of factor loading in principal component analysis and its application to metabolite set enrichment analysis. BMC Bioinformatics 15:51. https://doi.org/10.1186/1471-2105-15-51

57. Xia J, Broadhurst DI, Wilson M, Wishart DS (2013) Translational biomarker discovery in clinical metabolomics: an introductory tutorial. Metabolomics 9:280–299. https://doi.org/10.1007/s11306-012-0482-9
58. Burkov A (2019) The hundred-page machine learning book. ISBN 1999579518, 9781999579517. Published by Andriy Burkov
59. Marco-Ramell A, Palau-Rodriguez M, Alay A, Tulipani S, Urpi-Sarda M, Sanchez-Pla A, Andres-Lacueva C (2018) Evaluation and comparison of bioinformatic tools for the enrichment analysis of metabolomics data. BMC Bioinformatics 19:1. https://doi.org/10.1186/s12859-017-2006-0

Chapter 4
Immune, Inflammatory and Oxidative Stress-Related Biomarkers

4.1 Introduction

As outlined in Chap. 1, dysregulations in inflammatory processes, mitochondrial functions, energy expenditure, and oxidative stress have been implicated in the pathophysiology of SSD. A noteworthy proportion of SSD patients, ranging from 35% to 50%, manifest indications of either low-grade inflammation [1] or chronic inflammation [2–4]. This inflammatory profile is believed to play a significant role in inducing alterations in neurotransmission, neuronal signaling, synapse organization, and brain connectivity. These changes have the potential to influence cognitive and behavioral functions, as reviewed by [5].

In an effort to ascertain whether inflammatory markers could serve as discriminators between individuals at clinical high risk (CHR) for psychosis and healthy people, or differentiate between psychosis converters and nonconverters, Khoury and Nasrallah undertook a comprehensive analysis of 15 papers [6]. They observed a dearth of longitudinal studies assessing inflammatory marker measures linked to the transition to psychosis, leading them to propose a potential role for baseline plasma levels of interleukins IL-1β, IL-7, IL-8, matrix metalloproteinase (MMP)-8, cortisol, and albumin as predictors of psychotic transition. Notably, they discounted C-reactive protein (CRP) and IL-6 as state markers, drawing attention to confounding factors influencing IL-6 levels, such as alcohol and marijuana use, as well as the correlation between IL-10 and IL-6 receptors with antidepressant prescription. They also emphasized the limitations of underpowered studies with small sample sizes and advocated for large-scale studies to replicate these findings.

Feng et al. conducted an analysis of 10 long-term risperidone-based treatment studies and found evidence supporting the use of IL-6 blood levels as a marker of successful treatment with risperidone. However, they noted the presence of moderate bias, significant heterogeneity ($\chi^2 = 116.50$, $P < 0.001$), and large variability in effect size estimates across studies ($I^2 = 91.4\%$). Through meta-analysis in subgroups, they identified the influence of ethnicities, with Europeans exhibiting reduced levels of IL-6, while Asians did not, as well as dependence on SSD states and follow-up time [7]. Pinto et al. detected robust but similar patterns of variation in the levels of the five most frequently studied molecules—BDNF, TNF-α, IL-6, CRP, and cortisol—across the diagnoses of SSD, major depressive disorder (MDD), and bipolar disorder (BD). Their findings confirmed the promiscuity between these peripheral biomarkers, indicating an overlap in the pathophysiology of major psychiatric disorders [8].

Following a meticulous systematic review of 177 articles, De Simone and colleagues cautiously concluded that antipsychotic drugs play a role as antioxidants in the treatment of SSD patients. They proposed that this antioxidative function may contribute to the clinical efficacy of these drugs by influencing synaptic plasticity and functional connectivity processes [5]. Through their investigation, they identified several modified molecules by antipsychotic treatment, including various mitochondrial proteins associated with cell viability, energy metabolism, and the regulation of oxidative systems. Despite these insights, the researchers emphasized the need for additional studies to confirm the precise involvement of oxidants and mitochondria in the pathophysiology of SSD. Furthermore, they highlighted the importance of exploring the potential therapeutic benefits of antioxidants, either alone or in combination with antipsychotics, for the treatment of the disorder.

In a systematic review and meta-analysis, Fraguas and colleagues investigated inflammatory and oxidative markers in patients with early-onset (EO) psychosis compared to healthy controls. Their findings revealed no significant differences in the levels of catalase, glutathione, glutathione peroxidase, superoxide dismutase, total antioxidant status, and cell/DNA oxidative damage between the compared cohorts. Despite this, a qualitative analysis of individual studies suggested an association of inflammatory and oxidative markers with clinical, cognitive, and neurobiological outcomes, particularly in longitudinal assessments. The researchers acknowledged a high degree of heterogeneity among the studies [9]. Subsequently, the same research group conducted a systematic review encompassing a larger number of studies, comparing EO psychosis, first-episode (FE) psychosis, FE SSD, and EO SSD patients. Their meta-analysis revealed that total antioxidant status and docosahexaenoic acid levels were significantly lower in FE psychosis compared to controls. Conversely, levels of homocysteine, IL-6, and tumor necrosis factor alpha (TNF-α) were significantly higher, indicating a reduced antioxidant status and a pro-inflammatory imbalance in FEP patients [10].

4.2 Immune, Inflammatory, and Oxidative Stress-Related Biomarkers for SSD Assessed Through ROC Analysis

Given the complex biochemical interplay between inflammatory and oxidative stress processes and psychotic symptoms, numerous studies have delved into the exploration of these biomarkers to differentiate SSD from healthy individuals and other psychoses. In Table 4.1, we have compiled studies conducted over the past twelve years (2011–2023) that specifically aim to identify these biomarkers, assessing their diagnostic value using the receiver operating characteristic (ROC) curve analysis.

Protein kinase A catalytic subunit α (PKACα) and protein kinase G II (PKG II) have been linked to various inflammatory disorders [11–13], although they exhibited only moderate efficacy in distinguishing SSD patients from healthy individuals [14].

Shafiee-Kandjani et al. devised multiple models to differentiate adult-onset (AO) SSD patients based on interleukins and microRNAs, with IL-6 and miRNA-26a yielding the most robust models (AUC 95.8% and 84.3%, respectively) [15]. In contrast, other tested microRNAs (miRNA-106a and -125b) exhibited limited value as biomarkers (AUCs <55%) [16]. Another study identified IL-6 mRNA with moderate predictive ability (AUC 76%) in SSD patients versus healthy individuals [17]. Despite this, elevated IL-6 mRNA levels were correlated with an increased presentation of positive symptoms, as measured by the PANSS (Spearman's $\rho = 0.35$, p-value $= 0.014$). Furthermore, the combination of endotoxin, high mobility group box protein (HMGB1), and Complement 4 demonstrated the ability to differentiate FE SSD patients from controls [18].

IL-4 and IL-10, acting as anti-inflammatory interleukins, play a crucial role in brain repair by modulating microglia, and reduced levels are frequently associated with immune dysregulation [19, 20]. In a study by Liu et al., significant differences in the plasma levels of cytokines IL-4, IL-6, IL-10, IL-12, and TNF-α were observed in two groups of chronic SSD patients who had the onset of symptoms at different ages. Through a subsequent partial least square discriminant analysis (PLS-DA), only IL-4 and IL-6 emerged as the cytokines with variable importance in projection (VIP) values exceeding 1. Despite this, the researchers constructed various models, incorporating either a single cytokine or panels of them. The most favorable outcomes were achieved when considering IL-4 either alone or in combination with IL-6 and IL-12. This underscores the impact of considering the biological effects alongside statistical results to determine the most effective predictor [19, 20].

While S100B is associated with astrocyte activation and brain dysfunction, playing a crucial role in microglial activation in response to brain damage [21, 22], its performance as a diagnostic biomarker for SSD has been suboptimal, with an AUC of 67% [23, 24].

NADPH oxidase 1 (encoded by the NOX1 gene) serves as the primary generator of endogenous reactive oxygen species (ROS) in the central nervous system, and it is linked to inflammatory processes [25, 26]. Raftlin, a lipid-protein, stimulates excessive production of IL-17 in high concentrations, leading to neuroinflammation and damage to the blood-brain barrier [27]. In a study by Hurşitoğlu et al. [28], increased levels of NOX1 and Raftlin demonstrated strong potential as biomarkers for SSD (AUC 93% and 92%, respectively). Moreover, the correlation of these markers with patient symptoms underscores the significance of oxidative stress and neuroinflammation in the pathophysiology of SSD.

Table 4.1 Potential immune, inflammatory, and oxidative stress-related biomarkers for schizophrenia spectrum disorders (SSD) assessed by ROC curve analysis

Cohort (effect in concentration): biomarker	Sample Analytical platform Biomarker function	Cohort (n = male/female)/Age[a]	AUC (95%CI)/sensitivity/ specificity (%) [Predictor]
AO SSD vs HC (↑): IL-1β and IL-6 ($p < 0.001$), TFN-α ($p < 0.002$), miRNA-26a ($p < 0.003$), miR-106a ($p = 0.005$). Shafiee-Kandiani et al. 2021 and 2023 [15, 16]	Plasma Sandwich ELISA and RT-qPCR. Diagnostic	AO SSD (27/13)/29.78 (9.60). HC (23/17)/ 30.33 ± 8.58. Cohort nationality: Iran (Azeri ethnicity)	**AO SSD vs HC:** 95.8 (91.8–99.8) [IL-6] Cutoff: 9.3 pg/mL. 79.2 (69.4–89.2) [IL-1β] Cutoff: 125.2 pg/mL. 68.0 (55.8–80.2) [TFN-α] Cutoff: 24.5 pg/mL. 84.3 (76.9–93.8) [miRNA-26a] Cutoff: 0.1726
Non-DEF SSD vs HC: No significant differences in IgM-MDA, IgM-AzA and zsumIgM. **DEF SSD vs non-DEF + HC (↑): IgA-*P. putida*, zsumIgM.** **DEF vs non-DEF + HC (↓):** IgM-MDA, IgM-AzA. Note: $p \leq 0.005$ for each predictor. Maes et al. 2019 [34]	Serum. ELISA. Diagnostic. Note: Data were adjusted for the influence of age and sex on IgA and IgM responses	Low bacterial load (zsumIgA <0.077): HC/non-DEF/DEF 20/26/13 (29/30)/38.0 (11.7). High bacterial load (zsumIgA ≥0.077): HC/non-DEF/DEF 18/14/27. (37/22)/41.9 (11.6). Cohort nationality: Thailand (Bangkok)	**DEF vs non-DEF:** 96.3 (91.5–99.2/92.5/90 [IgM-MDA, IgM-AzA, IgA-*Hafnia alvei*, zsumIgM]. **DEF vs non-DEF + HC:** 92.7/70/89.7 [IgM-MDA, IgM-AzA, IgA-*P. putida*, zsumIgM]
SSD vs HC (↑): IL-6 mRNA ($p < 0.05$). Chase et al. 2016 [17]	PBMC. RT-qPCR. Diagnostic	SSD (34/19)/36.8 (11.8). HC (24/29)/35.5 ± 11.4. Cohort nationality: USA (Chicago)	**SSD vs HC:** 76 (66–85) [IL-6 mRNA]

EO and AO vs HC (↓): IL-4 (p < 1E-4, p = 0.07), IL-12 (p < 1E-4). **EO and AO vs HC (↑):** IL-6 (p < 1E-4, p < 0.005). **AO vs EO (↑):** IL-4 (p < 1E-4). **AO vs EO (↓):** IL-6 (p < 0.003). Liu et al. 2020 [20]	Plasma. ELISA. Diagnostic. Note: Cohort with chronic patients who presented first symptoms above 20 years old (AO) and below 20 years old (EO)	Chronic SSD (EO + AO) (142/68/41. EO (57/27)/38 (9). AO (18/8)/44 (9). HC (46/76)/41.1 (11.8). Cohort nationality: Taiwan	**Chronic SSD vs HC:** 71 (65–76)/66/57 [IL-4]. 75 (69–80)/72/65 [IL-4, IL-6, IL-12]. **EO vs HC:** 81 (74–88)/77/76 [IL-4]. 83 (77–90)/79/77 [IL-4, IL-6, IL-12]. **AO vs HC:** 64 (57–71)/61/0.50 [IL-4]. 70/(64–77)/67/64 [IL-4, IL-6, IL-12]
SSD vs HC (↑): hsCRP and hsCRP/IL-10 (p < 0.001). **SSD vs HC (↓):** IL-10 (p < 0.001). Zhang et al. 2017 [32]	Plasma. ELISA. Diagnostic for SSD patients with aggressive behavior	SSD (18/23)/37.00 (11.30). HC (11/29)/35.02 (13.36). Cohort nationality: China (Shanghai)	**SSD vs HC:** 78.3 (68.1–88.5)/85.4/67.5 [hsCRP/IL-10]. 27.5 (15.9–39.1) [IL-10]. 71.8 (60.5–83.2) [hsCRP]
SSD vs HC (↓): PKACα (p < 0.01). **SSD vs HC (↑):** PKG II (p < 0.001). Wu et al. 2022 [14]	Serum. ELISA. Diagnostic. Note: The gender distribution within cohorts and the average age of participants were not reported.	SSD (30) vs HC (20)/40–75 years old. Cohort nationality: China (Zhenjiang, Taizhou, Jiangsu)	**SSD vs HC:** 71.9 (55–89)/93/60 [PKACα] Cutoff <123 pg/mL. 77.5 (63–92)/77/75 [PKG II] Cutoff <137 pcg/L
SSD vs HC (↑): S100B (p < 0.05). Hong et al. 2016 [24]	Plasma. ELISA. Diagnostic	SSD (18/23)/37.00 (11.30). HC (10/23)/35.15 (13.72). Cohort nationality: China (Shanghai)	**SSD vs HC:** 66.6 (53.6–79.5)/97.6/36.4 [S100B] Cutoff: 144.46 pg/mL

(continued)

Table 4.1 (continued)

Cohort (effect in concentration): biomarker	Sample Analytical platform Biomarker function	Cohort (n = male/female)/Age[a]	AUC (95%CI)/sensitivity/ specificity (%) [Predictor]
SF-AP, SIP, SI-FEP, and SF-FEP vs HC (↑): WBC, neutrophil, monocyte, MLR (p < 0.003). **SF-AP and SF-FE vs HC** (↑): NLR (p ≤ 0.004). Onur et al. 2021 [29]	Blood cells. CBC and ELISA diagnostic. Note: The cohort nationality was not mentioned, but the study was conducted in Turkey	SF-AP (47/29)/27.9 [25.4–30.4]. SF-FEP (38/22)/28.3 [25.4–31.2]. SIP (93/5)/27.5 [26–28.9]. SI-FEP (46/2)/26.3 [24.4–28.2]. HC (17/15)/29.2 [27.2–31.2]	**SF-AP vs HC:** 70.4 (60–80)/52.6/12.5 [NLR] Cutoff: 2.47. **SIP+ SF-AP vs HC:** 84.2 (78–90)/70.1/12.5 [MLR] Cutoff: 0.26. **SF-FE + SI-FE vs HC:** 83.1 (76–89)/71.3/12.5 [MLR]
SSD vs HC (↑): CAR, NAR, NLR, PLR, MLR, RDW, MPV (p < 0.001). Balcioglu and Kirlioglu 2020 [31]	ELISA Diagnostic. Note: The sample was not reported, but it is implied that it is serum	SSD (310/308)/39.68 (10.56). HC (223/222)/31.22 (9.69). Cohort nationality: Turkey (Bakirköy, Istanbul)	**SSD vs HC:** 88.2 (86.3–90.2)/81/81 [CAR] Cutoff: 0.388 pg/mL. 74.1 (71.2–77)/68/67.5 [NAR] Cutoff = 0.885 pg/mL
bFE SSD vs HC (↑): Endotoxin (p = 3.6E-6), HMGB1 (p = 1.4E-9), C4 (p = 1.06E-4). **Drug-treated FE SSD vs bFE SSD** (↓): Endotoxin (p = 0.0503), HMGB1 (p = 0.022), C4 (p = 0.434). Chen et al. 2022 [18]	Plasma. ELISA. Diagnostic and theranostic	Baseline FE SSD (18/24)/25.21 (6.20). HC (20/15)/26.20 (4.15). 8-weeks drug-treated FE SSD (20). Cohort nationality: China (Beijing, Han ethnicity)	**bFE SSD vs HC:** 79.4 (69.3–89.5)/76.2/77.1 [endotoxin] Cutoff: 315.8 ng/L. 86.2 (78.3–94.1)/90.5/68.6 [HMGB1] Cutoff: 26.3 µg/L. 74.1 (63.1–85.1)/66.7/74.3 [C4] Cutoff: 467.2 µg/L. **Drug-treated FE SSD vs HC:** 93.3 (87.8–98.7)/85.7/85.7 [Endotoxin, HMGB1, C4] Cutoff: 0.495

4.2 Immune, Inflammatory, and Oxidative Stress-Related Biomarkers for SSD...

Drug-treated SSD vs HC (↑): ATP, ADP ($p \leq 0.0002$), Ado ($p = 0.042$), IL-1β, IL-6, IL-10, and IL-12 ($p < 0.05$). **Drug-treated SSD vs HC (NS)**: IL-8, TNF-α, and P2X7R. Kristóf et al. 2022 [33]	Serum HPLC/UV for ATP, ADP, Ado. MBA to others. Diagnostic	Drug-treated SSD (34/19)/39.7. HC (27/20)/35. Cohort nationality: Hungary (Budapest)	**Drug-treated SSD vs HC:** 76/90.6/54.6 [ADP] Cutoff: 743.4 nmol/L. 78.9/62.2/86.4 [IL-6] Cutoff: 7.8 pg/mL. 82.5/80/80 [IL-10] Cutoff: 7.4 pg/mL
Chronic SSD vs HC (↑): NOX1 and Raftlin ($p < 0.001$) Hurşitoğlu et al. 2022 [28]	Serum ELISA Diagnostic	Chronic SSD (24/21)/34.86 (6.74). HC (22/23)/34.35 (6.46). Cohort nationality: Turkey (Necip Fazıl, Istanbul)	**Chronic SSD vs HC:** 93.1/94.6/97.8 [NOX1] Cutoff: 18.12 ng/mL 91.5/92.5/97.8 [Raftlin] Cut-off: 19.46 ng/mL
SSD vs HC (↑): MDA, SOD, CAT ($p < 0.001$). Hurşitoğlu et al. 2021 [40]	Serum. TBA test for lipid peroxidation (MDA). Catalase (CAT) and superoxide dismutase (SOD) assay activity. Diagnostic	SSD (21/9)/34.77 (6.13). HC (19/11)/33.24 (9.64). Cohort nationality: Turkey (Kahramanmaras)	**SSD vs HC:** 97.9/93.3/90 [MDA] Cutoff: 2.72 ng/mL
SSD vs HC immune blood domain: HSV2, *Toxoplasma gondii* IgM, CMV, IgG2, IgG, ANCA. **SSD vs HC cognitive domain:** WAIS DSC, CVLT LDCR, CVLT SDCR, WAIS deterioration, WAIS DS, WAIS letter-number sequencing. **BD vs SSD multi-domain:** NART33 IQ, CVLT LDCR, WAIS DS, WAIS DSC, WAIS obj ass. IgG1. Fernandes et al., 2020 [36]	Serum. IT, nephelometry, ELISA, multiplex immunoassay, turbidimetry, immunofluorescence. CVLT and WAIS for cognitive assessment. Diagnostic and transdiagnostic.	SSD (43/15)/35.26 (10.97). HC (49/74)/39.19 (13.17). BD (50/48)/45.14 (12.93). Cohort nationality: France (Créteil)	**SSD vs HC:** 90/84.4/81.1 [cognitive domain] **SSD vs HC:** 71/71.5/61.3 [immune blood domain] **BD vs SSD:** 80/71.3/73.3 [immune blood + cognitive multi domain] **BD vs SSD:** 75/79.1/64.5 [immune blood domain: HSV2, IgG4, *T. gondii* IgM, IgM, anti-gliadin A (AGA—Gluten), anti-tissue transglutaminase (anti-tTG—Celiac disease)]

(continued)

Table 4.1 (continued)

Cohort (effect in concentration): biomarker	Sample Analytical platform Biomarker function	Cohort (n = male/female/Age[a])	AUC (95%CI)/sensitivity/ specificity (%) [Predictor]
TC (↑): Fatty acid translocase (CD36; $Q = 0.0004$) and insulin receptor (IR; $Q = 0.024$) in T-helper cells. **TC (↓)**: Glucose transporter 1 (GLUT 1; $Q = 0.0004$) in monocytes. Zaki et al., 2022 [35]	Serum, PBMC. Luminex MAGPIX multiplexed immunoassay for serum. Flow cytometry for PBMC. Diagnostic and transdiagnostic. Note: Data was adjusted for covariates. Cutoff, specificity, and sensitivity were not reported	**Training cohort (TC)**: Drug-treated SSD (28/8)/26.0 [22.0,30.2]. HC (22/4)/22.0 [21.0,24.8], the Netherlands (Groningen, Utrecht). **Testing cohort 1 (TeC1)**: Drug-naïve FE SSD (20/14)/29.8 [23.3,37.6]. HC (24/15) /33.4 [28.7,38.5]. Spain (Santander). **Testing cohort 2 (TeC2)**: Drug-treated Aut (14/11)/31.5 (8.3). UK (Cambridge). BD (12/13)/32.6 (9.5) (Bulgaria, Sofia + UK, Cambridge). MDD (12/13) ± 40.5 (10.0) Germany (Münster). Drug-naïve FE SSD (17/8)/247.3 (6.7) Spain (Santander). HC (49/51)/33.2 (10.2)	**TC:** 78 (66–92) [CD36, GLUT1]. 81 (69–92) [CD36, GLUT1, RI]. **TeC1:** 75 (64–86) [CD36, GLUT1]. 74 (62–86) [CD36, GLUT1, RI]. **TeC2** with [CD36, GLUT1]: **SSD vs HC:** 84 (77–92) **SSD vs all psychiatric diseases:** 83 (75–92). **SSD vs Aut:** 75 (61–89). **SSD vs BD:** 85 (74–97). **SSD vs MDD:** 90 (81–99)
SSD vs HC (↑): C1, C2, C3, C4, CH50 ($p < 0.001$). Cao Y. et al. 2023 [41]	Serum. ELISA. Diagnostic	SSD (38/51)/35.96 (12.65) HC (42/47)/35.78 (11.05). Cohort nationality: China (Hefei, Anhui province)	**SSD vs HC:** 85.7 (80.1–91.4)/79.8/79.8 [C1, C2, C3, C4, CH50] When each biomarker was considered alone, the AUC values were less than 70%

4.2 Immune, Inflammatory, and Oxidative Stress-Related Biomarkers for SSD... 85

CHR-t vs CHR-nt (↑): VEGF, IL-10/IL-6 ($p = 0.001$). **CHR-nt vs HC (↓):** VEGF ($p = 0.006$), IL-10/IL-6 ($p = 0.033$). The remaining 20 inflammatory biomarkers measured had no significant changes. However, IL-6 and IL-4 stood out with the highest coefficients associated with CHR status. Mondelli et al. 2023 [37]	Serum. Multiplex immunoassay. Prognostic. Note: Data was adjusted for multiple comparisons (Benjamini and Hochberg FDR < 0.05)	Baseline CHR (269). After 2 years of follow up: CHR-t (29/21)/22.56 (0.719). CHR-nt ((118/101)/22.63 (0.334). HC ((30/26)/22.93 (0.541). Cohort nationality: UK (London), the Netherlands (Amsterdam, the Hague), Austria (Vienna), Switzerland (Basel), Australia (Melbourne), France (Paris), Brazil (São Paulo), Spain (Barcelona) (Copenhagen), Denmark	**bCHR vs HC:** 82 (80–84)/75/71 [IFN-gamma, IL-1α, IL-5, IL-8, IL-2, IL-17α, IL-15, IL-16, GM-CSF, TNF-α, IL-12p70, IL-6, IL-4, age, sex, ethnicity, BMI, smoking, and site]. **bCHR vs CHR-t:** 57 (52–61)/48/64 [the same predictors of bCHR vs HC]
FE SSD vs HC (↑): IL-1β, IL-4 ($p < 0.0001$). **FE SSD vs HC (↓):** Tie-2 ($p = 0.029$). IL-6, IL-10, and TNF-α did not differ significantly. Yan et al. 2023 [42]	Serum. MSD platform. Diagnostic	FE SSD (12/28)/29.5 [23–46.5]. HC (14/22)/26.5 [24–31]. Cohort nationality: China (Hefei, Anhui province)	**FE SSD vs HC:** 83.6/97/85 [IL-1β]. 64.6/64/68 [angiopoietin-1 receptor (Tie-2)]
bCHR vs HC (↓): Complement factors C5, H, D, I, C3 ($p < 0.03$). **CHR-t vs CHR-nt (↓):** C5a ($p = 0.007$), C5 ($p = 0.012$). GM-CSF, IL-10, IL-1β, IL-6, IL-8, and TNF-α did not show significant changes in any cohort. Zhang et al. 2023 [43]	Serum. Luminex 200 for complement factors. ELISA for cytokines. Prognostic.	Baseline CHR (35/14)/18.2 (4.1). After 3 years of follow up: CHR-t (18/7)/18.1 (4.1). CHR-nt (17/7)/18.4 (4.2). Cohort nationality: China (Shanghai)	**bCHR vs HC:** AUCs 56–69% using individual complements. **CHR-t vs CHR-nt:** 73 (58–87)/63/76 [C5a] Cutoff: 4251. 71 (56–86)/83/64 [C5] Cutoff: 31169

(continued)

Table 4.1 (continued)

Cohort (effect in concentration): biomarker	Sample Analytical platform Biomarker function	Cohort (n = male/female)/Age[a]	AUC (95%CI)/sensitivity/ specificity (%) [Predictor]
bFE SSD vs HC (↓): SOD (p = 6.95E-10). **bFE SSD vs HC (↑): Indirect bilirubin** (IBIL; p = 1E-5). **Risp-treated FE vs bFE:** No significant difference. **Risp-treated FE vs HC (↓): SOD** (p = 2.2E-3). **Risp-treated FE vs HC (↑): IBIL** (p = 1.3E-2), glucose (p = 6.1E-4). Wang et al. 2023 [44]	Serum. Immune Turbidimetric test for HCY. Colorimetric assay for SOD. Theranostic	Baseline drug-naïve FE SSD (59/89)/24.2 (6.62). 6 months risperidone-treated FE SSD (39). HC (37/60)/23.1 (1.81). Note: A ratio of 9:1 was allocated to training and testing cohorts, respectively. Cohort nationality: China (Zhengzhou)	**Training cohort bFE SSD vs HC:** 83/79/77 [SOD, glucose, IBIL]. **Testing cohort bFE SSD vs HC:** 80/76/76 [SOD, glucose, IBIL]
FEP vs HC (↑): WBC (p = 0.004), neutrophil (Neu, p = 0.04), monocyte (Mono, p = 0.02), eosinophil (Eos, p = 0.002). **Drug-treated FEP vs HC (↑):** WBC (p = 0.008), lymphocyte (p = 0.01). Juchnowicz et al. 2023 [38]	Serum. Complete blood count (CBC). Diagnostic and theranostic	FEP (25/23)/21 [5]. HC (11/19)/24 [4]. Drug-treated FEP (21). Drug-naïve FEP (26). Cohort nationality: Poland (Lublin)	**FEP vs HC:** 77 (67–87) [Mono, Eos, Neu]. **Drug-treated FEP vs HC:** 87 (76–99) [WBC, Neu]. **Drug-naïve FEP vs HC:** 74 (64–90) [WBC, hematocrit]. **Drug-treated FEP vs drug-naïve FEP:** 86 (75–97) [RDW-SD, lymphocyte]
SSD vs HC (↓): Met-1, Cu (p < 0.001), Cu/Zn (p = 0.003). Not significant: Zn. Yilmaz et al. 2023 [45]	Plasma. ELISA for metallothionein-1 (Met-1). Spectrophotometry for Cu/Zn. Diagnostic	SSD (42/10)/36.1 (9.8). HC (31/7)/34.7 (9.2). Cohort nationality: Turkey (Elazığ)	**SSD vs HC:** 90 (82–95)/83/95 [Met-1] Cutoff <50.2.

4.2 Immune, Inflammatory, and Oxidative Stress-Related Biomarkers for SSD...

SSD vs HC (↓): Iridium reducing capacity (p = 9E-8). **Clozapine-treated SSD vs HC**: p = 3.6E-9. **Treated without clozapine vs HC**: p = 0.03. Kim et al. 2019 [46]	Serum. Spectrophotometry. Diagnostic	Clozapine-treated SSD (34/15)/42.3 (13.4). Treated without clozapine SSD (14/10)/44.6 (10.9). HC (25/20)/36.2 (13.6). Cohort nationality: USA (Maryland)	**SSD vs HC**: 89/88/75 [iridium-reducing capacity] Cutoff: 0.82.
SSD+SAD vs HC (↑): MDA, advanced oxidation protein products (AOPP), glutathione peroxidase (GSH-Px), protein carbonyls (PC) (p ≤ 0.003). **SSD vs HC** (↓): Total thiols (p = 0.05), reduced glutathione (GSH; p < 0.001). No significant ferric-reducing ability of plasma (FRAP), catalase (CAT). Guidara et al. 2020 [39]	Plasma. Spectrophotometry. Diagnostic	SSD (45/0)/37.6 (11.3). Schizoaffective-SAD (21/0)/28.8 (5.8). HC (101/0)/34.2 (9.8). Note: A ratio of 75:25 was allocated to training and testing cohorts, respectively. After permutation tests, all models exhibited AUCs <70%. Cohort nationality: Tunisia (Sfax)	**Training cohort SSD+SAD vs HC**: 79/67/70 [MDA] Cutoff: 25.5. 72/73/66 [AOPP] Cutoff: 48.9. 72/61/66 [PC] Cutoff: 0.65. 79.5/73/81 [GSH-Px] Cutoff: 2.57. 86/90/84 [all together] Cutoff: 0.33. **Testing cohort SSD+SAD vs HC**: 78/83/66 [MDA] Cutoff: 24.9. 89/83/79 [AOPP] Cutoff: 49.5 91/83/79 [PC] Cutoff: 0.84 78/72/69 [GSH-Px] Cutoff: 2.3. 91/84/96 [all together] Cutoff: 0.53

(continued)

Table 4.1 (continued)

Ado adenosine, *ANCA* antineutrophil cytoplasmic antibodies, *AO* adult-onset, *APD* acute psychotic disorder, *AUC* area under the curve, *Aut* drug-treated autism, *C4* complement component 4, *CAR* C-reactive protein/albumin ratio, *BD* bipolar disorder, *CBC* complete blood count, *CHR-t* patients at clinical high risk who have transitioned to psychosis, *CI* confidence interval, *CMV* cytomegalovirus, *CVLT* California Verbal Learning Test, *CVLT SDCR* total number of correct answers during list a short delay cued recall, *CVLT LDCR* total number of correct answers during list a long delay cued recall, *DEF SSD* deficit schizophrenia, *ELISA* enzyme linked immunosorbent assay, *EO* early-onset, *FEP* first-episode psychosis, *HC* healthy control, *HMGB1* high mobility group box 1 protein, *GM-CSF* granulocyte-macrophage colony-stimulating factor, *HPLC/UV* high performance liquid chromatography coupled to ultraviolet detector, *HSV2* Herpes simplex virus 2, *hsCRP* high-sensitivity C-reactive protein, *IgA-P. putida* immunoglobulin A against *Pseudomonas putida*, *IgGs* immunoglobulins, *IgM-AzA* immunoglobulin M against azelaic acid, *IgM-MDA* immunoglobulin M against malondialdehyde, *IL-6 mRNA* interleukin 6 messenger RNA, *IT* immunoturbidimetry, *MBA* multiplex bead array, *MDA* malondialdehyde, *MDD* major depressive disorder, *MLR* monocyte to lymphocyte ratio, *MPV* mean platelet volume, *miRNA-26a* microRNA-26a, *NAR* neutrophil/albumin ratio, *NART33* IQ National Adult Reading Test 33 items total Intelligence Quotient score, *NLR* neutrophil to lymphocyte ratio, *NOX-1* NADPH oxidase 1, *NS* not significant, *PBMC* peripheral blood mononuclear cells, *PLR* platelet/lymphocyte ratio, *PKACα* catalytic subunit C-alpha of protein kinase A, *PKG II* protein kinase G type II, *P2X7R* ATP-sensitive ligand-gated nonselective cation channel, a purinergic receptor, *RDW* red-cell distribution width, *RT-qPCR* reverse transcription quantitative polymerase chain reaction, *S100B* S100 calcium binding protein, *SIP* substance-induced psychosis, *SFAP* substance-free acute psychosis, *TBARS* thiobarbituric acid-reactive substances, *TNF-α* tumor necrosis factor-α, *VEGF* vascular endothelial growth factor, *WAIS* Wechsler Adult Intelligence Scale, *WAIS DSC* digit symbol coding total score, *WAIS DS* digit spam total score, *WAIS Obj Ass* object assembly total score, *WBC* white blood cells, *zsumIg* total IgM or IgA responses to the five enterobacteria tested

[a] Age is given as mean (standard deviation) or median [interquartile]

Onur et al. investigated various compound ratios for distinguishing patients with substance-induced psychosis (SIP) from those with acute psychotic disorder unrelated to substance abuse (SF-AP). The neutrophil-to-lymphocyte ratio (NLR) exhibited moderate predictive power against SF-AP, while the monocyte-to-lymphocyte ratio (MLR) was more effective in distinguishing substance-free or substance-induced FE patients (AUC 83%) [29]. Along the same lines, another research group observed the ability of the neutrophil/albumin ratio (NAR, AUC 74%) and C-reactive protein/albumin ratio (CAR, AUC 88%) to differentiate SSD patients from controls. Interestingly, mean platelet volume (MPV) showed a significant and negative association with a higher number of hospitalizations [30, 31].

Recognizing the need to predict aggressive behavior in SSD patients for the purpose of minimizing prolonged hospital stays, Zhang et al. developed a model using increased hsCRP to IL-10 ratio, but the model showed moderate results [32].

In light of the acknowledged involvement of purinergic signaling in SSD pathophysiology, particularly pointing to a general hypoadenosinergic state and an imbalance between A1 and A2A adenosine receptors, Kristóf et al. constructed diagnostic models using ADP, IL-6, and IL-10. However, the obtained AUC values were consistently within the range of 76–82% [33].

Deficit schizophrenia (DEF) is a distinct subtype within SSD, characterized by an immunoinflammatory response and a deficiency in IgM antibodies targeting specific oxidative epitopes (SOEs) such as malondialdehyde (MDA) and azelaic acid (AzA). Maes et al. proposed that diminished levels of IgM targeting SOEs could indirectly contribute to the negative symptoms and cognitive impairments observed in DEF SSD. They suggested these reduced IgM levels as potential classifiers for constructing a multistep mediation model to differentiate DEF from non-DEF phenotypes. Through a multivariate analysis involving plasma IgA and IgM responses to five enterobacteria (*Hafnia alvei, Pseudomonas aeruginosa, P. putida, Morganella morganii, Klebsiella pneumoniae*) and associated IgM responses to MDA and AzA, four models with good predictive ability were identified, achieving AUC values greater than 90%. These findings not only underscore the potential of specific immune markers in distinguishing DEF from non-DEF phenotypes but also provide compelling evidence that DEF represents a unique and identifiable phenotype within the broader spectrum of SSD [34].

Zaki et al. developed two diagnostic models utilizing insulin receptor (IR) and fatty acid translocase (CD36) expressions in T-helper cells, alongside glucose transporter 1 (GLUT1) expression in monocytes, for recent onset SSD patients, yielding favorable results. Subsequently, when applying these models to an independent drug-naïve FE SSD cohort, they achieved comparable outcomes using either two or three predictors. To mitigate concerns about potential overfitting and evaluate specificity to SSD, the researchers opted to validate the two-predictor model in an additional drug-naïve FE SSD cohort and against various psychiatric disorders, resulting in promising findings [35].

Fernandes et al. devised a probabilistic multi-domain data integration model that incorporated immune and inflammatory biomarkers from peripheral blood alongside cognitive biomarkers to forecast the diagnosis of SSD. Surprisingly, the multi-domain model demonstrated superior performance in distinguishing SSD from bipolar disorder (BD) patients. However, it was noteworthy that the traditional six-predictor model, which included only cognitive measurements, exhibited the highest AUC value at 90%, particularly effective in differentiating SSD patients from healthy controls [36].

Motivated to identify markers associated with the transition of patients at clinical high risk (CHR) to psychosis and seeking insights into the involvement of inflammatory markers in this process, Mondelli and colleagues initially employed in silico testing to determine the most effective classifier. Subsequently, they measured 20 potential biomarkers in the serum of CHR patients across 10 different countries. The markers comprised both pro-inflammatory and anti-inflammatory cytokines. While various classifiers exhibited limited predictive ability for the transition to psychosis, penalized logistic regression emerged as particularly effective in distinguishing CHR patients from normal controls, exhibiting an AUC of 82% [37]. In contrast, Zhang and colleagues demonstrated that complements C5 and C5a exhibit superior predictive power for CHR individuals who later transition to psychosis (AUCs >70%). This suggests that dysregulation of complements might occur at an earlier stage than the imbalance of inflammatory factors in the progression toward psychosis.

Juchnowicz et al. conducted a comprehensive analysis, assessing a complete blood count (CBC) along with oxidative stress-related parameters, including catalase, glutathione peroxidase, and superoxide dismutase, among others, in blood samples. Utilizing these data, the researchers constructed a Classification and Regression Trees (C&RT) model, incorporating lymphocyte levels and red cell distribution width. This model demonstrated notable accuracy (86%) in effectively distinguishing between treated and non-treated FEP patients [38]. Guidara et al. determined that oxidative stress-related markers play a role in the diagnosis of SSD patients. However, they found that these biomarkers were not particularly effective in predicting the occurrence of the disease [39].

4.3 Discussion

This comprehensive review encompassed 24 papers, involving approximately 2300 SSD patients from 20 different countries. Notably, China emerged as a leading contributor with 30% of the studies, followed by Turkey at 17%.

Several trends became evident in the analysis. Models using 3–7 biomarkers showed increased diagnostic potential compared to models relying on a single biomarker. Among the approximately 50 models developed with a single biomarker, only 22% exhibited an AUC greater than 80%. In contrast, 65% of models incorporating a panel of predictors achieved this notable result.

It is noteworthy that only since 2020 have some studies begun including training and testing cohorts in the development of diagnostic models. However, challenges persist due to insufficient description of data analysis methodologies and lack of standardization in technical language, making comparisons between studies difficult.

Two studies revealed the promising potential of IL-6 and IL-β1 in predicting the onset of SSD (AUCs higher than 80%) [16, 42]. Interestingly, a panel of cognitive tests demonstrated greater efficacy in distinguishing between SSD patients and both healthy individuals and those with bipolar disorder (BD) than immunoinflammatory biomarkers [36]. Additionally, stress-related biomarkers showed a fair diagnostic potential.

References

1. Kroken RA, Sommer IE, Steen VM, Dieset I, Johnsen E (2018) Constructing the immune signature of schizophrenia for clinical use and research; an integrative review translating descriptives into diagnostics. Front Psychiatry 9:753. https://doi.org/10.3389/fpsyt.2018.00753
2. Rajasekaran A, Venkatasubramanian G, Berk M, Debnath M (2015) Mitochondrial dysfunction in schizophrenia: pathways, mechanisms and implications. Neurosci Biobehav Rev 48:10–21. https://doi.org/10.1016/j.neubiorev.2014.11.005
3. Khandaker GM, Zimbron J, Lewis G, Jones PB (2013) Prenatal maternal infection, neurodevelopment and adult schizophrenia: a systematic review of population-based studies. Psychol Med 43:239–257. https://doi.org/10.1017/S0033291712000736
4. Trovão N, Prata J, VonDoellinger O, Santos S, Barbosa M, Coelho R (2019) Peripheral biomarkers for first-episode psychosis—opportunities from the neuroinflammatory hypothesis of schizophrenia. Psychiatry Investig 16:177–184. https://doi.org/10.30773/pi.2018.12.19.1
5. De Simone G, Mazza B, Vellucci L, Barone A, Ciccarelli M, de Bartolomeis A (2023) Schizophrenia synaptic pathology and antipsychotic treatment in the framework of oxidative and mitochondrial dysfunction: translational highlights for the clinics and treatment. Antioxidants (Basel) 12:975. https://doi.org/10.3390/antiox12040975
6. Khoury R, Nasrallah HA (2018) Inflammatory biomarkers in individuals at clinical high risk for psychosis (CHR-P): state or trait? Schizophr Res 199:31–38. https://doi.org/10.1016/j.schres.2018.04.017
7. Feng Z, Zhang Y, You X, Zhang W, Ma Y, Long Q, Liu Z, Hao W, Zeng Y, Teng Z (2020) Effects of risperidone on blood levels of interleukin-6 in schizophrenia: a meta-analysis. Medicine 99:e19694. https://doi.org/10.1097/MD.0000000000019694
8. Pinto JV, Moulin TC, Amaral OB (2017) On the transdiagnostic nature of peripheral biomarkers in major psychiatric disorders: a systematic review. Neurosci Biobehav Rev 83:97–108. https://doi.org/10.1016/j.neubiorev.2017.10.001
9. Fraguas D, Díaz-Caneja CM, Rodríguez-Quiroga A, Arango C (2017) Oxidative stress and inflammation in early onset first episode psychosis: a systematic review and meta-analysis. Int J Neuropsychopharmacol 20:435–444. https://doi.org/10.1093/ijnp/pyx015
10. Fraguas D, Díaz-Caneja CM, Ayora M, Hernández-Álvarez F, Rodríguez-Quiroga A, Recio S, Leza JC, Arango C (2019) Oxidative stress and inflammation in first-episode psychosis: a systematic review and meta-analysis. Schizophr Bull 45:742–751. https://doi.org/10.1093/schbul/sby125
11. Kang J-H, Mori T, Kitazaki H, Niidome T, Takayama K, Nakanishi Y, Katayama Y (2013) Serum protein kinase Cα as a diagnostic biomarker of cancers. Cancer Biomark 13:99–103. https://doi.org/10.3233/CBM-130340
12. Wu Y, Li Y, Liu J, Chen M, Li W, Chen Y, Xu M (2020) The diagnostic value of extracellular protein kinase A (ECPKA) in serum for gastric and colorectal cancer. Transl Cancer Res 9:3870–3878. https://doi.org/10.21037/tcr-20-764
13. Wu Y, Wu M, Wang Z et al (2022) PKG II secreted via the classical endoplasmic reticulum-Golgi apparatus secretory pathway blocks the activation of EGFR through phosporalting its threonine 406 and has an anti-tumor effect. Front Biosci 27:53. https://doi.org/10.31083/j.fbl2702053

14. Wu Y, Ye J, Zhao C, Pang J, Li Y, Lin X (2022) Extracellular protein kinase A and G are potential biomarkers of some inflammation-associated disorders. Eur J Inflam 20:1721727X2211051. https://doi.org/10.1177/1721727X221105132
15. Shafiee-Kandjani AR, Nejadettehad N, Farhang S, Bruggeman R, Shanehbandi D, Asadi M, Hassanzadeh M (2021) MicroRNAs and pro-inflammatory cytokines as candidate biomarkers for recent-onset psychosis. SSRN Electron J. https://doi.org/10.2139/ssrn.3840100
16. Shafiee-Kandjani AR, Nezhadettehad N, Farhang S, Bruggeman R, Shanebandi D, Hassanzadeh M, Azizi H (2023) MicroRNAs and pro-inflammatory cytokines as candidate biomarkers for recent-onset psychosis. BMC Psychiatry 23:631. https://doi.org/10.1186/s12888-023-05136-6
17. Chase KA, Cone JJ, Rosen C, Sharma RP (2016) The value of interleukin 6 as a peripheral diagnostic marker in schizophrenia. BMC Psychiatry 16:152. https://doi.org/10.1186/s12888-016-0866-x
18. Chen S, Gou M, Chen W, Xiu M, Fan H, Tan Y, Tian L (2022) Alterations in innate immune defense distinguish first-episode schizophrenia patients from healthy controls. Front Psych 13:1024299. https://doi.org/10.3389/fpsyt.2022.1024299
19. Şimşek Ş, Yıldırım V, Çim A, Kaya S (2016) Serum IL-4 and IL-10 levels correlate with the symptoms of the drug-naive adolescents with first episode, early onset schizophrenia. J Child Adolesc Psychopharmacol 26:721–726. https://doi.org/10.1089/cap.2015.0220
20. Liu J-Y, Chen H-Y, Lin J-J, Lu M-K, Tan H-P, Jang F-L, Lin S-H (2020) Alterations of plasma cytokine biomarkers for identifying age at onset of schizophrenia with neurological soft signs. Int J Med Sci 17:255–262. https://doi.org/10.7150/ijms.38891
21. Li R-L, Zhang Z-Z, Peng M, Wu Y, Zhang J-J, Wang C-Y, Wang Y-L (2013) Postoperative impairment of cognitive function in old mice: a possible role for neuroinflammation mediated by HMGB1, S100B, and RAGE. J Surg Res 185:815–824. https://doi.org/10.1016/j.jss.2013.06.043
22. Papuć E, Rejdak K (2020) Increased cerebrospinal fluid S100B and NSE reflect neuronal and glial damage in Parkinson's disease. Front Aging Neurosci 12:156. https://doi.org/10.3389/fnagi.2020.00156
23. Lara DR, Gama CS, Belmonte-de-Abreu P, Portela LV, Gonçalves CA, Fonseca M, Hauck S, Souza DO (2001) Increased serum S100B protein in schizophrenia: a study in medication-free patients. J Psychiatr Res 35:11–14. https://doi.org/10.1016/s0022-3956(01)00003-6
24. Hong W, Zhao M, Li H et al (2016) Higher plasma S100B concentrations in schizophrenia patients, and dependently associated with inflammatory markers. Sci Rep 6:27584. https://doi.org/10.1038/srep27584
25. Emiliani FE, Sedlak TW, Sawa A (2014) Oxidative stress and schizophrenia: recent breakthroughs from an old story. Curr Opin Psychiatry 27:185–190. https://doi.org/10.1097/YCO.0000000000000054
26. Sorce S, Krause K-H, Jaquet V (2012) Targeting NOX enzymes in the central nervous system: therapeutic opportunities. Cell Mol Life Sci 69:2387–2407. https://doi.org/10.1007/s00018-012-1014-5
27. Debnath M, Berk M (2014) Th17 pathway-mediated immunopathogenesis of schizophrenia: mechanisms and implications. Schizophr Bull 40:1412–1421. https://doi.org/10.1093/schbul/sbu049
28. Hurşitoğlu O, Kurutas EB, Strawbridge R, Uygur OF, Yildiz E, Reilly TJ (2022) Serum NOX1 and Raftlin as new potential biomarkers of interest in schizophrenia: a preliminary study. Neuropsychiatr Dis Treat 18:2519–2527. https://doi.org/10.2147/NDT.S385631
29. Onur D, Neslihan AK, Samet K (2021) A comparative study of complete blood count inflammatory markers in substance-free acute psychotic disorder and substance-induced psychosis. Early Interv Psychiatry 15:1522–1530. https://doi.org/10.1111/eip.13089
30. Tingle SJ, Severs GR, Goodfellow M, Moir JA, White SA (2018) NARCA: a novel prognostic scoring system using neutrophil-albumin ratio and Ca19-9 to predict overall survival in palliative pancreatic cancer. J Surg Oncol 118:680–686. https://doi.org/10.1002/jso.25209

31. Balcioglu YH, Kirlioglu SS (2020) C-reactive protein/albumin and neutrophil/albumin ratios as novel inflammatory markers in patients with schizophrenia. Psychiatry Investig 17:902–910. https://doi.org/10.30773/pi.2020.0185
32. Zhang Q, Hong W, Li H et al (2017) Increased ratio of high sensitivity C-reactive protein to interleukin-10 as a potential peripheral biomarker of schizophrenia and aggression. Int J Psychophysiol 114:9–15. https://doi.org/10.1016/j.ijpsycho.2017.02.001
33. Kristóf Z, Baranyi M, Tod P, Mut-Arbona P, Demeter K, Bitter I, Sperlágh B (2022) Elevated serum purine levels in schizophrenia: a reverse translational study to identify novel inflammatory biomarkers. Int J Neuropsychopharmacol 25:645–659. https://doi.org/10.1093/ijnp/pyac026
34. Maes M, Kanchanatawan B, Sirivichayakul S, Carvalho AF (2019) In schizophrenia, increased plasma IgM/IgA responses to gut commensal bacteria are associated with negative symptoms, neurocognitive impairments, and the deficit phenotype. Neurotox Res 35:684–698. https://doi.org/10.1007/s12640-018-9987-y
35. Zaki JK, Lago SG, Rustogi N et al (2022) Diagnostic model development for schizophrenia based on peripheral blood mononuclear cell subtype-specific expression of metabolic markers. Transl Psychiatry 12:457. https://doi.org/10.1038/s41398-022-02229-w
36. Fernandes BS, Karmakar C, Tamouza R et al (2020) Precision psychiatry with immunological and cognitive biomarkers: a multi-domain prediction for the diagnosis of bipolar disorder or schizophrenia using machine learning. Transl Psychiatry 10:162. https://doi.org/10.1038/s41398-020-0836-4
37. Mondelli V, Blackman G, Kempton MJ et al (2023) Serum immune markers and transition to psychosis in individuals at clinical high risk. Brain Behav Immun 110:290–296. https://doi.org/10.1016/j.bbi.2023.03.014
38. Juchnowicz D, Dzikowski M, Rog J, Waszkiewicz N, Karakuła KH, Zalewska A, Maciejczyk M, Karakula-Juchnowicz H (2023) The usefulness of a complete blood count in the prediction of the first episode of schizophrenia diagnosis and its relationship with oxidative stress. PLoS One 18:e0292756. https://doi.org/10.1371/journal.pone.0292756
39. Guidara W, Messedi M, Naifar M, Maalej M, Grayaa S, Omri S, Ben Thabet J, Maalej M, Charfi N, Ayadi F (2020) Predictive value of oxidative stress biomarkers in drug-free patients with schizophrenia and schizo-affective disorder. Psychiatry Res 293:113467. https://doi.org/10.1016/j.psychres.2020.113467
40. Hurşitoğlu O, Orhan FÖ, Kurutaş EB, Doğaner A, Durmuş HT, Kopar H (2021) Diagnostic performance of increased malondialdehyde level and oxidative stress in patients with schizophrenia. Noro Psikiyatr Ars 58:184–188. https://doi.org/10.2147/NDT.S385631
41. Cao Y, Xu Y, Xia Q, Shan F, Liang J (2023) Peripheral complement factor-based biomarkers for patients with first-episode schizophrenia. Neuropsychiatr Dis Treat 19:1455–1462. https://doi.org/10.2147/NDT.S420475
42. Yan F, Meng X, Cheng X, Pei W, Chen Y, Chen L, Zheng M, Shi L, Zhu C, Zhang X (2023) Potential role between inflammatory cytokines and Tie-2 receptor levels and clinical symptoms in patients with first-episode schizophrenia. BMC Psychiatry 23:538. https://doi.org/10.1186/s12888-023-04913-7
43. Zhang T, Zeng J, Ye J et al (2023) Serum complement proteins rather than inflammatory factors is effective in predicting psychosis in individuals at clinical high risk. Transl Psychiatry 13:9. https://doi.org/10.1038/s41398-022-02305-1
44. Wang S, Yuan X, Pang L, Song P, Jia R, Song X (2023) Establishment of an assistive diagnostic model for schizophrenia with oxidative stress biomarkers. Front Pharmacol 14:1158254. https://doi.org/10.3389/fphar.2023.1158254
45. Yılmaz S, Kılıç N, Kaya Ş, Taşcı G (2023) A potential biomarker for predicting schizophrenia: metallothionein-1. Biomedicines 11:590. https://doi.org/10.3390/biomedicines11020590
46. Kim E, Keskey Z, Kang M, Kitchen C, Bentley WE, Chen S, Kelly DL, Payne GF (2019) Validation of oxidative stress assay for schizophrenia. Schizophr Res 212:126–133. https://doi.org/10.1016/j.schres.2019.07.057

Chapter 5
Gut Microbiota-Brain Axis and Related Biomarkers

5.1 Introduction

Enthusiasm prevails in the scientific community as researchers endeavor to elucidate the etiology of various diseases through the lens of dysbiosis. The interest in this area is growing exponentially, to the extent that the gut microbiota is now regarded as the newest organ in the human body [1]. Although the exploration of this intricate symbiotic relationship dates back to the 1600s [2], it is only recently that we have begun to truly comprehend its complexity, allowing our imaginations to soar as we coin terms indicative of its multifaceted role in the human body. These terms include "metabolic machinery" [3], "a virtual endocrine organ" [4], "second human genome," referring to the gut microbiome [5], "psychobiotics," alluding to probiotic therapy [6], and "the quintessence of life" [2].

The influence of gut bacteria on human health is well established, impacting both the maintenance of health and the disruption of homeostatic regulation [7]. This microbial influence stems from the intricate interplay between bacterial diversity and the metabolites they produce [8]. At the core of this microbial influence is the gut-brain axis, a bidirectional communication system involving neural, endocrine, and immune pathways connecting the gut and the central nervous system. The gut microbiota plays a pivotal role in this axis by producing and modulating neurotransmitters and neuroactive compounds, including serotonin, dopamine, and gamma-aminobutyric acid (GABA). Additionally, the microbiota contributes to the development and function of the immune system.

Dysbiosis, characterized by alterations in microbiota composition, has been linked to an increase in systemic inflammation. Chronic inflammation, imbalances in neurotransmitters, and immune dysfunction are implicated in various psychiatric disorders, including SSD. The gut microbiota may contribute to these alterations, offering a potential avenue for understanding and addressing the pathophysiology of SSD [9].

A bibliometric study by Yang et al. in [10] identified a growing trend in research exploring the relationship between SSD and the gut microbiota, particularly since 2014 [10]. The focus of this research encompassed microbiota composition in SSD patients, investigations into the gut-brain axis, and microbial-based interventions. Consequently, numerous systematic reviews and meta-analyses have also delved into the subject, contributing to a more comprehensive understanding of the intricate interplay between the gut microbiota and SSD.

McGuinness et al. [3] conducted a systematic review aiming to understand the association between gut bacteria and mental disorders. Their findings suggested no substantial difference in the number or distribution (α-diversity) of bacteria among individuals with SSD, major depressive disorder (MDD), or bipolar disorder (BD). However, they observed differences in overall community composition (β-diversity). Specific bacterial taxa associated with mental disorders included lower levels of genera producing short-chain fatty acids (e.g., butyrate), elevated levels of lactic acid-producing bacteria, and increased abundance of bacteria linked to glutamate and GABA metabolism. Despite study heterogeneity, these associations were consistently identified.

In a broader meta-analysis involving nine mental disorders, Nikolova et al. reported depleted levels of *Faecalibacterium* and *Coprococcus*, known for anti-inflammatory butyrate production, and enriched levels of *Eggerthella* across the various disorders [11]. However, a review by Szeligowski et al. highlighted highly discrepant evidence regarding dysbiosis in SSD, making it challenging to determine whether microbiome changes are linked to an increased risk of the disorder or are a consequence of external factors or treatment [12]. They also found inconclusive evidence regarding the efficacy of pro/prebiotic supplementation in SSD treatment. The systematic review and meta-analysis carried out by the Minichino group provided additional insights by corroborating significant changes in alpha and beta diversity metrics following treatment with antipsychotics and antidepressants [13].

A recent meta-analysis by Murray et al. in 2023 indicated that the decrease in bacterial diversity in SSD patients did not reach statistical significance [14]. Additionally, Borkent et al. reported inconsistent findings in studies exploring the composition of the gastrointestinal microbiota in patients with SSD, BD, and MDD [15]. Despite the variability in results, a consistent pattern emerged with higher relative abundances of *Streptococcus*, *Lactobacillus*, and *Eggerthella*, concurrent with lower abundances of *Faecalibacterium*.

In summary, the current literature on the microbiome in psychiatric disorders, specifically SSD, portrays an intricate landscape with diverse findings. Nonetheless, the primary objective of this review was to examine whether the altered microbiota and its associated metabolites demonstrate biomarker potential in the construction of diagnostic models, a question assessed through ROC analysis.

5.2 Gut Microbiota and Associated Metabolites Assessed as Biomarkers for SSD Through ROC Analysis

After conducting a literature review spanning the past twelve years (2011–2023), some of the most pertinent studies were summarized in Table 5.1.

Driven by the urgency to identify reliable indicators for SSD and leveraging metagenomics technology to unravel microbial identities, Shen and colleagues pioneered the development of the first model. This model incorporated a panel of bacteria from diverse genera and species, successfully distinguishing SSD patients from healthy individuals [16]. They achieved promising results with an AUC of 84%, drawing significant conclusions about the diminished abundance of butyrate-producing bacteria and their regulatory role in the host's immune response. These bacteria demonstrated a negative correlation with glutathione and taurine/hypotaurine metabolic pathways, shedding light on their involvement in the stress-related and inflammatory pathophysiology of SSD.

Expanding upon this groundwork, 2 years later, Xu et al. opted to construct a model incorporating the microbiota dysbiosis index, in lieu of bacterial genera, alongside the increased levels of IgA and glutamate synthase. These molecules, pivotal in the regulation of intestinal mucosal immunity and glutamate synthesis, were selected as predictors for distinguishing SSD patients from controls. Their model demonstrated good performance, achieving an AUC of 86% [17]. Zhu et al. utilized a panel of 26 metagenomic operational taxonomic units (mOTUs) in their model, yielding an AUC of 90% in the training cohort, although it demonstrated a slight decrease to 76% in the testing cohort, indicative of potential overfitting [18]. In contrast, a study conducted in the United States encountered challenges in identifying specific bacteria or groups with significant predictive value. Consequently, the researchers opted for a comprehensive approach, building a model that incorporated all identified bacteria. This inclusive model demonstrated an AUC of 82% [19].

Yuan et al. explored the potential for following up on FE drug-naïve patients during risperidone treatment. Their model, incorporating 10 altered species, demonstrated an AUC of 88% [20]. In a parallel study, an Indian group obtained comparable results using four genera, achieving an AUC of 87% [21]. Liang's group took a distinctive approach to distinguish treated patients from controls, employing 22 aminoacyl-tRNA transferases with proven inflammatory cytokine activity to construct a diagnostic model, boasting an AUC of 91% [22].

Gao et al. delved into the analysis of altered fecal metabolites, uncovering 65 significant changes (including 11 with FDR ≤ 0.01) between paranoid and undifferentiated SSD patients. By constructing models that incorporated altered genera or phyla, correlated with pathways enriched for the identified metabolites, they pinpointed the Bacillota phylum as the most efficient model, achieving an AUC of 91% [23]. In a parallel exploration, Ling et al. observed imbalances in serum chemokines among chronic SSD patients, coupled with a reduced abundance of fecal Lachnospiraceae. To further understand these findings, they evaluated various

Table 5.1 Potential gut microbiota-brain axis biomarkers for schizophrenia spectrum disorders (SSD) assessed by ROC analysis

Cohort (effect in concentration): metabolite/concentration	Sample platform	Cohort (n = male/female)/Age[a] AUC (95% CI) [Predictor]	Altered bacteria (Phylum[b]/Family: Genus)	Correlations Bacterial diversity Differential abundance	Pathway/symptom
Drug-naïve SSD serum vs HC (↓): Cholic acid, 4,8-dimethylnonanoyl carnitine, prostaglandin A2, 3-hydroxycapric acid, PE(22:6), PE(20:4), PC(P-18:1) **Drug-naïve SSD serum vs HC (↑) pg/mL:** IL-1β/2.0 ($p = 0.35$), IL-4/3.2 ($p = 0.09$), IL-6/3.1 ($p = 0.06$), CCL2/377.9 ($p = 0.04$), TNF-α/3.3 ($p = 0.20$), IFN-γ/3.6 ($p < 0.001$) Note: VIP > 2 and FDR < 0.001 Fan et al. [26]	Feces: The same sequencing platform of Zhu et al. [18] Serum: UHPLC-Orbitrap/MS for untargeted metabolomics Luminex200 multiplex assay detection system for cytokines Note: Data were adjusted for gender, age, and BMI	**Training cohort (TC):** Drug-naïve SSD (34/29)/29 [25.0–32.0] HC (14/43)/35 [25.0–40.0] **Testing cohort (TeC):** SSD (23) vs HC (23) TC AUC: 99.2% [Cholic acid, 4,8-dimethylnonanoyl carnitine, prostaglandin A2, 3-hydroxycapric acid, PE (22:6)] TeC AUC: 99.5% Cohort nationality: China (Xi'an, Shaanxi Province)	**Pseudomonadota** Bacteroidaceae: *Bacteroides* Enterobacteriaceae: *Enterobacter, Klebsiella* Pasteurellaceae: *Haemophilus* **Bacteroidota** Tannerellaceae: *Parabacteroides* Prevotellaceae: *Prevotella* **Bacillota** Lachnospiraceae: *Blautia, Dorea, Coprococcus* Ruminococcaceae: *Anaerotruncus, Subdoligranulum* Veillonellaceae: *Dialister* Streptococcaceae: *Streptococcus, Lactococcus* Clostridiaceae: *Clostridium* **Actinomycetota** Coriobacteriaceae: *Collinsella, Eggerthella* Actinomycetaceae: *Actinomyces*	**Drug-naïve SSD feces vs HC (↑)** ($p = 0.001$): α- and β-diversity **Drug-naïve SSD feces vs HC (↑)** ($p ≤ 0.05$): Actinomycetota, *Bacteroides, Clostridium, Parabacteroides, Dialister, Haemophilus* Ruminococcaceae, Lachnospiraceae (SCFAs-producing) **Drug-naïve SSD feces vs HC (↓)** ($p ≤ 0.05$): *Prevotella,* (probiotic-producing), *Enterobacter, Klebsiella, Lactococcus, Streptococcus* The classifiers correlated negatively with α-diversity (Shannon index), except carnitine	The clusters fatty acyls & steroids (7), fatty acyls & glycerolipids (6), LysoPC & PC (6), PE & Pyridinecarboxylic acids & derivatives (23) and fatty acids & conjugates (3) were (↓) ($p < 0.05$) in SSD. They all correlated negatively with disease duration while (↑) *Clostridium symbiosum* and *Eggerthella lenta,* (↑) CCL2, and (↓) cholic acid correlated positively. PEs, PCs, and sphingolipids correlated negatively with inflammatory cytokines, while 7 (↑) amino acids correlated positively 1-methylnicotinamide (neuroprotective) decreased with disease progression

Relapsed/FE SSD serum vs HC (↓) µg/mL: Pro/20.82, Met/4.07, Arg/13.52, Gln/8.67, Asp/1.12, Trp/0.72 ($p < 0.001$ for each) **Relapsed/FE SSD serum vs HC (↑) µg/mL:** Gly/24.39 ($p = 0.005$), Leu/20.46 ($p = 0.002$) Ma et al. [25]	Feces: Illumina platform (paired-end library: 350-bp; read length: 150-bp; number of sequences and good's coverage: NI) Serum: LC/MS (Quattro Premier XE) for AAs Note: Data were adjusted for age, BMI, years of education, gender, smoking, and alcohol drinking	**Training cohort (TC):** Relapsed/FE SSD (40/36)/27.99 (9.57). Patients were divided according to serum [Met, Leu, Arg, Gln] levels into low, medium, and high HC (40/39)/33.47 (12.58) **Testing cohort (TeC):** SSD (48) vs HC (49) TC AUC: 99.0% [Met, Leu, Arg, Gln] TeC AUC: 89.8%. Cohort nationality: China (Xi'an, Shaanxi Province)	**Bacteroidota** Rikenellaceae: *Alistipes* **Bacillota** Veillonellaceae: *Mitsuokella multacida* Oscillospiraceae: *Oscillibacter* Clostridiaceae: *Clostridium*	**Low-level [predictors] vs mid-level and HC** (↑) ($p < 0.007$): α-diversity richness and evenness There was no difference in phyla abundance between cohorts **Low-level [predictors] vs HC** (↑) ($p = 0.03$): *Mitsuokella* and *Oscillibacter* Many species were not identified	Arginine correlated negatively with positive PANSS scores and with excitation, cognitive, anxiety, and depression factors. 15 amino acids were positively correlated with different aspects of cognitive function ($p < 0.05$). *Mitsuokella* correlated positively with Gln ($p = 0.0004$) *Oscillibacter* correlated negatively with Met ($p = 0.004$)

(continued)

Table 5.1 (continued)

Cohort (effect in concentration): metabolite/concentration	Sample platform	Cohort (n = male/female/Age[a] AUC (95% CI) [Predictor])	Altered bacteria (Phylum[b]/Family: Genus)	Correlations Bacterial diversity Differential abundance	Pathway/symptom
65 fecal metabolites showed significant changes between cohorts (11 with FDR ≤ 0.01) Gao et al. [23]	Feces Illumina NovaSeq PE250 platform (values of paired-end library, read length, number of sequences and good's coverage: NI) UPLC-Orbitrap/MS for untargeted metabolomics	Paranoid P-SSD (14)/41.64 (11.57) Undifferentiated U-SSD (14)/45.64 (13.35) HC (11)/40.27 (6.36) **P-SSD vs HC AUC:** 99.4% [Bacteroidota], 97.4% [Rhizobium]. **U-SSD vs HC AUC:** 94.2% [Pseudomonadota], 96.1% [Fusicatenibacter or Bacteroides] **P-SSD vs U-SDD AUC:** 90.8% [Bacillota], 85.7% [Rhizobium] Cohort nationality: China (Hainan Province)	**Pseudomonadota** Bacteroidaceae: *Bacteroides* Enterobacteriaceae: *Enterobacter* Comamonadaceae: *Delftia*, *Comamonas* Yersiniaceae: *Serratia* Rhizobiaceae: *Rhizobium* **Fusobacteriota** Fusobacteriaceae: *Fusobacterium* **Bacteroidota** Tannerellaceae: *Parabacteroides* Prevotellaceae: *Prevotella*, *Alloprevotella* **Bacillota** Lachnospiraceae: *Blautia*, *Roseburia*, *Fusicatenibacter*, *Coprococcus*, *Lachnospira*, *Agathobacter* Ruminococcaceae: *Faecalibacterium* Veillonellaceae: *Dialister* Oscillospiraceae: *Subdoligranulum*, *Ruminococcus* Selenomonadaceae: *Megamonas* **Actinomycetota** Bifidobacteriaceae: *Bifidobacterium* Coriobacteriaceae: *Collinsella*	α-Diversity and richness: NS between cohorts β-Diversity: DS between cohorts ($p < 0.001$) All bacteria below showed DS ($p \leq 0.005$): **HC vs P-SSD and U-SSD (↑):** *Bacteroides*, *Fusicatenibacter*, and *Lachnospira* **P-SSD vs HC (↑):** *Serratia*, *Comamonas*, *Delftia*, *Rhizobium* **U-SSD (↑):** *Enterobacter*, *Intestinibacter*, *Haemophilus*, and *Comamonas*	Pathway enrichment analysis (KEGG) for differential metabolites showed altered His, Ala, Asp and Glu metabolism; protein digestion and absorption; secondary bile acid biosynthesis; aminoacyl-tRNA biosynthesis; neuroactive ligand-receptor interaction among others ($p <$ 1e-4, FDR: NI). Pentadecanoic acid was negatively correlated with *Rhizobium*, *Delftia* and *Comamonas* ($r = -0.797$, -0.774 and -0.752, respectively) Lachnospiraceae_UCG-010 correlated with the highest number of metabolites (12) and correlated negatively with menatetrenone, a form of vitamin K2 ($r = -0.662$)

Drug-treated SSD feces vs HC($\uparrow$): , DA ($p < 0.001$); BSCFAs (isobutyric, isovaleric), Phe, Tyr, Ser, IAA, spermidine, putrescine ($p < 0.05$); decarboxylase, transaminase, deaminase ($p < 1e-4$); KYN, total AAs, Gln, indole, urea ($p < 0.01$) **Drug-treated SSD feces vs HC ($\downarrow$):** GABA ($p < 0.05$) **Drug-treated SSD plasma vs HC ($\downarrow$):** 8 fatty acids ($p < 0.05$) Liang et al. [22]	Feces: Illumina HiSeq X ten platform (paired-end library: 150-bp; read length: 150-bp; number of sequences per sample: 10Gb; good's coverage: NI) Plasma & Feces: GC/MS for untargeted metabolomics of SCFAs, MCFAs, LCFAs UPLC-QTOF/MS for targeted metabolomics of AA, KYNA, KYN etc.	Drug-treated SSD (66/34)/43.11 (9.81) HC (32/21)/48.92 (12.95) **Drug-treated SSD vs HC AUC:** 85% [38 amino acid-biosynthesis pathways] **AUC:** 91% [22 aminoacyl-tRNA transferases] This result indicates a shift in intestinal metabolism from the typical protein synthesis observed under normal conditions to protein catabolism in SSD patients Cohort nationality: China (Gaizhou, Liaoning Province)	**Pseudomonadota** Bacteroidaceae: *Bacteroides* Sutterellaceae: *Sutterella* **Fusobacteriota** Fusobacteriaceae: *Fusobacterium* **Thermodesulfobacteriota** Desulfovibrionaceae: *Desulfovibrio* **Bacteroidota** Rikenellaceae: *Alistipes* Tannerellaceae: *Parabacteroides* Prevotellaceae: *Prevotella* **Bacillota** Lachnospiraceae: *Dorea, Roseburia, Coprococcus* Streptococcaceae: *Streptococcus* Eubacteriaceae: *Eubacterium* Acidaminococcaceae: *Phascolarctobacterium* Clostridiaceae: *Clostridium* **Actinomycetota** Eggerthelaceae: *Eggerthella*	α- and β-diversity: NS between cohorts 4 of the 22 SSD-enriched species ($p < 0.05$, FDR < 0.015) were unable to metabolize carbohydrates: *Fusobacterium mortiferum* (FC = 140), *Desulfovibrio piger* (FC = 10), *Phascolarctobacterium succinatutens* (FC = 4), and *Sutterella wadsworthensis* (FC = 3). Lachnospiraceae ($\uparrow$) ($p < 0.05$) Saccharolytic species ($\downarrow$) ($p < 0.05$): *Clostridium, Dorea, Eubacterium, Coprococcus, Bacteroides, Roseburia, Prevotella, Parabacteroides, Streptococcus*	Intestinal microbial fermentation changed from carbohydrates to proteins, with significantly reduced anaerobic glycolysis and starch degradation Thr, Trp, indole, IAA, and choline correlated positively with negative PANSS scores while Ser, Thr, Trp, Lys, and KYNA correlated positively with positive PANSS scores ($p < 0.05$ for both). There was a significant increase in fermentation products of Trp (KYN, IAA, indole)

(continued)

Table 5.1 (continued)

Cohort (effect in concentration): metabolite/concentration	Sample platform	Cohort (n = male/female)/Age[a] AUC (95% CI) [Predictor]	Altered bacteria (Phylum[b]/Family: Genus)	Correlations Bacterial diversity Differential abundance	Pathway/symptom
SSD serum vs HC (↑) pg/mL: Eotaxin, IL-1β, IL-4, IL-6, IL-8, MIP-1α, TNF-α ($p < 0.05$ for each) **SSD serum vs HC (↓) pg/mL:** IFN-γ, IL-9, IL-1Rα, IL-13, MCP-1, MIP-1β and RANTES ($p < 0.05$ for each) Ling et al. [24]	Feces: Illumina MiSeq (paired-end library and read length: NI; number of sequences per sample: 33.9 Mb; good's coverage: 98.80%) Serum: Bio-Plex 200 system for immune-inflammatory factors and cytokines	SSD (90)/> 62 years old HC (71) Authors did not report demographic data AUC: 95.7% [*Prevotella, Faecalibacterium, Roseburia, Actinomyces, Butyricicoccus*] AUC: 81.8% [*Faecalibacterium*] AUC: 76.4% [*Butyricicoccus*], AUC: 71.3% [*Roseburia*] Cohort nationality: China (Lishui, Zhejiang Province)	**Pseudomonadota** Sutterellaceae: *Sutterella* **Fusobacteriota** Fusobacteriaceae: *Fusobacterium* **Thermodesulfobacteriota** Desulfovibrionaceae: *Desulfovibrio* **Verrucomicrobiota** Akkermansiaceae: *Akkermansia* **Bacteroidota** Prevotellaceae: *Prevotella, Alloprevotella* **Bacillota** Lachnospiraceae: *Roseburia, Fusicatenibacter, Coprococcus* Ruminococcaceae: *Faecalibacterium* Streptococcaceae: *Streptococcus* Oscillospiraceae: *Gemmiger, Butyricicoccus* Peptostreptococcaceae: *Romboutsia* Selenomonadaceae: *Mitsuokella* **Actinomycetota** Actinomycetaceae: *Actinomyces* Bifidobacteriaceae: *Bifidobacterium*	α-Diversity and richness: NS between cohorts β-Diversity: DS between cohorts ($p < 0.01$) **SSD vs HC (↑)** ($p < 0.05$ for each): Bacteroidota, Actinomycetota, Verrucomicrobiota **SSD vs HC (↓):** Lachnospiraceae and Ruminococcaceae **SSD vs HC (↓):** *Roseburia, Butyricicoccus, Faecalibacterium, Coprococcus, Gemmiger, Fusicatenibacter* **SSD vs HC (↑):** *Sutterella, Prevotella, Actinomyces, Akkermansia, Romboutsia, Streptococcus, Mitsuokella, Desulfovibrio*	SSD metabolism (↑) ($p < 0.05$ for each): cofactors, vitamins, transport, catabolism Microbiota metabolism (↑): Lipoic acid metabolism, folate, lipopolysaccharide biosynthesis, Microbiota metabolism (↓): Bile acid and fatty acid biosynthesis *Actinomyces* & *Prevotella* showed inverse correlations with altered cytokines and chemokines. Butyrate-producing genera, such as *Roseburia, Faecalibacterium,* and *Butyricicoccus* correlated negatively with altered cytokines

SSD blood vs HC (↑): Hs-CRP/3.18 mg/L ($p = 0.18$), 33% CVD risk TNF-α/2.88 pg/mL ($p = 0.93$), IL-6/1.10 pg/mL ($p = 0.16$), IL-10/0.42 pg/mL ($p = 0.68$) Nguyen et al. [19]	Feces: Illumina MiSeq and HiSeq 4000 platforms (paired-end library: NI; read length: 150-bp), number of sequences *per* sample: minimum of 7.9 Mb (Median 47.6 Mb; good's coverage: NI) Blood: SECTOR Imager 2400 for cytokines; ELISA for CRP	SSD (83% drug-treated) (29/19)/53.2 (10.3) HC (29/19)/54.1 (12.6) **AUC:** 82% [all bacteria] Note: The feature importance score for each bacterium was <2%, indicating that no single feature(s) contributed to model performance in any significant way Cohort nationality: USA (San Diego)	**Pseudomonadota** Bacteroidaceae: *Bacteroides* **Thermodesulfobacteriota** Desulfovibrionaceae: *Desulfovibrio* **Verrucomicrobiota** Akkermansiaceae: *Akkermansia* **Bacteroidota** Tannerellaceae: *Parabacteroides* Prevotellaceae: *Prevotella* Odoribacteraceae: *Butyricimonas* Porphyromonadaceae: *Porphyromonas* **Bacilota** Lachnospiraceae: *Coprococcus, Lachnospira* Veillonellaceae: *Veillonella, Dialister* Ruminococcaceae: *Anaerotruncus, Ruminococcus, Oscillospira* **Actinomycetota** Micrococcaceae: *Rothia*	α-Diversity: NS between cohorts β-Diversity: DS between cohorts ($p = 0.001$), for KEGG orthologs ($p = 0.008$) and for EC numbers ($p = 0.007$) **SSD vs HC (↑)** ($p < 0.002$): Lachnospiraceae (butyrate-producing) No phylum or genus was found with significant abundance (880 mOTUs)	Cytokines and hs-CRP: NS between cohorts SSD (↓) ($q < 0.001$): TMAO reductase and Kdo2-lipid A biosynthesis (Bottom-20 pathways) SSD (↑) ($q < 0.001$): Glycerol degradation to 1,3-propanediol (Top-20 pathways). TMAO reductase correlated positively with CVD risk. Kdo2-lipid A correlated negatively with anti-inflammatory IL-10

(continued)

Table 5.1 (continued)

Cohort (effect in concentration): metabolite/concentration	Sample platform	Cohort (n = male/female/Age[a] AUC (95% CI) [Predictor]	Altered bacteria (Phylum[b]/Family: Genus)	Correlations Bacterial diversity Differential abundance	Pathway/symptom
Drug-naïve SSD vs HC (↑): Hs-CRP/0.28 mg/L ($p = 0.04$), HCY/14.81 mmol/L ($p = 1.49E-6$) **Risp-treated vs HC** (↑): Hs-CRP ($p < 0.05$) **Risp-treated vs HC** (↓): HCY ($p < 0.05$) Yuan et al. [20]	Feces: Ion S5 XL platform (single-end library, read length, number of sequences and good's coverage: NI) Serum: Immunoturbidimetric assay for HCY and hs-CRP Note: Data were adjusted for gender, age, and BMI	FE drug-naïve SSD (51/56)/19 [19.0-25.0] HC (37/70)/23 [22.0-25.0] Risp-treated SSD for 6 months (60) Training cohort (TC): SSD (82) vs HC (86) Testing cohort (TeC): SSD (25) vs HC (21) TC AUC: 87.9 (83–93) % [Romboutsia, Blautia, Streptococcus, Weissella, Anaerostipes, Alistipes, Terrisporobacter, Haemophilus, Dorea, Lachnoclostridium] TeC AUC: 86.7 (74–99) % Cohort nationality: China (Zhengzhou, Henan Province)	**Pseudomonadota** Pasteurellaceae: *Haemophilus* Enterobacteriaceae: *Citrobacter* **Bacteroidota** Rikenellaceae: *Alistipes* **Bacillota** Lachnospiraceae: *Lachnoclostridium, Blautia, Dorea, Fusicatenibacter, Anaerostipes* Streptococcaceae: *Streptococcus* Peptostreptococcaceae: *Romboutsia, Terrisporobacter* Streptococcaceae: *Streptococcus, Lactococcus* Lactobacillaceae: *Weissella* Enterococcaceae: *Enterococcus* **Actinomycetota** Bifidobacteriaceae: *Bifidobacterium*	**Drug-naïve SSD vs HC** α-diversity: Shannon (↓), $p = 1.21e-9$; PhyD (↑), $p = 0.005$; Simpson (↓), $p = 1.23e-8$ Chao1, ACE, observed species: NS between cohorts MDI (↑), $p = 1.3E-15$ **Drug-naïve vs HC** β-diversity: (↑) ($p < 0.05$) **Drug-naïve vs HC** (↑): *Lachnoclostridium*, adjusted $p = 0.001$ **Drug-naïve vs HC** (↓): 8 genera, adjusted $p < 1E-8$, e.g., *Streptococcus, Blautia, Romboutsia, Dorea, Lactococcus* **Risp-treated vs HC** (↓): *Lachnoclostridium*, $p = 0.02$	Hs-CRP at baseline correlated positively with (↓) PANSS total scores after risp treatment ($p = 0.02$), HCY at baseline correlated positively with positive (↓) PANSS scores after risp treatment ($p = 0.01$) *Lachnoclostridium* and Bacillotta levels correlated positively with negative PANSS scores ($p = 0.002$). (↓) *Lachnoclostridium* correlated positively with PANSS scores ($p = 0.002$). (↑) *Romboutsia* ($p = 0.067$) correlated negatively with PANSS scores ($p = 0.014$)

FE/relapsed SSD vs HC (↓): Trp/0.73 μM ($p < 2.2E-16$) FE/relapsed SSD vs HC (↑): KYNA/0.03 μM ($p < 3.7E-8$) Zhu et al. [18]	Feces: Illumina platform for TC cohort (paired end library: 350-bp; read length: 150-bp); BGISEQ-500 platform for TeC cohort (single-end library and read length: 100 bp). Number of sequences per sample: 11.46 to 13.13 Gb; good's coverage: NI Serum: ELISA for DA, GABA, KYNA, KYN, Ser. HPLC for Trp	Training cohort (TC): FE/relapsed drug-free SSD (90) HC (81) Age was not reported Testing cohort (TeC): Drug-treated SSD (45) vs HC (45) TC AUC: 89.6% TeC AUC: 76.5% [26 mOTUs] Note: Data were adjusted for gender, age, diet, and BMI Cohort nationality: China (Xi'an, Shaanxi Province)	**Pseudomonadota** Enterobacteriaceae: *Cronobacter* Bacteroidaceae: *Bacteroides* **Verrucomicrobiota** Akkermansiaceae: *Akkermansia* **Bacillota** Lachnospiraceae: *Dorea* Ruminococcaceae: *Anaerotruncus* Veillonellaceae: *Veillonella*, *Dialister*, *Megasphaera* Streptococcaceae: *Streptococcus* Clostridiaceae: *Clostridium*, *Alkaliphilus* Acidaminococcaceae: *Acidaminococcus* Lactobacillaceae: *Lactobacillus* Eubacteriaceae: *Eubacterium* Enterococcaceae: *Enterococcus* **Actinomycetota** Bifidobacteriaceae: *Bifidobacterium*	**FE/relapsed vs HC**: α- and β-diversity increased at the genus level ($p = 0.027$ and < 0.001, respectively) **FE/relapsed vs HC** (↑): *Veillonella atypica*, *Veillonella dispar*, *Bifidobacterium dentium*, *Dialister invisus*, *Lactobacillus oris*, and *Streptococcus salivarius* Trp level correlated negatively with the abundances of 38 bacterial species; KYNA level correlated positively with 10 bacterial species	*Streptococcus vestibularis*, one of the 26 mOTUs, was associated with Glu and isovaleric acid synthesis and GABA degradation and was correlated negatively with BACS. Transplantation into mice caused deficits in sociability and social novelty **SSD depleted microbial functional modules:** Glu/Asp transport system, lipopolysaccharide biosynthesis, beta-carotene biosynthesis **SSD enriched functional modules:** GABA shunt, Mn, Zn, and Fe transport system

(continued)

Table 5.1 (continued)

Cohort (effect in concentration): metabolite/concentration	Sample platform	Cohort (n = male/female)/Age[a] AUC (95% CI) [Predictor]	Altered bacteria (Phylum[b]/Family: Genus)	Correlations Bacterial diversity Differential abundance	Pathway/symptom
SSD vs HC (↑): GOGAT/1300 ± 850 nmol/min/g ($p = 0.002$), IgA/320 ± 170 OD 450 nm ($p = 1.16e-8$) Xu et al. [17]	Feces Illumina Hiseq2500 (paired end library: 150-bp: read length: NI; number of sequences *per* sample: 2 Gb; good's coverage: NI) ELISA for GOGAT and IgA Note: Data were adjusted for gender	**Training cohort (TC):** SSD (20/20)/35 (11) HC (20/20) 34 (9) **Testing cohort (TeC):** SSD (28/16)/35 (11) HC (21/16) 35 (11) TeC AUC: 86% [IgA, GOGAT, and MDI] Cohort nationality: China (Shenzhen city)	**Verrucomicrobiota** Akkermansiaceae: *Akkermansia* **Bacillota** Veillonellaceae: *Megasphaera* Clostridiaceae: *Clostridium* Lactobacillaceae: *Lactobacillus* Enterococcaceae: *Enterococcus* **Actinomycetota** Bifidobacteriaceae: *Bifidobacterium* Eggerthellaceae: *Eggerthella*	**SSD vs HC (↓):** Richness (Chao 1) ($p < 0.05$) and MDI ($p < 0.001$) MDI correlated negatively with decreased richness ($p = 0.02$) **SSD vs HC (↑):** ($p <$ E-3) *Eggerthella, Megasphaera, Clostridium, Lactobacillus, Akkermansia, Bifidobacterium* SSD (↓): *Enterococcus* ($p < 0.001$)	IgA correlated positively with MDI. GOGAT activity correlated positively with (↑) Deltaproteobacteria class and correlated negatively with (↓) Rhodocyclales order. These last two correlations were associated with increased levels of IgA, indicating that gut glutamate metabolism in SSD patients may be associated with intestinal immune alterations

No metabolites were investigated. Shen et al. [16]	Feces Illumina HiSeq 2500 platform (paired-end library: 250-bp; read length: 453-bp; number of sequences per sample: 21.19 to 107.17 Mb; good's coverage: 99.79%)	SSD (36/28)/42 (11) HC (35/13)/39 (14) **SSD vs HC AUC: 83.7%** [Gammaproteobacteria (class level), Enterobacteriales (order level), Alcaligenaceae, Enterobacteriaceae and Lachnospiraceae (family level), *Acidaminococcus*, *Phascolarctobacterium*, *Blautia*, *Desulfovibrio* and *Megasphaera* (genus level) and *Plebeius fragilis* (species level)] Cohort nationality: China (Huludao, Liaoning Province)	**Pseudomonadota** Bacteroidaceae: *Bacteroides* Succinivibrionaceae: *Succinivibrio* Enterobacteriaceae: *Klebsiella* **Bacteroidota** Prevotellaceae: *Prevotella* **Bacillota** Clostridiaceae: *Clostridium*, Lachnospiraceae: *Coprococcus*, *Roseburia*, *Blautia* Veillonellaceae: *Megasphaera* Ruminococcaceae: *Faecalibacterium* **Actinomycetota** Coriobacteriaceae: *Collinsella*	α-Diversity: NS between cohorts β-Diversity: DS between cohorts ($p = 0.016$) **SSD (↑):** Pseudomonadota phylum ($p = 0.004$) **SSD (↑)** ($p < 0.05$): *Megasphaera*, *Klebsiella*, *Clostridium*, *Succinivibrio*, *Methanobrevibacter* **SSD species (↑)** ($p < 0.05$): *Collinsella aerofaciens*, *Bacteroides fragilis* **SSD species (↓)** ($p < 0.05$): *Roseburia faecis*, *Blautia producta*, *Coprococcus* sp.	*Blautia*, *Coprococcus*, and *Roseburia* were negatively associated with vitamin B6, and with the glutathione and taurine/hypotaurine metabolic pathway. They positively correlated with the methane metabolic pathway

(continued)

Table 5.1 (continued)

Cohort (effect in concentration): metabolite/concentration	Sample platform	Cohort (n = male/female)/Age[a] AUC (95% CI) [Predictor]	Altered bacteria (Phylum[b]/Family: Genus)	Correlations Bacterial diversity Differential abundance	Pathway/symptom
Drug-naïve vs HC (↑): Neutrophil lymphocyte ratio (NLR; $p < 0.01$) **Risp-treated vs HC (↑):** NLR ($p < 0.001$) No significantly altered metabolites Gokulakrishnan et al. [21]	Feces Illumina MiSeq (paired-end library: 2 x 250-bp; read length: 500-bp; number of sequences per sample and good's coverage: No information was provided	Drug-naïve SSD (9/11)/33 (8.1) Risp-treated SSD (8/12)/34 (9.3) HC (11/9)/32 (5.6) All models used [*Ruminococcus*, UCG005, *Clostridium-sensu-stricto-1*, *Bifidobacterium*] **SSD vs HC** AUC: 79 (69–91) % **Drug-naïve SSD vs HC** AUC: 68 (52–85) % **Risp-treated SSD vs HC** AUC: 93 (84–100) % **Drug-naïve SSD vs Risp-treated SSD** AUC: 87 (75–98) % Cohort nationality: India (Bengaluru)	**Pseudomonadota** Bacteroidaceae: *Bacteroides* Moraxellaceae: *Acinetobacter* **Bacteroidota** Prevotellaceae: *Prevotella* **Bacillota** Lachnospiraceae: *Dorea* Ruminococcaceae: *Ruminococcus*, *Faecalibacterium*, *Subdoligranulum*, Clostridiaceae: *Clostridium* Veillonellaceae: *Dialister* Eubacteriaceae: *Eubacterium* **Actinomycetota** Bifidobacteriaceae: *Bifidobacterium*	α-Diversity: NS between cohorts β-Diversity: DS between cohorts ($p = 0.02$) **Risp-treated SSD vs HC** (↑) ($p < 0.01$): UCG005, *Clostridium-sensu-stricto-1* **Risp-treated SSD vs drug-naïve SSD** (↑) ($p < 0.01$): *Ruminococcus*, UCG00, *Eubacterium siraeum* **Drug-naïve SSD vs risp-treated SSD** (↑) ($p < 0.01$): *Acinetobacter*	Pathway-symptom associations were not investigated. Authors discussed relevant findings from the literature. Specifically, it was reported that an increase in negative symptoms correlated with a decrease in the abundance of the Ruminococcaceae. The treatment with Risp improved the abundance of Ruminococcaceae. Previous studies have demonstrated that *Eubacterium siraeum* is associated with glutamate metabolism and is negatively correlated with the severity of symptoms in SSD

AA amino acids, *ACE* abundance-based coverage estimator, *Ala* alanine, *Arg* arginine, *Asp* aspartate, *AUC* area under the curve, *BACS* Brief Assessment of Cognition Scores, *BSCFA* branched short-chain fatty acid, *BMI* body mass index, *CCL2* chemokine (C-C motif) ligand 2, *Chao1* richness estimator, *CVD* cardiovascular disease, *DA* dopamine, *DS* differed significantly, *ELISA* Enzyme Linked Immunosorbent Assay, *FA* fatty acid, *FC* fold change, *FDR* false discovery rate, *FE* first episode, *GABA* gamma-aminobutyric acid, *GOGAT* glutamate synthase, *Gln* glutamine, *Glu* glutamate, *Gly* glycine, *HC* healthy control, *HCY* homocysteine, *His* histidine, *HPLC* high performance liquid chromatography, *hs-CRP* high-sensitivity C-reactive protein, *IAA* indole-3-acetic-acid, *IgA* immunoglobulin A, *IQR* interquartile range, *KEGG* Kyoto Encyclopedia of Genes and Genomes, *KYNA* kynurenic acid, *KYN* kynurenine, *LCFA* long-chain fatty acid, *Leu* leucine, *LC/MS* liquid chromatography coupled to mass spectrometer, *Lys* lysine, *lysoPC* lysophosphatidylcholine, *MCFA* medium-chain fatty acid, *MCP1* monocyte chemoattractant protein-1, *MDI* microbial dysbiosis index, *Met* methionine, *MIP-1α* macrophage inflammatory protein-1α, *mOTU* metagenomic operational taxonomic unit, *NS* not significant, *NI* not informed, *Orbitrap* ion trap mass analyzer, *PANSS* Positive and Negative Syndrome Scale, *PC* phosphatidylcholine, *PE* phosphatidylethanolamine, *Phe* phenylalanine, *PhyD* phylogenetic diversity index, *Pro* proline, *QTOF* quadrupole time of flight mass spectrometer, *r* Pearson correlation, *RANTES* regulated on activation normal T-cell expressed and secreted, *Risp-treated* risperidone-treated patient, *SCFA* short-chain fatty acid, *SD* standard deviation, *Ser* serine, *Thr* threonine, *TMAO* trimethylamine N-oxide, *TNF* tumor necrosis factor, *Trp* tryptophan, *Tyr* tyrosine, *UHPLC* ultra-high performance liquid chromatography

[a]Age is given as mean (standard deviation) or median [interquartile]

[b]Phyla are named according to the newest rules of the International Code of Nomenclature of Prokaryotes (ICNP) [27]: Firmicutes (Bacillota), Proteobacteria (Pseudomonadota), Bacteroidetes (Bacteroidota), Actinobacteria (Actinomycetota), Fusobacteria (Fusobacteriota)

models employing butyrate-producing genera. The most effective model emerged, incorporating a set of five genera, achieving an outstanding AUC of 96% [24].

Ma et al. adopted a serum-based approach, constructing a model using the four most altered serum amino acids, which exhibited strong correlations with some reduced fecal genera. This approach successfully distinguished SSD patients from controls [25]. Conversely, Fan's group focused on five significantly altered metabolites correlated with a panel of altered cytokines and gut bacteria to distinguish drug-free SSD patients from controls [26]. Both studies showcased excellent predictive performances (AUCs >90%) in training and testing cohorts, emphasizing the efficacy of integrating multi-field information for achieving more robust models. Fan's group provided a compelling insight into the weak correlation between gut microbes and inflammatory cytokines, suggesting that serum metabolites may play a more direct role in the inflammatory process of SSD, while the intestinal microbiota might primarily modulate immune activation indirectly.

5.3 Discussion

This review aimed to assess the potential of gut microbiota and associated metabolites as biomarkers for the development of diagnostic models for SSD, utilizing ROC analysis to evaluate model performance. The exploration of using gut microbiota for this purpose is a relatively recent development. A survey conducted on the PubMed database revealed that attention to this topic began in 2018, marked by the study conducted by Shen et al. [16]. This initial focus has since evolved, culminating in four studies in 2022 and three publications in 2023. Among the 11 articles analyzed, 9 originated from China, involving a total of 836 SSD patients, while 1 study was from the United States and another from India.

Based on the findings outlined in Table 5.1, the 30 metabolites displaying the most significant alterations were subjected to overrepresentation analysis for chemical structures using MetaboAnalyst 6.0 (December 2023). The results revealed that the most significantly enriched classes were amino acids and analogues (FDR 7.2E-21) and indolyl carboxylic acids and derivatives (FDR 0.01), as illustrated in Fig. 5.1.

Enrichment analysis serves as a crucial step in validating the relevance of identified metabolites, assessing whether they are overrepresented in pathways or functions associated with a particular disease and thereby enhancing confidence in their role as potential disease signatures [28]. When the 30 altered metabolites underwent enrichment analysis for disease signature, considering blood as the origin of the metabolites, MetaboAnalys, SSD emerged as the second-highest enriched condition (FDR 4.7E-30), as depicted in Fig. 5.2. Notably, diseases closely related to the SSD classification, such as colorectal cancer, epilepsy, and Alzheimer's disease, became easily distinguishable in a clinical diagnosis. This suggests that the identified signature holds promise for advancing the understanding of the disorder's underlying mechanisms and for developing more targeted diagnostic approaches. As research

5.3 Discussion

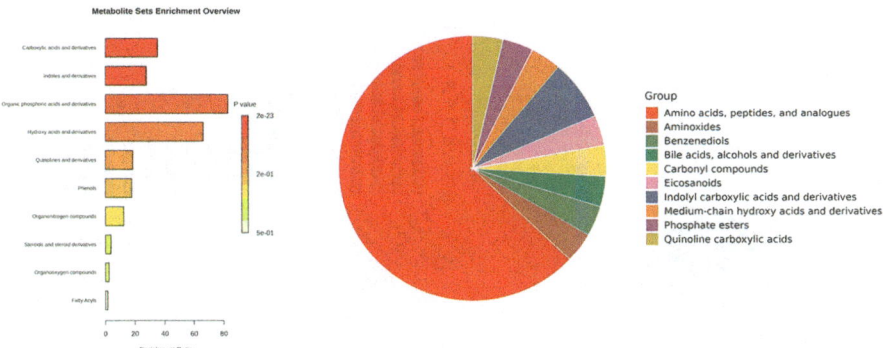

Fig. 5.1 Enrichment of chemical classes and subclasses of altered compounds in blood and feces of SSD patients

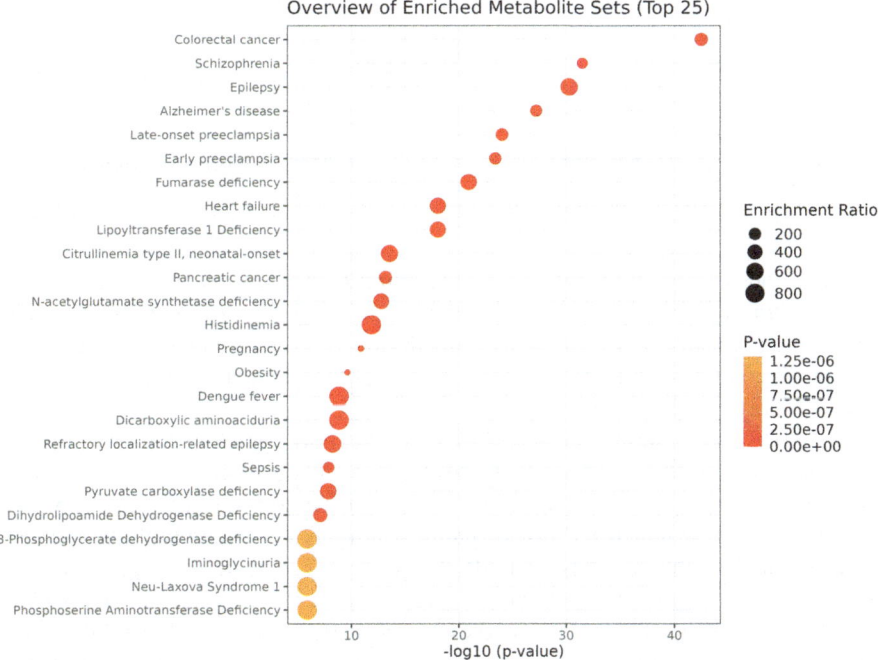

Fig. 5.2 Enrichment analysis in the disease signature using MetaboAnalyst 6.0 and the thirty altered metabolites identified from SSD patients across the eleven studies analyzed

in this field progresses, these metabolites may become integral components in the early detection and personalized management of SSD.

Moreover, this study identified significant alterations in 7 phyla, 34 families, and 62 genera of bacteria. As depicted in Fig. 5.3, Bacillota emerged as the most prevalent phylum across all studies (100%), followed by Actinomycetota (91%) and Pseudomonadota (82%). Notably, Lachnospiraceae (80%) stood out among

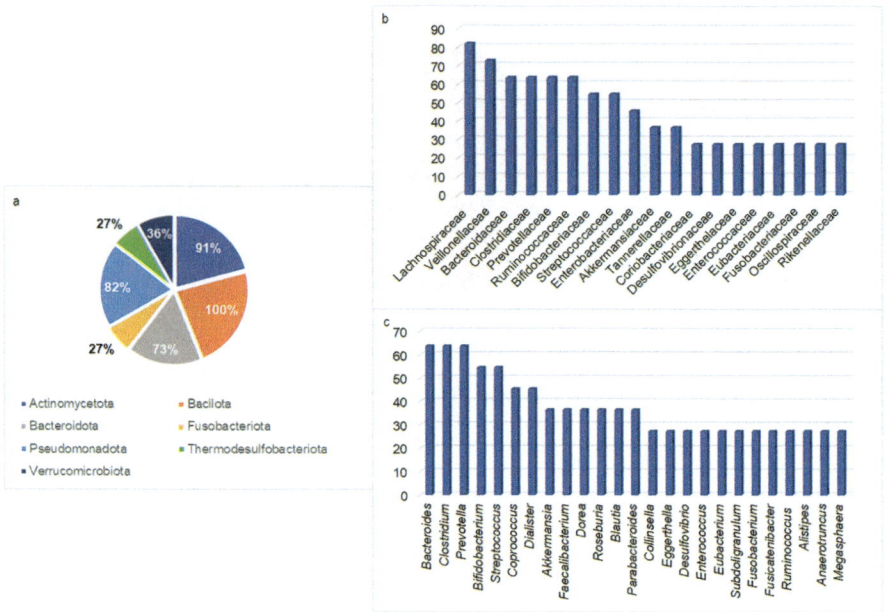

Fig. 5.3 (**a**) Most significantly altered phyla, (**b**) families, and (**c**) genera of gut microbiota in SSD patients. Results in percentages relative to the 11 studies analyzed

families, and its most frequent genera, including *Coprococcus*, *Dorea*, *Roseburia*, and *Blautia*, exhibited reduced abundance in the feces of SSD individuals. Interestingly, Lachnospiraceae has been identified as one of the most prevalent families in 10 countries globally, underscoring its potential as a biomarker that transcends racial and ethnic barriers [29].

Guided by these findings, researchers developed diagnostic models using diverse strategies, ranging from the exclusive utilization of metabolites to incorporating individual phyla, genera, or panels of altered bacteria. Some studies encompassed all identified dysregulated bacteria or pathways highly correlated with these bacteria as predictors in the models. Overall, the performance of these models was consistently deemed good or excellent, reflected in AUC values ranging from 80% to 99%. Notably, nearly 50% of the studies incorporated training and testing cohorts, representing a notable advancement compared to the methodologies discussed in preceding chapters.

References

1. Baquero F, Nombela C (2012) The microbiome as a human organ. Clin Microbiol Infect 18(Suppl 4):2–4. https://doi.org/10.1111/j.1469-0691.2012.03916.x
2. Dhanaraju R, Rao DN (2022) The human microbiome: an acquired organ? Reson 27:247–272. https://doi.org/10.1007/s12045-022-1312-7

3. McGuinness AJ, Davis JA, Dawson SL et al (2022) A systematic review of gut microbiota composition in observational studies of major depressive disorder, bipolar disorder and schizophrenia. Mol Psychiatry 27:1920–1935. https://doi.org/10.1038/s41380-022-01456-3
4. Clarke G, Stilling RM, Kennedy PJ, Stanton C, Cryan JF, Dinan TG (2014) Minireview: gut microbiota: the neglected endocrine organ. Mol Endocrinol 28:1221–1238. https://doi.org/10.1210/me.2014-1108
5. Zhu B, Wang X, Li L (2010) Human gut microbiome: the second genome of human body. Protein Cell 1:718–725. https://doi.org/10.1007/s13238-010-0093-z
6. Munawar N, Ahsan K, Muhammad K, Ahmad A, Anwar MA, Shah I, Al Ameri AK, Al Mughairbi F (2021) Hidden role of gut microbiome dysbiosis in schizophrenia: antipsychotics or psychobiotics as therapeutics? Int J Mol Sci 22:7671. https://doi.org/10.3390/ijms22147671
7. Hou K, Wu Z-X, Chen X-Y et al (2022) Microbiota in health and diseases. Signal Transduct Target Ther 7:1–28. https://doi.org/10.1038/s41392-022-00974-4
8. Rahman S, O'Connor AL, Becker SL, Patel RK, Martindale RG, Tsikitis VL (2023) Gut microbial metabolites and its impact on human health. Ann Gastroenterol Hepatol 36:360–368. https://doi.org/10.20524/aog.2023.0809
9. Nemani K, Hosseini Ghomi R, McCormick B, Fan X (2015) Schizophrenia and the gut-brain axis. Prog Neuro-Psychopharmacol Biol Psychiatry 56:155–160. https://doi.org/10.1016/j.pnpbp.2014.08.018
10. Yang C, Lin X, Wang X, Liu H, Huang J, Wang S (2022) The schizophrenia and gut microbiota: a bibliometric and visual analysis. Front Psych 13:1022472. https://doi.org/10.3389/fpsyt.2022.1022472
11. Nikolova VL, Smith MRB, Hall LJ, Cleare AJ, Stone JM, Young AH (2021) Perturbations in gut microbiota composition in psychiatric disorders: a review and meta-analysis. JAMA Psychiatry 78:1343–1354. https://doi.org/10.1001/jamapsychiatry.2021.2573
12. Szeligowski T, Yun AL, Lennox BR, Burnet PWJ (2020) The gut microbiome and schizophrenia: the current state of the field and clinical applications. Front Psych 11:156. https://doi.org/10.3389/fpsyt.2020.00156
13. Minichino A, Preston T, Fanshawe JB, Fusar-Poli P, McGuire P, Burnet PWJ, Lennox BR (2023) Psycho-pharmacomicrobiomics: a systematic review and meta-analysis. Biol Psychiatry 9:S0006-3223(23)01486-5. https://doi.org/10.1016/j.biopsych.2023.07.019
14. Murray N, Al Khalaf S, Bastiaanssen TFS, Kaulmann D, Lonergan E, Cryan JF, Clarke G, Khashan AS, O'Connor K (2023) Compositional and functional alterations in intestinal microbiota in patients with psychosis or schizophrenia: a systematic review and meta-analysis. Schizophr Bull 49:1239–1255. https://doi.org/10.1093/schbul/sbad049
15. Borkent J, Ioannou M, Laman JD, Haarman BCM, Sommer IEC (2022) Role of the gut microbiome in three major psychiatric disorders. Psychol Med 52:1222–1242. https://doi.org/10.1017/S0033291722000897
16. Shen Y, Xu J, Li Z, Huang Y, Yuan Y, Wang J, Zhang M, Hu S, Liang Y (2018) Analysis of gut microbiota diversity and auxiliary diagnosis as a biomarker in patients with schizophrenia: a cross-sectional study. Schizophr Res 197:470–477. https://doi.org/10.1016/j.schres.2018.01.002
17. Xu R, Wu B, Liang J et al (2020) Altered gut microbiota and mucosal immunity in patients with schizophrenia. Brain Behav Immun 85:120–127. https://doi.org/10.1016/j.bbi.2019.06.039
18. Zhu F, Ju Y, Wang W et al (2020) Metagenome-wide association of gut microbiome features for schizophrenia. Nat Commun 11:1612. https://doi.org/10.1038/s41467-020-15457-9
19. Nguyen TT, Kosciolek T, Daly RE, Vázquez-Baeza Y, Swafford A, Knight R, Jeste DV (2021) Gut microbiome in schizophrenia: altered functional pathways related to immune modulation and atherosclerotic risk. Brain Behav Immun 91:245–256. https://doi.org/10.1016/j.bbi.2020.10.003
20. Yuan X, Wang Y, Li X et al (2021) Gut microbial biomarkers for the treatment response in first-episode, drug-naïve schizophrenia: a 24-week follow-up study. Transl Psychiatry 11:422. https://doi.org/10.1038/s41398-021-01531-3

21. Gokulakrishnan K, Nikhil J, Viswanath B et al (2023) Comparison of gut microbiome profile in patients with schizophrenia and healthy controls – a plausible non-invasive biomarker? J Psychiatr Res 162:140–149. https://doi.org/10.1016/j.jpsychires.2023.05.021
22. Liang Y, Shi X, Shen Y, Huang Z, Wang J, Shao C, Chu Y, Chen J, Yu J, Kang Y (2022) Enhanced intestinal protein fermentation in schizophrenia. BMC Med 20:67. https://doi.org/10.1186/s12916-022-02261-z
23. Gao Y, Liu X, Pan M et al (2022) Integrated untargeted fecal metabolomics and gut microbiota strategy for screening potential biomarkers associated with schizophrenia. J Psychiatr Res 156:628–638. https://doi.org/10.1016/j.jpsychires.2022.10.072
24. Ling Z, Jin G, Yan X, Cheng Y, Shao L, Song Q, Liu X, Zhao L (2022) Fecal dysbiosis and immune dysfunction in Chinese elderly patients with schizophrenia: an observational study. Front Cell Infect Microbiol 12:886872. https://doi.org/10.3389/fcimb.2022.886872
25. Ma Q, Gao F, Zhou L et al (2022) Characterizing serum amino acids in schizophrenic patients: correlations with gut microbes. J Psychiatr Res 153:125–133. https://doi.org/10.1016/j.jpsychires.2022.07.006
26. Fan Y, Gao Y, Ma Q et al (2022) Multi-omics analysis reveals aberrant gut-metabolome-immune neetwork in schizophrenia. Front Immunol 13:812293. https://doi.org/10.3389/fimmu.2022.812293
27. Oren A, Garrity GM (2021) Valid publication of the names of forty-two phyla of prokaryotes. Int J Syst Evol Microbiol 71. https://doi.org/10.1099/ijsem.0.005056
28. Qiu S, Cai Y, Yao H, Lin C, Xie Y, Tang S, Zhang A (2023) Small molecule metabolites: discovery of biomarkers and therapeutic targets. Signal Transduct Target Ther 8:132. https://doi.org/10.1038/s41392-023-01399-3
29. Dhakan DB, Maji A, Sharma AK, Saxena R, Pulikkan J, Grace T, Gomez A, Scaria J, Amato KR, Sharma VK (2019) The unique composition of Indian gut microbiome, gene catalogue, and associated fecal metabolome deciphered using multi-omics approaches. Gigascience 8:giz004. https://doi.org/10.1093/gigascience/giz004

Chapter 6
Neuroendocrine Biomarkers

6.1 Introduction

The relationship between the neuroendocrine system (NES) and SSD is intricate and multi-faceted. The NES, encompassing the hypothalamus, pituitary gland, and adrenal glands, plays a pivotal role in regulating various physiological processes through hormone release. Notably, NES is implicated in the regulation of dopamine release, thyroid function, and the body's stress response. The activation of the hypothalamic-pituitary-adrenal (HPA) axis during stress leads to the release of cortisol, a stress hormone. The HPA axis serves to maintain homeostasis in response to environmental and internal stressors. Evidence suggests that individuals with SSD may experience hyper- or hypofunction of the HPA axis, potentially contributing to the premature mortality observed in this population [1]. Furthermore, HPA axis dysregulation is associated with elevated rates of cardiovascular diseases, metabolic syndrome, and alterations in immune responses. NES also interacts with the immune system, and there is a growing body of evidence pointing to immune system dysregulation in SSD, potentially contributing to the inflammatory processes associated with the disorder.

A systematic review of 16 studies highlighted elevated cortisol levels in saliva, particularly in individuals at genetic or clinical high risk (CHR) of psychosis [2]. A subsequent meta-analysis, incorporating 22 studies, demonstrated that blood cortisol levels were significantly increased in individuals experiencing their first episode (FE) of psychosis compared to controls [3]. In contrast, a recent study did not identify differences in serum aldosterone and cortisol concentrations measured at the time of admission, discharge, and one-year follow-up in FE non-affective psychosis patients [4].

In summary, the connection between the neuroendocrine system and SSD encompasses intricate interactions among hormonal regulation, neurotransmitter systems, stress response, and immune function. Dysregulation in these systems not only may contribute to the development and progression of SSD but also presents a potential source of valuable biomarkers for the disorder.

6.2 Neuroendocrine Biomarkers for SSD Assessed Through ROC Analysis

Assuming the time frame covered by this review (2011–2023), there has been a limited exploration of a biochemical panel of neuroendocrine markers for diagnosing SSD, a question assessed through ROC analysis. Models based on discriminatory analyses for diagnostic purposes have become notably scarce over the years. Most studies have predominantly concentrated on assessing the potential of neuroendocrine biomarkers for monitoring treatment response or for prognosis, as outlined in Table 6.1.

Given that major psychiatric diseases share several biochemical aspects involved in neuroendocrine changes, it is unlikely that a diagnostic test based on a single analyte will be able to effectively differentiate these disorders. Recognizing this challenge, Lyu et al. extracted demographic and routine biochemical data from over 3500 psychiatric inpatients to construct a model considering only cortisol and subsequently a panel containing 5 analytes [5]. However, regardless of the number of metabolites tested, the results demonstrated only moderate transdiagnostic potential to differentiate between SSD, bipolar disorder (BD), and major depressive disorder (MDD) patients.

In a distinct approach, Cai et al. [6] explored the potential of progesterone—a modulator of the dopaminergic pathway [7]—and inosine—a signaling molecule in the central nervous system [8]—to distinguish between good and poor responders during the treatment of SSD patients in both cross-sectional and longitudinal cohorts simultaneously. The study yielded satisfactory results, with AUCs ranging from 82% to 91%. Conversely, a model based on estradiol and other gonadal and pituitary hormones did not achieve the same success, despite approximately 80% of patients receiving estradiol showing a reduction in symptoms after two months [9].

Chan et al. conducted an extensive, multi-site, and multi-approach discriminatory analysis in approximately a thousand patients, encompassing those at CHR, pre-onset SDD and BD. They utilized a panel of 26 serum compounds, primarily proteins, and incorporated Comprehensive Assessment of At-Risk Mental State (CAARMS) positive subscale symptom scores into the model. While they achieved favorable results, it is essential to acknowledge the potential influence of confounding factors, such as hormonal variations related to sex [10]. Four years later, Cooper et al. obtained similar results (AUCs between 80% and 90%) employing a different approach based on multi-model inference, with a much smaller set of proteins. Additionally, they assessed first-order interactions between protein abundance and sex, adding a nuanced layer to their analysis [11].

6.2 Neuroendocrine Biomarkers for SSD Assessed Through ROC Analysis

Table 6.1 Potential neuroendocrine biomarkers for schizophrenia spectrum disorders (SSD) assessed by ROC analysis

Cohort (effect): biomarker/ concentration level	Sample platform biomarker function	Cohort (n = male/ female)/Age[a]	AUC (95%CI)/ sensitivity/ specificity (%) [Classifier]
SSD vs BD (↑): CRP, C4, HCY, ACTH, cortisol. **SSD vs BD or vs MDD (↓):** Folate, vitamin B12. **SSD vs MDD (↑):** CRP, C3, C4, HCY. Note: $p < 0.05$ for each compound. Lyu et al., 2021 [5]	Blood collected before 10 a.m. Data collected from inpatients' medical records. Transdiagnostic.	SSD (742/917)/35 [27, 49]. BD (719/1182)/33 [26, 47]. MDD (470/1051)/44 [26, 57]. Cohort nationality: China (Beijing).	**SSD vs BD:** 55 (53–57)/74/36 [cortisol] Cutoff: 13.67 µg/dL. **SSD vs BD:** 64 (62–66)/57/64 [cortisol, CRP, HCY, folate, vit. B12] Cutoff: 0.47 **SSD vs MDD:** 68 (66–70)/ 67/61, [age, gender, CRP, C4, HCY, folate, vitamin B12] Cutoff: 0.51.
Non-responder vs responder (↓): Progesterone, $p < 0.001$. **Non-responder vs responder (↑):** Inosine, $p < 0.001$. Cai et al., 2022 [6]	Plasma collected at 7 a.m. UPLC/ MS-MS. Theranostic.	Responder SSD (203/212)/30.2 (9.5). Non-responder SSD (123/169)/32.8 (10.3). Cohort nationality: China (Changsha).	**Non-responder vs responder:** 82.5 (79.5–85.6)/75/77, [progesterone] Cutoff: 55.8 ng/mL. **Non-responder vs responder:** 91.4 (89–94)/85/84, [inosine] Cutoff: 29.1 ng/mL.
Responder SSD 56 days (↓): FSH (FC = 0.9, $p = 0.008$). **Responder SSD 56 days (↑):** Estradiol (FC = 1.7, $p = 0.047$). LH, progesterone, testosterone, DHEA, and prolactin were not significantly different. Thomas et al., 2021 [9]	Serum. ECLIA and CMIA. Theranostic. Note: The blood collection time was not reported.	Baseline SSD (56 females)/35.29. After treatment with estradiol for 28 or 56 days: Responder SSD (44)/36.4 (8.2). Non-responder SSD (12)/33.3 (8.6). Cohort nationality: Australia (Melbourne).	**28 days-treat vs baseline:** 52 [estradiol]. **56 days-treat vs baseline:** 55 [estradiol]. Note: CI, sensitivity, specificity, and cutoff were not reported.

(continued)

Table 6.1 (continued)

Cohort (effect): biomarker/ concentration level	Sample platform biomarker function	Cohort (n = male/female)/Age[a]	AUC (95%CI)/ sensitivity/ specificity (%) [Classifier]
SMI vs HC (↑): Leptin (L) ($p < 0.01$). Adiponectin (A) level had no significant difference between cohorts. Reponen et al., 2021 [18]	Plasma collected between 8–11 a.m. RIA. Assessment of CVD Risk due to treatment. Diagnostic. Note: Data were adjusted for multiple confounding variables.	**SMI** [SSD (n = 701) + BD (n = 391)] (576/516))/32 (11). HC (113/63) 32 (8). **SMI subgroups:** Non-treat (139/162) 33 (12). APY (141/128) 30 (10). AP O/C/Q (296/226) 31 (10). Cohort nationality: Norway (Oslo, Trondheim, Lillehammer).	**Model 1:** L/A ratio identifying high TC/HDL AUCs 66–70%. **Model 2:** L/A ratio identifying high TG/HDL, AUCs 65–77%. **Model 3:** Leptin-TG/HDL, AUCs 54–67%. **Model 4:** Leptin-TC/HDL, AUCs 53–62%. **Model 5:** A-TG/HDL, AUCs 24–33%. **Model 6:** A-TC/HDL, AUCs 22–30%.
bFE SSD vs drug-treated FE SSD (↓): TSH (1–2.5 mU/L). **Drug-treated FE SSD vs bFE SSD (↓):** BDI ($p < 0.01$). Smierciak et al., 2022 [12]	Plasma collected between 6–8 a.m. Routine (20) and antioxidant potential (4) lab tests. MRI and ^{1}H-MRS of brain. Theranostic.	Baseline FE SSD (22/18)/22.68 (7.39). 12-weeks drug-treated FE SSD subgroups: TSH 1–2.5 mU/L (20), TSH 2.5–4.2 mU/L (17). Cohort nationality: Poland (Krakow).	**bFE SSD vs drug-treated FE SSD:** 80 [TSH at baseline, BDI-after 12 weeks treatment] Cutoff of TSH ≤ 2.5 mU/L. Note: CI, sensitivity, and specificity were not informed.

(continued)

6.2 Neuroendocrine Biomarkers for SSD Assessed Through ROC Analysis

Table 6.1 (continued)

Cohort (effect): biomarker/ concentration level	Sample platform biomarker function	Cohort (n = male/ female)/Age[a]	AUC (95%CI)/ sensitivity/ specificity (%) [Classifier]
Analytes tested: Testosterone (total), IgA, eotaxin, IL-10, IL-1RA, IL-8, leptin, FSH, TSH, macrophage inhibitory factor, APOA1, APOH, haptoglobin, tenascin C, PPP, alpha-2 macroglobulin, beta-2 macroglobulin, tyrosine protein kinase receptor, SCF complex, serum glutamic oxaloacetic transaminase, coagulation factor VII, carcinoembryonic antigen, von Willebrand factor, insulin-like growth factor binding *protein*-2, RAGE, VCAM-1, angiotensin converting enzyme, chromogranin-A. Chan et al., 2015 [10]	Serum. Multiplexed immunoassay platform. Prognostic and transdiagnostic. Note: Data were adjusted for age and sex.	**Training cohort (TC):** SSD (79/48)/31. HC (124/80)/31. **Testing cohort 1 (TeC1):** SSD (63/30)/29. HC (51/37)/33. **Testing cohort 2 (TeC2):** Pre-onset SSD (78/15)/22. HC (136/48)/22. CHR (33/25)/22. Pre-onset BD (70/40)/21. Cohort nationality: Germany (Mannheim, Magdeburg), Netherlands (Rotterdam).	**TC:** 96 (93.7–97.6)/90/90 [26 analytes]. **TeC1:** 97 (95.2–99.6) /87/97 [22 analytes]. **Subgroups of TeC2: Pre-onset SSD vs HC:** 90 (85.6–95.2)/88/81 [24 analytes]. **Pre-onset BD vs HC:** 53 (45.7–61.1)/25/86 [24 analytes]. **Pre-onset SSD vs pre-onset BD:** 91 (86.5–94.9)/88/83 [24 analytes]. **CHR vs pre-onset SSD:** 82 (70.6–92.5)/89/66 [22 analytes]. **CHR vs pre-onset SSD:** 90 (81.6–97.8)/89/79 [22 analytes + CAARMS scores].
Analytes tested: APOA4, APOC3, APOA2, APOH, histidine phosphotransferase (HPT), protease C1 inhibitor (IC1), inter-alpha-trypsin inhibitor heavy chain 4 (ITIH4), adenine nucleotide translocase 3 (ANT3). Cooper et al., 2019 [11]	Serum. LC/TQ-MS. Diagnostic. Note: Models were built with SSD-associated proteins according to data reported in the literature.	**Training cohort (TC):** FE drug-naïve SSD (31/29)/30.95. HC (43/34)/31.90. Germany (Cologne) **Testing cohort (TeC):** FE drug-naïve SSD (9/0)/28.3 (5.9). HC (12/0)/28.3 (7.7). Cohort nationality: Netherlands (Rotterdam).	**TC (both sexes):** 86 [APOA4, APOC3, APOA2, APOH, HPT, IC1, ITIH4, ANT3]. **TC (males):** 89. **TC (females):** 82. **TeC (males):** 89. **TC (both sexes):** 85 [APOA4, APOC3, APOA2, HPT, IC1, ITIH4]. **TC (males):** 89. **TC (females):** 82. **TeC (males):** 82. Note: CI, sensitivity, and specificity were not reported.

(continued)

Table 6.1 (continued)

Cohort (effect): biomarker/ concentration level	Sample platform biomarker function	Cohort (n = male/ female)/Age[a]	AUC (95%CI)/ sensitivity/ specificity (%) [Classifier]
Chronic SSD vs HC (↑): IL-2, cortisol. **Chronic SSD vs HC (↓):** IL-8. **Remitting vs non-remitting (↑):** IL-2, IL-6. Note: $p < 0.05$ for each compound. Shangguan et al., 2023 [16]	Serum collected between 7–9 a.m. RIA for cortisol. ELISA for interleukins. Diagnostic.	Chronic SSD (162) including 27 remitting patients. HC (62). Cohort nationality: China (Haidian-Beijing), Han ethnicity.	**Chronic SSD vs HC:** 60 [cortisol]. 67 [IL-2]. 75 [cortisol, IL-2]. **Remitting vs non-remitting:** 71 [IL-6]. Note: CI, sensitivity, and specificity were not reported.
Agitated SSD vs non-agitated (↑): CRP, FT4, FT3, uric acid. **Agitated SSD vs non-agitated (↓):** TSH, creatinine. Note: $p < 0.05$ for each compound. Li et al., 2021 [17]	Serum collected between 7–7:30 a.m. Immunoturbidimetry for CRP. Architect i2000sr for hormones. Cobas C702 for other compounds. Diagnostic.	Agitated SSD (36/35) /30 [22–39]. Non-agitated SSD (37/46)/34 [29–47] HC (35/40)/35 [30–46]. Cohort nationality: China (Jinan).	**Agitated SSD vs non-agitated:** 63 (54–72)/32/90 [CRP] Cutoff: 3.47 mg/mL. 73 (65–81)/75/68 [FT3] Cutoff: 2.79 pg/mL. 62 (53–71)/79/43 [TSH] Cutoff: 2.72 µIU/mL. 66 (58–75)/58/71 [creatinine] Cutoff: 75.76 µmol/mL.

ACTH adrenocorticotropic hormone, *APO* apolipoprotein, *AP O/C/Q* patients receiving olanzapine, clozapine, or quetiapine, *APY* patients under treatment other than O/C/Q, *AUC* area under the curve, *BD* bipolar disorder, *BDI-12* Beck Depression Inventory—the scale of severity of depressive symptoms after 12 weeks of treatment, *CAARMS* Comprehensive Assessment of At-Risk Mental State, *CHR* clinical high risk for psychosis, *C4* complement protein 4, *CI* confidence interval, *CMIA* chemiluminescent microparticle immunoassay, *CRP* C-reactive protein, *CVD* cardiovascular disease, *DHEA* dehydroepiandrosterone, *ELISA* enzyme linked immunosorbent assay, *FIA* fluorescent bead immunoassay, *FSH* follicle-stimulating hormone, *FT3* triiodothyronine, *FT4* free thyroxine, *HC* health control, *^{1}H–MRS* proton magnetic resonance spectroscopy, *HYC* homocysteine, *IQR* interquartile range, *LC/TQ-MS* liquid chromatography coupled to triple quadrupole mass spectrometer, *LH* luteinizing hormone, *MDD* major depressive disorder, *MRI* magnetic resonance image, *non-responder* patients unresponsive to treatment, *Post/Dex* blood collection after 9 hours of dexamethasone administration, *PPP* pancreatic polypeptide, *Pre/Dex* blood collection one day before dexamethasone administration, *RAGE* receptor for advanced glycosylation end products, *RIA* radioimmunoassay, *SMI* severe mental illnesses, *TC/HDL* total cholesterol/ high-density lipoprotein ratio, *TG/HDL* triglyceride/HDL ratio, *TSH* thyroid-stimulating hormone or thyrotropin, *VCAM-1* vascular cell adhesion molecule-1

[a]Age is given as the mean (standard deviation) or median [interquartile]

To comprehend the relationship between SSD and the hypothalamic, pituitary, and thyroid (HPT) axis, Śmerciak et al. developed a model that considered thyroid functioning, thyroid-stimulating hormone (TSH) levels, and their correlation with various biochemical, clinical, and imaging tests. Through unsupervised cluster analysis, two patient groups were identified based on the presented TSH levels (1–2.5 mU/L and 2.5–4.2 mU/L) [12]. The most notable association was found between the baseline TSH level and scores on the depression symptom severity scale after 12 weeks of treatment, as measured by the Beck Depression Inventory (BDI-12). Higher TSH levels were correlated with lower BDI scores, indicative of an improvement in the severity of illness (BDI scores ranging from 0–13 to 29–63, signifying minimal to severe depression). Consequently, these variables were incorporated into the model to predict treatment response, yielding an AUC of 80%.

Leptin (L) plays a role in suppressing cortisol secretion during HPA axis activation and may undergo alterations in SSD patients due to a desensitization mechanism induced by antipsychotics [13]. Conversely, low levels of adiponectin (A) have been associated with psychiatric illness and impaired HPA axis function [14]. Motivated by these observations, Reponen's group sought to explore the potential of dysregulated levels of adipokines and the ratio between them in predicting cardiovascular disease (CVD) risk among BD and SSD patients, irrespective of antipsychotic treatment. Correlating these adipokines with elevated atherogenic lipid ratios, the group found that predictive models based solely on adiponectin or leptin alone demonstrated poor performance (AUCs less than 30% and 60%, respectively). However, the combination of the L/A ratio with total cholesterol/HDL or triglycerides/HDL exhibited slightly more promising results, with AUCs around 70%. Notably, this trend was particularly evident among patients treated with antipsychotics other than olanzapine, chlorpromazine, and quetiapine [15].

Shangguan et al. assessed models constructed with cortisol and IL-2 to differentiate chronic and remitting patients from healthy controls. However, the obtained AUCs fell below 75% [16].

Following the observation of a robust association between C-reactive protein (CRP), FT3 (triiodothyronine), TSH, and creatinine with agitation in SSD, Li et al. developed a model utilizing these compounds as predictors to distinguish agitated SSD patients from those without agitation. However, the AUCs obtained were not highly promising, with a maximum value of 73% [17].

6.3 Discussion

The purpose of this review was to examine the utilization of neuroendocrine biomarkers over the past 12 years (2011–2023), employing ROC curve analysis. A limited number of articles, only 9 in total, were identified, with a notable concentration between 2021 and 2022. The research spanned seven countries and encompassed the analysis of over 3000 patients with SSD. The primary focus of inquiry was the potential of neuroendocrine biomarkers to track improvements during treatment or predict disease progression in individuals at CHR for the disorders.

In general, the overall performance of the models proved to be modest, with AUC values ranging from 22% to 80%. However, a noteworthy exception was identified—Chan et al. demonstrated a robust AUC of 90% [10]. Their model incorporated a diverse array of compound classes and integrated scores derived from the Comprehensive Assessment of At-Risk Mental States (CAARMS) clinical tool. CAARMS is designed to assess early signs and symptoms of psychosis, suggesting that the inclusion of a comprehensive set of factors may enhance predictive accuracy.

Similarly, Fernandes et al. achieved a comparable outcome by combining cognitive test scores with inflammatory biomarkers, as elaborated in Chap. 4 [18]. This pattern in findings implies a potential trend that merits further exploration. Notably, several studies have consistently indicated a strong correlation between hormonal and inflammatory factors and the trajectory of the disease, as well as the exacerbation of cognitive symptoms [19–24]. This convergence of findings underscores the importance of considering a broad spectrum of biomarkers and clinical assessments in understanding and predicting the course of SSD. Future research efforts may benefit from a comprehensive approach, integrating various classes of biomarkers and clinical tools, to enhance the accuracy and reliability of predictive models.

References

1. Bradley AJ, Dinan TG (2010) A systematic review of hypothalamic-pituitary-adrenal axis function in schizophrenia: implications for mortality. J Psychopharmacol 24:91–118. https://doi.org/10.1177/1359786810385491
2. Karanikas E, Garyfallos G (2015) Role of cortisol in patients at risk for psychosis mental state and psychopathological correlates: a systematic review. Psychiatry Clin Neurosci 69:268–282. https://doi.org/10.1111/pcn.12259
3. Hubbard DB, Miller BJ (2019) Meta-analysis of blood cortisol levels in individuals with first-episode psychosis. Psychoneuroendocrinology 104:269–275. https://doi.org/10.1016/j.psyneuen.2019.03.014
4. Obdržálková M, Ustohal L, Hlaváčová N, Mayerová M, Češková E, Kašpárek T, Ježová D (2023) Selected neuroendocrine factors as potential molecular biomarkers of early nonaffective psychosis course in relation to treatment outcome: a pilot study. Heliyon 9:e21173. https://doi.org/10.1016/j.heliyon.2023.e21173
5. Lyu N, Xing G, Yang J, Zhu X, Zhao X, Zhang L, Wang G (2021) Comparison of inflammatory, nutrient, and neurohormonal indicators in patients with schizophrenia, bipolar disorder and major depressive disorder. J Psychiatry Res 137:401–408. https://doi.org/10.1016/j.jpsychires.2021.03.010
6. Cai H, Zeng C, Zhang X et al (2022) Diminished treatment response in relapsed versus first-episode schizophrenia as revealed by a panel of blood-based biomarkers: a combined cross-sectional and longitudinal study. Psychiatry Res 316:114762. https://doi.org/10.1016/j.psychres.2022.114762
7. Sun J, Walker AJ, Dean B, van den Buuse M, Gogos A (2016) Progesterone: the neglected hormone in schizophrenia? A focus on progesterone-dopamine interactions. Psychoneuroendocrinology 74:126–140. https://doi.org/10.1016/j.psyneuen.2016.08.019
8. Nascimento FP, Macedo-Júnior SJ, Lapa-Costa FR, Cezar-Dos-Santos F, Santos ARS (2021) Inosine as a tool to understand and treat central nervous system disorders: a neglected actor? Front Neurosci 15:703783. https://doi.org/10.3389/fnins.2021.703783
9. Thomas N, Gurvich C, Hudaib AR, Gavrilidis E, de Castella RA, Thomas EH, Kulkarni J (2021) Serum estradiol as a blood-based biomarker predicting hormonal treatment outcomes in women with schizophrenia. Psychoneuroendocrinology 126:105165. https://doi.org/10.1016/j.psyneuen.2021.105165

10. Chan MK, Krebs M-O, Cox D et al (2015) Development of a blood-based molecular biomarker test for identification of schizophrenia before disease onset. Transl Psychiatry 5:e601. https://doi.org/10.1038/tp.2015.91
11. Cooper JD, Han SYS, Tomasik J, Ozcan S, Rustogi N, van Beveren NJM, Leweke FM, Bahn S (2019) Multimodel inference for biomarker development: an application to schizophrenia. Transl Psychiatry 9:83. https://doi.org/10.1038/s41398-019-0419-4
12. Śmierciak N, Szwajca M, Popiela TJ, Bryll A, Karcz P, Donicz P, Turek A, Krzyściak W, Pilecki M (2022) Redefining the cut-off ranges for TSH based on the clinical picture, results of neuroimaging and laboratory tests in unsupervised cluster analysis as individualized diagnosis of early chizophrenia. J Pers Med 12:247. https://doi.org/10.3390/jpm12020247
13. Stieg MR, Sievers C, Farr O, Stalla GK, Mantzoros CS (2015) Leptin: a hormone linking activation of neuroendocrine axes with neuropathology. Psychoneuroendocrinology 51:47–57. https://doi.org/10.1016/j.psyneuen.2014.09.004
14. Wędrychowicz A, Zając A, Pilecki M, Kościelniak B, Tomasik PJ (2014) Peptides from adipose tissue in mental disorders. World J Psychiatry 4:103–111. https://doi.org/10.5498/wjp.v4.i4.103
15. Reponen EJ, Tesli M, Dieset I et al (2021) Adiponectin is related to cardiovascular risk in severe mental illness independent of antipsychotic treatment. Front Psych 12:623192. https://doi.org/10.3389/fpsyt.2021.623192
16. Shangguan F, Chen Z, Lv Y, Zhang X-Y (2023) Interaction between high interleukin-2 and high cortisol levels is associated with psychopathology in patients with chronic schizophrenia. J Psychiatr Res 165:255–263. https://doi.org/10.1016/j.jpsychires.2023.07.039
17. Li C, Shi Z, Ji J, Niu G, Liu Z (2021) Associations of C-reactive protein, free triiodothyronine, thyroid stimulating hormone and creatinine levels with agitation in patients with schizophrenia: a comparative cross-sectional study. Neuropsychiatr Dis Treat 17:2575–2585. https://doi.org/10.2147/NDT.S322005
18. Fernandes BS, Karmakar C, Tamouza R et al (2020) Precision psychiatry with immunological and cognitive biomarkers: a multi-domain prediction for the diagnosis of bipolar disorder or schizophrenia using machine learning. Transl Psychiatry 10:162. https://doi.org/10.1038/s41398-020-0836-4
19. Kim H, Baek S-H, Kim J-W, Ryu S, Lee J-Y, Kim J-M, Chung Y-C, Kim S-W (2023) Inflammatory markers of symptomatic remission at 6 months in patients with first-episode schizophrenia. Schizophrenia (Heidelb) 9:68. https://doi.org/10.1038/s41537-023-00398-1
20. Thomas M, Rakesh D, Whittle S, Sheridan M, Upthegrove R, Cropley V (2023) The neural, stress hormone and inflammatory correlates of childhood deprivation and threat in psychosis: a systematic review. Psychoneuroendocrinology 157:106371. https://doi.org/10.1016/j.psyneuen.2023.106371
21. de Bartolomeis A, Barone A, Vellucci L, Mazza B, Austin MC, Iasevoli F, Ciccarelli M (2022) Linking inflammation, aberrant glutamate-dopamine interaction, and post-synaptic changes: translational relevance for schizophrenia and antipsychotic treatment: a systematic review. Mol Neurobiol 59:6460–6501. https://doi.org/10.1007/s12035-022-02976-3
22. Martínez AL, Brea J, Rico S, de Los Frailes MT, Loza MI (2021) Cognitive deficit in schizophrenia: from etiology to novel treatments. Int J Mol Sci 22:9905. https://doi.org/10.3390/ijms22189905
23. Gogos A, Ney LJ, Seymour N, Van Rheenen TE, Felmingham KL (2019) Sex differences in schizophrenia, bipolar disorder, and post-traumatic stress disorder: are gonadal hormones the link? Br J Pharmacol 176:4119–4135. https://doi.org/10.1111/bph.14584
24. Schwarz E, van Beveren NJM, Ramsey J, Leweke FM, Rothermundt M, Bogerts B, Steiner J, Guest PC, Bahn S (2014) Identification of subgroups of schizophrenia patients with changes in either immune or growth factor and hormonal pathways. Schizophr Bull 40:787–795. https://doi.org/10.1093/schbul/sbt105

Chapter 7
Neurotrophic Biomarkers

7.1 Introduction

Neurotrophins, a group of neuroprotective proteins including brain-derived neurotrophic factor (BDNF), nerve growth factor (NGF), neurotrophin-3 (NT-3), neurotrophin-4/5 (NT-4/5), and many others, exert profound effects on cell survival and activity within the central nervous system (CNS). These polypeptides play a crucial role in supporting the survival, growth, and differentiation of neurons [1].

Among neurotrophins, BDNF stands out as the most extensively studied due to its involvement in neuronal plasticity, apoptosis, modulation of neurotransmitters, and the survival of key neuron types such as dopaminergic, cholinergic, and serotonergic neurons. These functions have been implicated in the memory and cognitive changes observed in SSD [2].

Cognitive impairment is a prominent feature in SSD, particularly in first episode (FE) and clinical high-risk (CHR) patients [3]. Neurotrophins emerge as potential biomarkers offering insights into the association between SSD and cognitive deficits, given their fundamental roles in the survival, development, regeneration, maintenance, and plasticity of neurons [4]. A meta-analysis revealed moderately reduced serum and plasma BDNF levels in SSD compared to controls, with a notable accentuation of this decrease with disease duration [5]. However, its utility as a biomarker for cognitive dysfunction and differentiation between drug-treated and drug-naïve SSD patients was inconclusive in a subsequent meta-analysis [6]. Nevertheless, it proved effective as a state marker differentiating between chronic and FE subgroups, supporting the neurobiological/neurodevelopmental hypothesis in SSD pathophysiology and presenting itself as a novel treatment target [7].

NGF, another important neurotrophin, plays a key role in the behavioral response to social challenges and is associated with anxiety-related behaviors. Its involvement in memory impairments in SSD is noteworthy, with antipsychotic drugs demonstrating the ability to modulate NGF production [1, 8]. Changes in NGF levels in

SSD patients further support its role in the pathophysiology of the disorder [9]. A meta-analysis affirmed low NGF levels in plasma or blood of SSD patients, irrespective of age and gender [10]. In addition, certain researchers have noted that the manifestation of negative or depressive symptoms in individuals with SSD is associated with an elevation in peripheral NT-3 concentration and a concurrent decrease in BDNF levels [11, 12].

In conclusion, understanding the intricate roles of some neurotrophins across different stages and origins of psychotic conditions in SSD necessitates longitudinal studies and extended follow-up periods [13].

7.2 Neurotrophic Biomarkers for SSD Assessed Through ROC Analysis

This comprehensive review spans a 12-year period (2011–2023) and aims to examine studies dedicated to the development of diagnostic models for SSD utilizing neurotrophic biomarkers, with a specific focus on the assessment through ROC curve analysis. Table 7.1 succinctly presents an overview of the studies identified during the literature mining process. The subsequent paragraphs delve into comprehensive elaboration and detailed descriptions of these studies, facilitating an in-depth examination of the research findings.

Xiong et al. demonstrated effective discrimination between FE SSD patients and healthy individuals by utilizing serum levels of NGF and interleukin IL-2 (AUC 85%). However, the combined predictive power of these two biomarkers performed marginally less successfully than IL-2 alone (AUC 87%). Notably, this combination did not exhibit robust transdiagnostic utility between SSD and major depressive disorder (MDD) [14].

Diagnostic models employing BDNF presented challenges. Chiou and Huang found that a model based on BDNF exhibited poor diagnostic efficacy for both SSD (AUC 59%) and MDD (AUC 56%) patients compared to healthy individuals. Even when stratified by sex, the differentiation between men with MDD (AUC 65%) and women with SSD (AUC 62%) remained only moderately discernible from healthy subjects [15].

In a different context, Pillai et al. explored the potential of BDNF as a marker for relapse prediction, specifically for first relapse, first hospitalization, and first exacerbation after treatment with second-generation oral antipsychotics or long-acting risperidone microspheres. Regrettably, none of the three adverse outcomes were successfully predicted [16].

Djordjević et al. investigated chronically treated patients to unravel the intricate relationship between BDNF production and nitric oxide (NO), the latter being indirectly measured by the nitrite/nitrate ratio in the blood. Although they observed low levels of BDNF predicting Positive and Negative Syndrome Scale (PANSS) total scores with high sensitivity (100%), statistical significance was not reached [17].

7.2 Neurotrophic Biomarkers for SSD Assessed Through ROC Analysis

Table 7.1 Potential neurotrophic biomarkers for schizophrenia spectrum disorders (SSD) assessed through ROC analysis

Cohort (effect in concentration): biomarker	Sample platform biomarker function	Cohort (n = male/female) /Age[a]	AUC (95%CI)/sensitivity/specificity (%) [Predictor]
SSD vs HC or MDD vs HC (↓): NGF, IL-2 ($p < 0.001$ for both) **SSD vs MDD:** no results have presented significant changes Xiong et al. [14]	Serum ELISA. Diagnostic and transdiagnostic	Drug-naïve FE SSD (17/13)/22.83 (4.07) HC (15/13)/23.11 (3.21) MDD (17/13)/25.63 (6.41) Cohort nationality: China (Kunming, Yunnan Province)	**SSD vs HC:** 78.1 (66.1–90.1) /60.0/89.3 [NGF] Cutoff: 55.3 pg/mL 86.6 (76.1–97.1)/ 76.7/96.4 [IL-2] cutoff: 73.9 pg/mL 84.5/79.3 [NGF, IL-2] **SSD vs MDD:** 79.4 (66.2–92.7)/90.0/76.7 [IL-2] Cutoff: 72.7 pg/mL 58/86.7/40 [NGF] Cutoff: 47.9 pg/mL
BDNF **SSD vs HC (↓):** FC = 0.84, $p = 0.007$ **Male:** FC = 0.91, $p = 0.269$ **Female:** FC = 0.77, $p = 0.008$ **MDD vs HC (↓):** FC = 0.87, $p = 0.145$ **Male:** FC = 0.78, $p = 0.083$ **Female:** FC = 0.89, $p = 0.368$ Chiou and Huang [15]	Serum ELISA Diagnostic	SSD (116/108) 33.8 (10.1) HC (143/247) 31.3 (6.7) MDD (66/207) 39.4 (11.8) Cohort nationality: Taiwan (Kaohsiung)	**SSD vs HC:** 59.1 (54.3–63.9)/68.5/54.9 [BDNF]. Cutoff: 6.6 ng/mL **Male MDD vs male HC:** 65.2/81.1/48.5 [BDNF] Cutoff: 5.1 ng/mL **Female SSD vs female HC:** 62.3/74.1/57.4 [BDNF] Cutoff: 5.9 ng/mL
FGF10 **FE vs HC (↓):** Female ($p < 0.005$), and male ($p < 0.013$) **Chronic vs HC (↓):** Female ($p < 0.004$) and male ($p < 0.001$) **Female SSD vs male SSD:** Not significant Yu et al. [20]	Serum ELISA Diagnostic	HC (60/51) 28.68 (4.59) SSD (62/68)/29.65 (7.29) Subgroups: Drug-naïve FE (25/32)/29.02 (6.91) Chronic (treated >6 months) (36/37) 30.14 (7.58) Cohort nationality: China (Foshan)	**SSD vs HC:** 69.5 (63–76.1)/67.6/64.6 [FGF10] Cutoff: 149.4 pg/mL **FE vs HC:** 66.5 (57.7–75.4)/65.8/64.9 [FGF10] Cutoff: 152.3 pg/mL

(continued)

Table 7.1 (continued)

Cohort (effect in concentration): biomarker	Sample platform biomarker function	Cohort (n = male/female) / Age[a]	AUC (95%CI)/sensitivity/specificity (%) [Predictor]
FGF9 **Drug-naïve FE vs HC (↓)** **Chronic vs drug-naïve FE (↑)** **8-week drug-treated FE vs drug-naïve FE (↑)** Note: FGF10 levels had no significant difference between female and male Li et al. [21]	Serum ELISA Diagnostic, theranostic, and predictive	HC (60/51) 28.68 (4.59) SSD (62/68)/29.65 (7.29) Subgroups: Drug-naïve FE (25/32)/29.02 (6.91) 8-week drug-treated FE (9/10)/26.79 (5.95) Chronic (drug-treated >6 months) (36/37) 30.14 (7.58) Cohort nationality: China (Foshan)	**Drug-naïve FE vs HC:** 97.3 (95.4–99.3)/95.5/86 [FGF9] Cutoff: 166.4 pg/mL **Chronic SSD vs drug-naïve FE:** 96.8 (94.4–99.1)/86/91.9 [FGF9] Cutoff: 165.0 pg/mL
FE SSD vs HC (↓): ErbB4 (FC = 0.7, p = 0.002, male p = 0.005, female p = 0.098) BDNF (FC = 0.7, p = 0.000, male p = 0.000, female p = 0.005) **FE SSD vs HC (↑):** TET1 (FC = 1.3, p = 0.000, male p = 0.000, female p = 0.049) Li et al. [22]	Blood ELISA Diagnostic	Drug-naïve FE SSD (27/26)/28.15 (10.42) HC (29/28) 31.33 (10.69) Cohort nationality: China (Changsha)	**FE SSD vs HC:** 76.4 (66.9–85.9)/79.2/75.4 [BDNF] Cutoff: 26.4 ng/mL 75.5 (66.2–84.8)/ 83.0/66.7 [TET1] Cutoff: 65.8 pg/mL 65.4 (55.3–75.6)/75.5/50.9 [ErB4] Cutoff: 10.8 ng/mL 82.5 (74.7–90.3)/81.1/73.7 [ErB4, BDNF, TET1] Cutoff: 0.461 ng/mL
VEGF **Baseline SSD vs HC (↓):** (FC = 0.8, p < 0.001) **Responder vs non-responder SSD (↑):** (FC = 1.4, p < 0.001) Xiao et al. [23]	Serum ELISA Theranostic	HC (47/36) 34.7 (10.9) Baseline SSD (62/56) 34.2 (11.9) Subgroups after 6-week drug treatment: Responder SSD (85) Non-responder (33). Cohort nationality: China (Yangzhou)	**Responder vs non-responder SSD:** 77.4 (68.8–84.6)/63.5/78.8 [VEGF] Cutoff: 130.0 pg/mL Likelihood ratio (+): 2.99% Likelihood ratio (−): 0.46%

7.2 Neurotrophic Biomarkers for SSD Assessed Through ROC Analysis

Chronic SSD vs HC (↓): BDNF (FC = 0.79, p = 1.339E-05) **Male:** FC = 0.73, p = 0.003 **Female:** FC = 0.86, p = 0.008 **Chronic SSD vs HC (↑):** NO_2^-/NO_3^- (FC = 1.4, p = 4.357E-8). **Male:** FC = 1.35, p = 0.001 **Female:** FC = 1.25, p = 9.361E-05 Djordjević et al. [17]	Serum ELISA Diagnostic and predictive Note: BDNF levels varied according to age of onset, duration of illness, and antipsychotic drug	Chronic (drug-treated SSD) (22/16)/34 HC (18/21)/33 Cohort nationality: Serbia (Niš)	**Chronic SSD vs HC:** 83.0 (70.6–91.7)/91.7/70.0 [BDNF] Cutoff ≤24.84 ng/mL Likelihood ratio (+): 3.06% Likelihood ratio (−): 0.12% 79.1 (67.7–88.0)/69.4/84.9 [NO_2^-/NO_3^-] Cutoff >76.8 μmol/L Likelihood ratio (+): 3.33% Likelihood ratio (−): 0.36%
SSD vs HC (↑): TBARS, PCC, NO_2^-, CRP, IL-6, TNF-α, B2M, C3, MMP-9, GG genotype of BDNF196GA SNP Note: Data were adjusted for confounding variables. p < 0.001 for each metabolite, FC = 1.2 for all, except for CRP (2.2), IL-6 (1.5), MMP-9 (27.0) Ali et al. [18]	Serum ADVIA Chemistry XPT system Colorimetric assay. IMMULITE 2000. ELISA PCR-RFLP for genotyping SNPs Diagnostic	SSD (34/10)/25.1 (4) HC (32/18)/26.1 (3.9) Cohort nationality: Egypt (Cairo)	**SSD vs HC:** 99.8 (99.3–100)/94.4/100 [CRP] 99.5 (98.5–100)/97.5/97.5 [MMP-9] 86.1 (78.3–93.9)/70.0/87.5 [IL-6] 82.6 (71.8–93.3)/66.7/83.3 [B2M] 75.7 (64.0–87.4)/58.3/83.3 [C3] 70.7 (59.9–81.5)/56.8/82.0 [NO_2^-] 70.5 (59.6–81.4)/52.3/92.0 [TBARS] 69.7 (58.9–80.4)/61.4/78.0 [PCC] 66.7 (54.6–78.7)/60.0/72.5 [TNF-α] 61.0 (47.4–74.6)/41.7/83.3 [C4]

(continued)

Table 7.1 (continued)

Cohort (effect in concentration): biomarker	Sample platform biomarker function	Cohort (n = male/female) /Age[a]	AUC (95%CI)/sensitivity/specificity (%) [Predictor]
Baseline SSD vs HC (↑): NT-3 (FC = 4.16), MMP-9 (FC = 4.97), $p < 0.001$ **Risp-treated vs baseline (↑): NT-3** (FC = 1.57, $p = 0.036$) **Risp-treated vs baseline (↓): MMP-9** (FC = 0.45, $p = 0.022$) **Responder vs baseline (↑): NT-3** (FC = 1.36, $p = 0.135$) **Responder vs baseline (↓): MMP-9** (FC = 0.96, $p = 0.476$) **Non-responder vs baseline (↑): NT-3** (FC = 1.14, $p = 0.182$) **Non-responder vs baseline (↓): MMP-9** (FC = 0.33, $p = 0.017$) Chenniappan et al. [24]	Serum ELISA Theranostic	Baseline SSD (31/31)/32.21 (8.86) HC (31/31)/38.7 (8) 6-week Risp-treated subgroups: Responder (25) Non-responder (37) Cohort nationality: India (Puducherry)	**Risp-treated vs baseline SSD:** 65.9/64.0/62.0 [MMP-9] 60.6/60.0/62.0 [NT-3] Cutoff: 1225 ng/mL Cutoff: 957 pg/mL
BDNF levels were not reported Pillai et al. [16]	Plasma ELISA Theranostic	Baseline SSD (157/64)/38.69 (11.76) 28-week drug-treated SSD subgroups: OSGA (80/33)/39.29 (12.21) LASGA (77/31)/38.06 (11.29) Relapse (34 LASGA, 26 OSGA) Hospitalization (47 LASGA, 35 OSGA) Exacerbation (22 LASGA, 19 OSGA) Cohort nationality: USA	**Drug-treated SSD vs bSSD:** The estimated AUC (standard error) for BDNF as a predictor of first relapse was 49.5 (4.2) %; for hospitalization, it was 50.5 (4.0) %, and for exacerbation, it was 49.9 (4.9) % Note: CI, sensitivity, and specificity were not reported

7.2 Neurotrophic Biomarkers for SSD Assessed Through ROC Analysis

BDNF (↓): Patients with low scores in semantic fluency ($p = 0.003$) and CGI negative symptoms ($p = 0.039$) No significant association with social functioning, positive symptoms subscale of CGI-SCH, and three-subscale PANSS Isayeva et al. [25]	Serum ELISA Step One Plus with TaqMan probes for genotyping SNPs Transdiagnostic Note: data were adjusted for age and sex	Psychotic patients (74/31)/48.85 (10.45) Subgroups: SSD (64) SAD (41) Cohort nationality: Italy (Cagliari, Sardinian ethnicity)	**SSD vs SAD:** 57.1 (51.8–62.4)/62/50 [BDNF] Cutoff: 2.83 ng/mL
SSD vs HC (↑): FINS, HOMA-IR **SSD vs HC (↓):** BDNF, GDNF Note: $p < 0.001$ for each compound Pang et al. [26]	Serum Multiplatform assays Diagnostic	SSD (70/72)/25 (4) HC (72/68)/26 (4) Cohort nationality: China (Zhengzhou)	**SSD vs HC:** 89 (83–94)/94/82 [BDNF, GDNF, HOMA-IR]

AUC area under the curve, *B2M* beta-2 microglobulin, *BDNF* brain-derived neurotrophic factor, *C3/C4* complement components 3 and 4, *CGI-SCH* Clinical Global Impression Scale for Schizophrenia, *CI* confidence interval, *CRP* C-reactive protein, *ELISA* enzyme linked immunosorbent assay, *ErbB4* receptor tyrosine protein kinase 4, *FC* fold change, *FE* first episode, *FGF9/FGF10* fibroblast growth factors 9 and 10, *FINS* fasting insulin, *GDNF* glial cell-derived neurotrophic factor, *HC* healthy control, *HOMA-IR* homeostatic model assessment for insulin resistance, *IL* interleukin, *LASGA* patients treated with long-acting risperidone microspheres, *MDD* major depressive disorder, *MMP-9* matrix metallopeptidase 9, *NGF* nerve growth factor, *NT-3* neurotrophin-3, *OSGA* patients treated with second-generation oral antipsychotics, *PCC* protein carbonyl content, *PCR-RFLP* polymerase chain reaction-restriction fragment length polymorphism, *Risp-treated* patients treated with risperidone, *SAD* schizoaffective disorders, *SNP* single nucleotide polymorphism, *TBARS* thiobarbituric acid reactive substances, *TET1* ten-eleven translocation methylcytosine dioxygenase 1, *TNF-α* tumor necrosis factor alpha, *VEGF* vascular endothelial growth factor

[a]Age is given as mean (standard deviation) or median [interquartile]

Ali et al. supported this finding at the genetic level. Their examination of matrix metallopeptidase 9 (MMP-9) and BDNF polymorphisms through multivariate logistic regression analysis identified CC homozygotes of MMP-9 −1562C>T SNP (single nucleotide polymorphism) as the most robust predictor of higher PANSS total scores, outperforming the GG genotype of BDNF 196G>A SNP. Intriguingly, among all markers tested, only immunoinflammatory markers exhibited sensitivity, specificity, and AUC values exceeding 70% [18]. In a study by Asadi et al., post-transcriptional regulators of BDNF expression, including tristetraprolin (AUC 77%) and miR-16 (not significant), did not emerge as strong diagnostic markers between female SSD patients and healthy individuals, in addition to exhibiting low sensitivity (57%) [19].

Low levels of fibroblast growth factor 10 (FGF10) demonstrated performance comparable to BDNF in predicting patients at different disease stages [20]. Subsequently, the same research group developed a model using FGF9 to differentiate FE drug-naïve patients from chronically treated patients, achieving an outstanding result with an AUC of 97% [21].

Li and colleagues developed distinct models employing receptor tyrosine-protein kinase 4 (ErbB4), BDNF, and the enzyme ten-eleven translocation methylcytosine dioxygenase 1 (TET1) to differentiate individuals with SSD from healthy controls, yielding moderate to good results [22].

Xiao et al. explored the utility of vascular endothelial growth factor (VEGF) as a theranostic marker to distinguish responder from non-responder patients to antipsychotic treatment, achieving a moderate result with an AUC of 77% [23]. Similar outcomes were observed with models constructed using NT-3 or MMP-9, both considered markers of synaptic plasticity [24].

The temporal decline of BDNF serum levels exhibited limited transdiagnostic power in distinguishing between SSD and schizoaffective patients in a 24-month observational prospective cohort study of Sardinian psychotic patients. Furthermore, in this study, the rs7934165 polymorphism (but not the Val66Met polymorphism) of the BDNF gene showed only a moderating effect on the relationship between the Brief Assessment of Cognition in Schizophrenia (BACS) subscale for semantic fluency and BDNF serum levels in psychotic patients ($p = 0.047$) [25].

7.3 Discussion

This review encompasses 11 articles reporting results from over 1000 patients with SSD over a 12-year period (2011–2023). Notably, 50% of the studies were conducted in China, with the remaining studies distributed across Italy, Egypt, India, Serbia, and the United States.

From the aggregated findings, neurotrophic factors generally exhibited moderate performance in most models, as evidenced by most AUC values falling below 70%. Interestingly, certain models employing a panel of three biomarkers, as opposed to a singular one, demonstrated AUCs surpassing 80%. This diversity in diagnostic efficacy underscores the nuanced nature of utilizing neurotrophic biomarkers in SSD. While some markers show promise in specific contexts, others face challenges, emphasizing the complexity of their role in SSD diagnosis.

A recent review underscored that the levels of BDNF, NGF, NT-3, and glial cell line-derived neurotrophic factor (GDNF) display variability that is contingent upon factors such as symptom severity, disease relapse or remission, and treatment modalities in individuals with SSD [27]. However, based on the data extracted from the studies included in this review, the evaluation of serum levels of these proteins has a limited likelihood of evolving as a component that contributes significantly to the enhancement of the diagnosis and prognosis of SSD.

References

1. Gioiosa L, Iannitelli A, Aloe L (2009) Stress, anxiety and schizophrenia and neurotrophic factors: the pioneer studies with nerve growth factor. Riv Psichiatr 44:88–94. https://doi.org/10.1708/420.4978
2. Gören JL (2016) Brain-derived neurotrophic factor and schizophrenia. Ment Health Clin 6:285–288. https://doi.org/10.9740/mhc.2016.11.285
3. Bora E (2019) Peripheral inflammatory and neurotrophic biomarkers of cognitive impairment in schizophrenia: a meta-analysis. Psychol Med 49:1971–1979. https://doi.org/10.1017/S0033291719001685
4. Bathina S, Das UN (2015) Brain-derived neurotrophic factor and its clinical implications. Arch Med Sci 11:1164–1178. https://doi.org/10.5114/aoms.2015.56342
5. Fernandes BS, Steiner J, Berk M, Molendijk ML, Gonzalez-Pinto A, Turck CW, Nardin P, Gonçalves C-A (2015) Peripheral brain-derived neurotrophic factor in schizophrenia and the role of antipsychotics: meta-analysis and implications. Mol Psychiatry 20:1108–1119. https://doi.org/10.1038/mp.2014.117
6. Dombi ZB, Szendi I, Burnet PWJ (2022) Brain derived neurotrophic factor and cognitive dysfunction in the schizophrenia-bipolar spectrum: a systematic review and meta-analysis. Front Psych 13:827322. https://doi.org/10.3389/fpsyt.2022.827322
7. Rodrigues-Amorim D, Rivera-Baltanás T, Bessa J et al (2018) The neurobiological hypothesis of neurotrophins in the pathophysiology of schizophrenia: a meta-analysis. J Psychiatr Res 106:43–53. https://doi.org/10.1016/j.jpsychires.2018.09.007
8. Martinez-Cengotitabengoa M, MacDowell KS, Alberich S et al (2016) BDNF and NGF signalling in early phases of psychosis: relationship with inflammation and response to antipsychotics after 1 year. Schizophr Bull 42:142–151. https://doi.org/10.1093/schbul/sbv078
9. Neugebauer K, Hammans C, Wensing T et al (2019) Nerve growth factor serum levels are associated with regional gray matter volume differences in schizophrenia patients. Front Psych 10:275. https://doi.org/10.3389/fpsyt.2019.00275
10. Qin X-Y, Wu H-T, Cao C, Loh YP, Cheng Y (2017) A meta-analysis of peripheral blood nerve growth factor levels in patients with schizophrenia. Mol Psychiatry 22:1306–1312. https://doi.org/10.1038/mp.2016.235
11. Arabska J, Łucka A, Strzelecki D, Wysokiński A (2018) In schizophrenia serum level of neurotrophin-3 (NT-3) is increased only if depressive symptoms are present. Neurosci Lett 684:152–155. https://doi.org/10.1016/j.neulet.2018.08.005
12. Wysokiński A (2016) Serum levels of brain-derived neurotrophic factor (BDNF) and neurotrophin-3 (NT-3) in depressed patients with schizophrenia. Nord J Psychiatry 70:267–271. https://doi.org/10.3109/08039488.2015.1087592

13. Nieto RR, Carrasco A, Corral S, Castillo R, Gaspar PA, Bustamante ML, Silva H (2021) BDNF as a biomarker of cognition in schizophrenia/psychosis: an updated review. Front Psych 12:662407. https://doi.org/10.3389/fpsyt.2021.662407
14. Xiong P, Zeng Y, Wan J, Xiaohan DH, Tan D, Lu J, Xu F, Li HY, Zhu Z, Ma M (2011) The role of NGF and IL-2 serum level in assisting the diagnosis in first episode schizophrenia. Psychiatry Res 189:72–76. https://doi.org/10.1016/j.psychres.2010.12.017
15. Chiou Y-J, Huang T-L (2019) Accuracy of brain-derived neurotrophic factor levels for differentiating between Taiwanese patients with major depressive disorder or schizophrenia and healthy controls. PLoS One 14:e0212373. https://doi.org/10.1371/journal.pone.0212373
16. Pillai A, Schooler NR, Peter D et al (2018) Predicting relapse in schizophrenia: is BDNF a plausible biological marker? Schizophr Res 193:263–268. https://doi.org/10.1016/j.schres.2017.06.059
17. Djordjević VV, Lazarević D, Ćosić V, Knežević MZ, Djordjević VB, Stojanović I (2016) Diagnostic accuracy of brain-derived neurotrophic factor and nitric oxide in patients with schizophrenia: a pilot study. J Med Biochem 35:7–16. https://doi.org/10.1515/jomb-2015-0010
18. Ali FT, Abd El-Azeem EM, Hamed MA, Ali MAM, Abd Al-Kader NM, Hassan EA (2017) Redox dysregulation, immuno-inflammatory alterations and genetic variants of BDNF and MMP-9 in schizophrenia: pathophysiological and phenotypic implications. Schizophr Res 188:98–109. https://doi.org/10.1016/j.schres.2017.01.016
19. Asadi MR, Gharesouran J, Sabaie H et al (2022) Assessing the expression of two posttranscriptional BDNF regulators, TTP and miR-16 in the peripheral blood of patients with schizophrenia. BMC Psychiatry 22:771. https://doi.org/10.1186/s12888-022-04442-9
20. Yu Y, Xie G-J, Hu Y, Li X-S, Chen G-Y, Zheng G-E, Chen X, Cheng Y (2019) Dysregulation of fibroblast growth factor 10 in the peripheral blood of patients with schizophrenia. J Mol Neurosci 69:69–74. https://doi.org/10.1007/s12031-019-01331-x
21. Li X-L, Yu Y, Hu Y, Wu H-T, Li X-S, Chen G-Y, Cheng Y (2022) Fibroblast growth factor 9 as a potential biomarker for schizophrenia. Front Psych 13:788677. https://doi.org/10.3389/fpsyt.2022.788677
22. Li C, Tao H, Yang X, Zhang X, Liu Y, Tang Y, Tang A (2018) Assessment of a combination of serum proteins as potential biomarkers to clinically predict schizophrenia. Int J Med Sci 15:900–906. https://doi.org/10.7150/ijms.24346
23. Xiao W, Zhan Q, Ye F, Tang X, Li J, Dong H, Sha W, Zhang X (2018) Baseline serum vascular endothelial growth factor levels predict treatment response to antipsychotic medication in patients with schizophrenia. Eur Neuropsychopharmacol 28:603–609. https://doi.org/10.1016/j.euroneuro.2018.03.007
24. Chenniappan R, Nandeesha H, Kattimani S, Goud AC, Thiagarajan D (2022) Risperidone reduces matrix metalloproteinase-9 and increases neurotrophin-3 in schizophrenia spectrum of disorder. Indian J Clin Biochem 37:342–348. https://doi.org/10.1007/s12291-021-00985-y
25. Isayeva U, Manchia M, Collu R et al (2022) Exploring the association between brain-derived neurotrophic factor levels and longitudinal psychopathological and cognitive changes in Sardinian psychotic patients. Eur Psychiatry 65:e71. https://doi.org/10.1192/j.eurpsy.2022.2333
26. Pang LJ, Li X, Yuan XX, Hei GR, Zhang LY, Wang SY, Chen YS, Song PL, Song XQ (2023) Establishment of diagnostic model for schizophrenia based on neurotrophic factor and other biomarkers. Zhonghua Yi Xue Za Zhi 103:1310–1315. https://doi.org/10.3760/cma.j.cn112137-20221212-02631
27. Sikorski P, Sikorska E, Michalska M, Stec A, Kasarełło K, Fudalej S (2023) Neuroprotective factors in schizophrenia: BDNF, NGF, NT3, GDNF and their connection to the pathogenesis of schizophrenia. A narrative review. Curr Probl Psychiatr 24:105–113. https://doi.org/10.12923/2353-8627/2023-0010

Chapter 8
Neurotransmitter Biomarkers

8.1 Introduction

The conceptualization of schizophrenia spectrum disorders (SSD) as a neurodevelopmental disorder underscores synaptic alterations and aberrant cortical-subcortical connections. Over time, researchers have diligently sought biomarker molecules associated with these systems, recognizing their profound implications for the disorder's pathophysiology. Systematic reviews and meta-analyses have been instrumental in supporting the potential use of these molecules for early diagnosis, monitoring, treatment response assessment in SSD patients, and the identification of novel drug targets.

Lai et al. [1] highlighted the potential limitations of monoamines, including dopamine and epinephrine, as diagnostic markers for SSD. They noted that these monoamines may function more as state markers, given that most antipsychotic treatments block their receptors throughout the body. This blockade induces dynamic changes in the peripheral profile of monoamine-related receptors and metabolites. On the contrary, glutamatergic dysregulation plays a pivotal role in the pathogenesis of SSD and seems to remain unaffected by current therapeutic interventions. Individuals with SSD exhibit altered peripheral levels of various molecules within the glutamatergic system, including glutamate, D- and L-serine, glycine, and agmatine. These molecules have the capacity to activate N-methyl-D-aspartate (NMDA) receptors and modulate glutamatergic neuronal pathways. A meta-analysis highlighted elevated blood levels of glutamate in SSD patients [2]. Additionally, Zahid et al.'s systematic review evidenced disruptions in the normal relationships between glutamate or GABA levels and neurophysiological activity during rest and task performance [3]. Consistently, both peripheral and central concentrations of kynurenine and 3-hydroxykynurenine, but not tryptophan and kynurenic acid, were found to be linked to psychosis [4].

The systematic review conducted by Li et al. in [5] identified a metabolic biomarker signature consisting of 10 compounds associated with psychosis, with a primary focus on SSD. These compounds include tryptophan, kynurenine, glutamate, D-serine, glutathione, lactate, creatine, 3-hydroxybutyrate, linoleic acid, and N-acetyl aspartate. Collectively, these biomarkers represent pathways most significantly impacted in SSD, encompassing energy metabolism, neural structure and function, neurotransmitter biosynthesis, neuroinflammation, synaptic plasticity, mitochondrial oxidative state, a marker of neuron health, and a neuroprotective compound. This compilation serves as a valuable starting point for a targeted metabolomic approach aimed at developing a diagnostic method for SSD based on neurotransmitter biomarkers.

In a recent study, Luo et al. introduced a novel framework combining multimodal neuroimaging meta-analysis and genetic/molecular architecture to identify reliable SSD biomarkers [6]. Their findings revealed disease-related aberrant regions significantly positively correlated with dopamine receptor D2, dopamine transporter (DAT), noradrenaline transporter (NAT), 5-hydroxytryptamine receptor 4 (5-HT4), and cannabinoid receptor type 1 (CB1). This confirmed the involvement of dopaminergic, serotonergic, GABAergic, and glutamatergic systems in SSD's entire pathophysiological process. Furthermore, the authors delved into the dysregulation of cannabinoid receptor 1 (CB1) across various brain regions, highlighting its role in the progression of SSD by impacting intricate circuits governing cognition and memory. The findings collectively portrayed SSD as a multidimensional serious mental illness, involving abnormalities at the genetic, molecular, brain structure, and function levels. While no single factor played a predominant role in the onset and aggravation of SSD, the interactions among them jointly contributed to the disorder's complex pathophysiology.

8.2 Neurotransmitter Biomarkers for SSD Assessed Through ROC Analysis

This comprehensive review spans a 12-year period (2011–2023) and aims to scrutinize studies dedicated to developing diagnostic models for SSD utilizing neurotransmitter biomarkers, with a specific emphasis on assessment through ROC curve analysis. Table 8.1 concisely provides an overview of the studies identified during the literature mining process. The following paragraphs offer more detailed descriptions of these studies, facilitating an in-depth examination of the research findings.

Dysfunction in dopaminergic, glutamatergic, serotonergic, and GABAergic signaling can give rise to a broad spectrum of anomalies in both macro and microcircuits, leading to cognitive, behavioral, and social impairments [7]. Additional abnormalities, such as NMDA receptor hypofunction and low-grade brain inflammation, may also contribute to cognitive impairment and negative symptoms. This has prompted many researchers to explore neurotransmitters as promising biomarkers for diagnosing SSD.

8.2 Neurotransmitter Biomarkers for SSD Assessed Through ROC Analysis

Table 8.1 Potential neurotransmitter biomarkers for schizophrenia spectrum disorders (SSD) assessed by ROC analysis

Cohort (effect in concentration): biomarker	Sample Analytical platform Biomarker function	Cohort (n = male/female)/Age[a]	AUC (95%CI)/sensitivity/specificity (%) [predictor]
SSD vs HC ($\uparrow$): Hydroxyhemopyrrolin-2-one (HPL)/creatinine ratio, DA, NE, ADR, visual and auditory speed of processing discrepancy, competing words discrepancy ($p < 0.0001$ for each); Copper/zinc ratio ($p = 0.01$), serum vit. B12 ($p = 0.09$) **SSD vs HC ($\downarrow$):** Vit. B6 activated ($p = 0.0009$), red cell folate ($p = 0.0005$), vit. D ($p = 0.0026$), visual spam, vision on right, reverse digit span ($p = 0.0001$) Fryar-Williams and Strobel [8]	Urine MS for DA, NE, ADR, and creatinine Colorimetric method for HPL Serum Liaison platform for vit. D; c-ECL for vit. B12; c-ECLPB for red cell folate; HPLC/fluorescence detector for vit. B6; FAAS for free copper; ICP-MS for red cell zinc Diagnostic	SSD (37/30)/40.5 (1.3) HC (33/34)/45.7 (1.4) Cohort nationality: Australia (Adelaide and Woodville), with multi-ethnic backgrounds	**SSD vs HC:** 92/89/85 [Visual cognitive domain: visual span, visual speed of processing discrepancy, distance vision on right] 89/87/82 [Auditory cognitive domain: reverse digit span, auditory speed of processing discrepancy, competing words discrepancy] 86/84/75 [Catecholamine domain: DA, NE, ADR] 70/70/64 [Oxidative stress domain: HPL/creatinine] 80/55/88 [Nutrition domain: Cu/Zn, Vit. B12, B6, D, folate]
FE SSD vs HC ($\downarrow$): GABA DSC and Glx levels were not significantly different Sigvard et al. [9]	In vivo analysis of brains ^{18}F-DOPA PET for DSC ^{1}H–MRS with PRESS for Glx. MEGAPRESS for GABA Diagnostic	Drug-naïve FE SSD (9/14)/22.30 (3.9) HC (8/12)/22.35 (4.3) Cohort nationality: Denmark (Copenhagen)	**FE SSD vs HC:** 83.8/83.7 AIC = 48.5 weight 93.5%. [Dopamine synthesis capacity (DSC) in the nucleus accumbens, GABA in the anterior cingulate cortex, Glx in the thalamus]
Non-responder vs responder SSD ($\uparrow$): Glx and glutamate in anterior cingulate cortex Note: $p \leq 0.05$ for each compound; data were adjusted for age and gender Egerton et al. [10]	In vivo analysis of brains ^{18}F-DOPA PET for DSC ^{1}H–MRS with PRESS for glutamate, Glx Theranostic	Responder SSD (41/7)/29.9 (9.8) Non-responder SSD (36/8)/28.9 (7.5) Cohort nationality: UK (Manchester, London, Edinburgh, Cardiff), multiethnic patients	**Non-responder vs responder SSD:** 61/73/49 [glutamate in the anterior cingulate cortex] 59 [DSC in the striatum]

(continued)

Table 8.1 (continued)

Cohort (effect in concentration): biomarker	Sample Analytical platform Biomarker function	Cohort (n = male/ female)/Age[a]	AUC (95%CI)/sensitivity/ specificity (%) [predictor]
FE SSD vs HC (↑): Choline, Glx ($p < 0.001$ for each) **FE SSD vs HC (↓):** GABA ($p = 0.026$), *myo*-inositol ($p < 0.003$ for each) Note: FDR < 0.05 for each metabolite. Data were adjusted for age and tissue portion Chiu [11]	*Postmortem* analysis of brains: Anterior cingulate cortex ^{1}H–MRS Diagnostic	FE SSD (11/8)/29.11 (6.68) HC (9/5)/27.71 (5.88) Cohort nationality: China (Hong Kong)	**FE SSD vs HC:** 72.7 (55.6–89.9)/73.7/64.3 [GABA], Cutoff <2.22 mM 90.6 (80–100)/84.2/92.9 [Glx], Cutoff >9.58 mM 90.6 (80.9–100)/78.9/85.7 [choline], Cutoff >2.00 mM 79.7 (63.8–95.6)/78.9/78.6 [*myo*-inositol], Cutoff <7.11 mM 94 (84.8–100)/89.5/100 [GABA, Glx], Cutoff < GABA conc. 95.5 (88.3–100)/94.7/92.9 [GABA, *myo*-inositol, choline, Glx], Cutoff (mM): 1.98 GABA, 8.19 Glx, 2.05 choline, 9.0 *myo*-inositol
bSSD vs HC (↑): 11,12-DHET (FC = 1.1), 20-carboxy-ArA (FC = 1.4), AEA (FC = 1.5), OEA (FC = 1.5) **bSSD vs HC (↓):** ArA (FC = 0.8) **Post-treat vs bSSD (↓):** 11,12-DHET (FC = 0.8), 20-carboxy-ArA (FC = 0.5), AEA (FC = 0.8), OEA (0.7) **Post-treat SSD vs HC (↑):** OEA (1.1) **Post-treat SSD vs HC (↓):** ArA (FC = 0.7), 11,12-DHET (0.9), 20-carboxy-ArA (0.8) Wang et al. [12]	Serum LC-TQ/MS Diagnostic and theranostic Note: $p < 0.05$ and FDR < 0.02, VIP > 1 for each metabolite	Baseline SSD (115)/29.0 [25.0,33.2] HC (108)/30.0 [26.0, 33.0] 8-weeks post-treat SSD (109) Cohort nationality: China (Shandong Province)	**bSSD vs HC:** 94 (90.9–97.1)/91/84 [ArA, AEA, OEA] **Post-treat vs bSSD:** 76 (69.8–82.4)/76/67 [AEA, OEA, 20-carboxy-ArA, 11,12-DHET]

(continued)

8.2 Neurotransmitter Biomarkers for SSD Assessed Through ROC Analysis

Table 8.1 (continued)

Cohort (effect in concentration): biomarker	Sample Analytical platform Biomarker function	Cohort (n = male/female)/Age[a]	AUC (95%CI)/sensitivity/specificity (%) [predictor]
FE SSD, BD, MDD vs HC or vs Siblings (↓): NR2. **FE SSD, MDD vs HC or vs Siblings (↓): NR1** **Siblings vs HC (↑): NR2** Other cohorts did not show significant differences Loureiro et al. [13]	Plasma ELISA Diagnostic Note: results were not influenced by age, antipsychotic treatment, gender, duration of disorder, drug use	HC (106/60)/31.4 (12.0) Siblings (23/53)/31.5 (11.0) FE psychoses (106/60)/30.3 (12.2) Subgroups: FE SSD (84)/28.5 (12.0) FE BD (51)/24.8 (8.6) FE MDD (31)/39.2 (12.4) Cohort nationality: Brazil (Ribeirão Preto-SP)	**FE psychosis vs HC:** 62/42.8/84.3 [NR1] Cutoff: 17.65 pg/mL **FE psychosis vs HC:** 80/71.9/80.6 [NR2] Cutoff: 2.92 ng/mL
SSD vs HC (↑): agmatine (FC = 4.4; $p < 0.0001$), S100B (FC = 1.2; not significant) Note: data were influenced by age and gender Uzbay et al. [14]	Plasma HPLC/fluorescence detector for agmatine Immunoassay for S100B Diagnostic	SSD (11/7)/37.39 (2.97) HC (11/8)/33.89 (1.42) Cohort nationality: Turkey (Bursa)	**SSD vs HC:** 96.9 (92.0–101.7)/83.3/93.8 [agmatine] Cutoff: 20.71 ng/mL
SSD vs depression or HC: No significant change regarding age, total platelet count, and platelet serotonin concentration (PSC) Peitl et al. [15]	Plasma (platelet) ELISA Diagnostic and transdiagnostic	SSD (221/118)/39.3 (14.0) Depression (200/129)/38.3 (14.5) HC (160/125)/37.7 (11.8) Cohort nationality: Croatia (Zagreb)	**SSD vs Depression:** 51.5 (47.1–55.8) [PSC], $p = 0.514$. SSD vs HC: 51.0 (46.5–55.6) [PSC], $p = 0.655$

(continued)

Table 8.1 (continued)

Cohort (effect in concentration): biomarker	Sample Analytical platform Biomarker function	Cohort (n = male/female)/Age[a]	AUC (95%CI)/sensitivity/specificity (%) [predictor]
1st endophenotype vs 2nd (↑): Glutamate ($p < 0.001$), lymphocytes ($p < 0.003$) **1st endophenotype vs 2nd (↓):** neutrophils ($p < 0.003$) Bryll et al. [16]	Serum Multiplatform assay (routine lab tests, MDA, FRAP, CRP) MRS Predictive	SSD (70% FE) in acute psychotic decompensation (22/18)/22.68 (7.39) Patients were compared based on the levels of glutamate in the anterior cingulate cortex Cohort nationality: Poland (Krakow)	**1st endophenotype vs 2nd:** 71 [Glutamate in ACC] Note: CI, sensitivity, specificity, and *p*-value were not reported
SSD DLPFC vs HC (↑): D-Ser (> 200 nmol/g tissue) **SSD DLPFC vs HC (↓):** D-asp/total asp (<1), VGLuT2, EAAT2, GAD67 **SSD hippo vs HC (↑):** L-Asn **SSD hippo vs HC (↓):** EAAT2, SLC17A7, synapsin-1 De Rosa et al. [17]	*Post-mortem* brains: DLPFC and hippocampus HPLC for amino acids. RT-PCR and Western blotting for mRNA and proteins Diagnostic	SSD (12/8)/52.5 [39–61] HC (16/4)/73.5 [66–80] Cohort nationality: USA (Los Angeles)	**SSD vs HC (Random Forest):** AUC for DLPFC: 80 (65–92) [VGluT2, EAAT2, GAD67, D-asp/total asp, D-Ser, PSD-95, GRIA1, GRM5] **AUC for hippocampus:** 85 (71–95) [EAAT2, L-Asn, CAMK2A, Synapsin-1, GRIA2, GRIA4, and SLC17A7] **SSD vs HC (CART):** [CAMK2A, GRIA2] **DLPFC:** 73 (60–85) **Hippo:** 92 (83–99)
FE SSD, BD, MDD vs HC (↓): α-MSH, neurotensin, orexin-A, oxytocin, substance P ($p < 0.001$) **BD vs HC (↓):** β-endorphin ($p < 0.05$) Yu et al. [18]	Plasma Immunoassay in Flexmap 3D Diagnostic and transdiagnostic	FE SSD (26/28)/22.0 (7.3) BD (21/31)/28.9 (11.5) MDD (15/20)/30.3 (10.3) HC (23/31)/27.9 (9.6) Cohort nationality: China (Sichuan)	**FE SSD vs BD or MDD:** 40–66% **FE SSD vs HC:** 83 (74–91) [neurotensin] 80 (68–89) [oxytocin] **MDD vs HC:** 87 (77–94) [neurotensin] 85 (74–92) [oxytocin] 80 (68–89) [α-MSH]

(continued)

8.2 Neurotransmitter Biomarkers for SSD Assessed Through ROC Analysis

Table 8.1 (continued)

Cohort (effect in concentration): biomarker	Sample Analytical platform Biomarker function	Cohort (n = male/female)/Age[a]	AUC (95%CI)/sensitivity/specificity (%) [predictor]
bSSD vs HC (↓): GABA (p = 0.002) **bSSD vs HC (↑): CE (20:3), CA** (p = 0.004), **GCA** (p = 0.034) **Treat SSD vs bSSD (↓):** CA (p = 0.009), GCA (p = 0.07) **Treat SD vs bSSD (↑): CE (20:3)** (p < 0.001), **GABA** (p = 0.032) Wang et al. [19]	Serum FIA-MS/MS and LC-MS/MS Diagnostic and theranostic	Baseline SSD (0/38)/39.7 (9.8) HC (0/19)/40.0 (9.1) 8-week drug-treated SSD Cohort nationality: China (Tianjin)	**Treated SSD vs bSSD:** 89.1 [GABA, CE (20:3), cholic acid (CA), glycocholic acid (GCA)] Note: CI, sensitivity, specificity, and FDR were not reported

ADR adrenaline, *AEA* anandamide, *AIC* Akaike information criterion, *ArA* arachidonic acid, *Asn* asparagine, *Asp* aspartate, *AUC* area under the curve, *BD* bipolar disorder, *CAMK2A* calcium-calmodulin-dependent protein kinase II alpha, *CART* classification and regression tree, *c-ECL* competitive electrochemiluminescent immunoassay, *c-ECLPB* competitive electrochemiluminescent protein binding assay, *CE* cholesteryl ester, *CI* confidence interval, *CRP* C reactive protein, *DA* dopamine, *DEF* deficit SSD, *DHET* dihydroxy-5Z,8Z,14Z-eicosatrienoic acid, *DLPFC* dorsolateral prefrontal cortex, *DSC* dopamine synthesis capacity, *EAAT2* glutamate transporter, *ELISA* enzyme linked immunosorbent assay, *FAAS* flame atomic absorption spectrophotometry, *FC* fold change, *FDR* false discovery rate, *FE* first episode, *[18]F-DOPA PET* ^{18}F-fluorodopa positron emission tomography, *FRAP* ferric reducing ability of plasma *GABA* gamma-aminobutyric acid, *GAD67* glutamic acid decarboxylase 67, *Glx* glutamate + glutamine, *GRIA1* glutamate ionotropic receptor AMPA type subunit 1, *GRM5* glutamate metabotropic receptor 5, *HC* health control, *^{1}H–MRS* proton magnetic resonance spectroscopy, *HPLC* high performance liquid chromatography, *ICP-MS* inductively coupled plasma mass spectroscopy, *IQR* interquartile range, *IT* immunoturbidimetry, *LC/TQ-MS* liquid chromatography coupled to triple quadrupole mass spectrometer, *MDA* malondialdehyde, *MDD* major depressive disorder, *MEGAPRESS* Meshcher-Garwood point resolved spectroscopy, *MRS* magnetic resonance spectroscopy, *MS* mass spectrometry, *NART33 IQ* National Adult Reading Test 33 items total IQ score, *NE* noradrenaline or norepinephrine, *NR1 and NR2* N-methyl-D-aspartate receptors (NMDAR) subunits, *OEA* oleoylethanolamide, *PSD-95* postsynaptic density protein 95, *Ser* serine, *VGLuT2* vesicular glutamate transporter encoded by the *SLC17A7* gene

[a]Age is given as mean (standard deviation) or median [interquartile range]

Fryar-Williams and Strobel constructed different models aimed at distinguishing SSD patients from healthy individuals, achieving the most favorable outcome with visual cognitive scores (AUC 92%) rather than peripheral catecholamine levels (AUC 86%) [8].

Sigvard et al. opted for a multimodal prediction model that included neurotransmitters measured in *in vivo* brain tissues [9]. Notably, none of the seven models built with individual neurotransmitters exhibited significant predictive performance, with AUC values consistently below 60%. Among the models constructed with a three-predictor panel, only one achieved an AUC greater than 80%. The authors concluded that the diagnosis of psychosis relies more on the disruption of the indirect pathway in brain macrocircuitry than on the dopaminergic or GABAergic abnormalities themselves. Similar results were obtained by Egerton et al., who employed glutamate and dopamine synthesis capacity in separate models [10]. In contrast, a postmortem analysis of the anterior cingulate cortex revealed an exceptional four-biomarker panel—GABA, choline, glutamate + glutamine, and myo-inositol—exhibiting an AUC of 95% in discriminating FE SSD from healthy individuals. This not only provided strong discriminatory power but also supported the hypothesis of dysfunction in GABAergic/glutamatergic interneurons [11].

Cannabinoid neurotransmitters, along with arachidonic acid, emerged as effective predictors for distinguishing untreated and treated SSD patients from healthy controls [12]. In contrast, the NMDA receptor subunits R1 and R2, when employed in a model, exhibited low sensitivity in distinguishing FE psychoses encompassing SSD, bipolar disorder (BD), and major depressive disorder (MDD) [13].

The polyamine agmatine, acting as an agonist of NMDA, α2-adrenergic, and imidazoline receptors, resulted in a model with an impressive AUC of 97%, and 94% accuracy for negative cases, as detailed in the study by Uzbay et al. [14].

Peitl et al. explored the potential of platelet serotonin concentration in distinguishing depressed patients from those with SSD. However, their investigation revealed no significant differences between these groups, and neither group exhibited significant distinctions when compared with healthy individuals [15].

In pursuit of a personalized approach to diagnose and treat patients with SSD, Bryll et al. conducted an evaluation of peripheral and cerebral biochemical indices in correlation with the clinical condition of the patients. This involved neuroimaging and numerous laboratory tests, encompassing assessments of thyroid hormones, metabolic, immunoinflammatory, and oxidative-stress markers. The study primarily focused on forty FE SSD patients, all admitted with acute psychotic decompensation. Through K-means cluster analysis, two distinct endophenotypes were identified. Glutamate concentration in the anterior cingulate cortex (ACC) emerged as the sole biochemical parameter that exhibited statistical significance between the identified endophenotypes. Specifically, patients in phenotype (II) demonstrated reduced glutamate levels, along with higher negative and total PANSS scores ($p < 0.05$). The authors concluded that glutamate concentration in the ACC could potentially serve as a marker indicative of a less favorable course of SSD, progressing toward more severe functional deterioration and the need for more intensive treatment [16].

In a glutamatergic hypothesis-driven study of postmortem brains, De Rosa et al. [17] investigated various components involved in the functioning of the glutamatergic synapse, including amino acids, mRNA, and proteins. While univariate analysis revealed no significant differences in SSD patients and healthy controls, multivariable analysis, utilizing two types of classifiers (Random Forest and Classification and Regression Tree), indicated pertinent differences in amino acid levels, vesicular glutamate transporters (EAAT2, VGluT2), enzymes involved in glutamate-dependent GABA neurotransmitter synthesis (GAD67), and proteins associated with postsynaptic NMDA receptor signaling, such as CAMK2α and the presynaptic marker synapsin-1. Using different combinations of these predictors, the models exhibited AUC values ranging from 80 to 92%, suggesting that these biomarkers could be good diagnostic predictors for SSD. Both studies collectively contribute to supporting the involvement of the glutamatergic hypothesis in the pathophysiology of SSD.

Yu et al. [18] attempted to construct single-biomarker models based on neuropeptides to differentiate FE SSD patients from healthy controls, MDD, and BD. However, they achieved AUC values less than 83%, indicating that these biomarkers possess moderate or poor diagnostic potential. On the other hand, Wang et al. [19] achieved a favorable AUC of 89% with a model incorporating GABA, a cholesterol ester, and two bile acids. However, like the study by Yu et al., crucial parameters for a more comprehensive evaluation of the model, such as sensitivity and specificity, were not reported.

8.3 Discussion

The primary objective of this review was to investigate the diagnostic potential of neurotransmitter biomarkers through ROC analysis. A total of 12 studies, spanning 8 countries and involving 1,000 SSD patients, revealed promising models. The key focus was on glutamatergic pathways and in the related molecules, such as glutamate and GABA, although their individual performance was moderate. Models that incorporated multiple factors and diverse chemical classes of compounds proved to be the most successful. Notably, a model utilizing three components of the endocannabinoid system demonstrated an impressive AUC of 94% [12], and the model using standalone agmatine yielded an AUC of 97% [14]. Furthermore, the Fryar-Williams model [8], which incorporated a cognitive domain as a predictor, exhibited greater effectiveness with an AUC of 92%, surpassing a panel of catecholamines (AUC 86%) in distinguishing SSD patients from healthy controls. This result anticipated a similar outcome achieved by Fernandes' model 5 years later, which also utilized a cognitive domain (AUC 90%) and outperformed an immunoinflammatory blood domain (AUC 71%) [20], highlighting the importance of cognitive changes that occur in patients affected by SSD.

These diverse findings underscore the intricate involvement of neurotransmitters in SSD and emphasize the significance of considering various factors, including the selection of biomarkers and the context of their measurement. Such considerations are crucial in comprehending the neurobiological underpinnings of the disorder and in developing more efficient diagnostic models.

References

1. Lai C-Y, Scarr E, Udawela M, Everall I, Chen WJ, Dean B (2016) Biomarkers in schizophrenia: a focus on blood-based diagnostics and theranostics. World J Psychiatry 6:102–117. https://doi.org/10.5498/wjp.v6.i1.102

2. Song J, Viggiano A, Monda M, De Luca V (2014) Peripheral glutamate levels in schizophrenia: evidence from a meta-analysis. Neuropsychobiology 70:133–141. https://doi.org/10.1159/000364828
3. Zahid U, Onwordi EC, Hedges EP, Wall MB, Modinos G, Murray RM, Egerton A (2023) Neurofunctional correlates of glutamate and GABA imbalance in psychosis: a systematic review. Neurosci Biobehav Rev 144:105010. https://doi.org/10.1016/j.neubiorev.2022.105010
4. Skorobogatov K, De Picker L, Verkerk R, Coppens V, Leboyer M, Müller N, Morrens M (2021) Brain Blood: a systematic review on the concordance between peripheral and central kynurenine pathway measures in psychiatric disorders. Front Immunol 12:716980. https://doi.org/10.3389/fimmu.2021.716980
5. Li C, Wang A, Wang C, Ramamurthy J, Zhang E, Guadagno E, Trakadis Y (2018) Metabolomics in patients with psychosis: a systematic review. Am J Med Genet B Neuropsychiatr Genet 177:580–588. https://doi.org/10.1002/ajmg.b.32662
6. Luo Y, Dong D, Huang H et al (2023) Associating multimodal neuroimaging abnormalities with the transcriptome and neurotransmitter signatures in schizophrenia. Schizophr Bull 49:1554–1567. https://doi.org/10.1093/schbul/sbad047
7. Yang AC, Tsai S-J (2017) New targets for schizophrenia treatment beyond the dopamine hypothesis. Int J Mol Sci 18:1689. https://doi.org/10.3390/ijms18081689
8. Fryar-Williams S, Strobel JE (2015) Biomarkers of a five-domain translational substrate for schizophrenia and schizoaffective psychosis. Biomark Res 3:3. https://doi.org/10.1186/s40364-015-0028-1
9. Sigvard AK, Bojesen KB, Ambrosen KS et al (2023) Dopamine synthesis capacity and GABA and glutamate levels separate antipsychotic-naïve patients with first-episode psychosis from healthy control subjects in a multimodal prediction model. Biol Psychiatry Glob Open Sci 3:500–509. https://doi.org/10.1016/j.bpsgos.2022.05.004
10. Egerton A, Murphy A, Donocik J et al (2021) Dopamine and glutamate in antipsychotic-responsive compared with antipsychotic-nonresponsive psychosis: a multicenter positron emission tomography and magnetic resonance spectroscopy study (STRATA). Schizophr Bull 47:505–516. https://doi.org/10.1093/schbul/sbaa128
11. Chiu PW (2018) In vivo gamma-aminobutyric acid and glutamate levels in people with first-episode schizophrenia: a proton magnetic resonance spectroscopy study. Schizophr Res 193:295–303. https://doi.org/10.1016/j.schres.2017.07.021
12. Wang D, Sun X, Yan J et al (2018) Alterations of eicosanoids and related mediators in patients with schizophrenia. J Psychiatr Res 102:168–178. https://doi.org/10.1016/j.jpsychires.2018.04.002
13. Loureiro CM, Shuhama R, Fachim HA, Menezes PR, Del-Ben CM, Louzada-Junior P (2018) Low plasma concentrations of N-methyl-d-aspartate receptor subunits as a possible biomarker for psychosis. Schizophr Res 202:55–63. https://doi.org/10.1016/j.schres.2018.06.037
14. Uzbay T, Goktalay G, Kayir H, Eker SS, Sarandol A, Oral S, Buyukuysal L, Ulusoy G, Kirli S (2013) Increased plasma agmatine levels in patients with schizophrenia. J Psychiatr Res 47:1054–1060. https://doi.org/10.1016/j.jpsychires.2013.04.004
15. Peitl V, Getaldić-Švarc B, Karlović D (2020) Platelet serotonin concentration is associated with illness duration in schizophrenia and chronological age in depression. Psychiatry Investig 17:579–586. https://doi.org/10.30773/pi.2020.0033
16. Bryll A, Krzyściak W, Karcz P, Pilecki M, Śmierciak N, Szwajca M, Skalniak A, Popiela TJ (2021) Determinants of schizophrenia endophenotypes based on neuroimaging and biochemical parameters. Biomedicines. https://doi.org/10.3390/biomedicines9040372
17. De Rosa A, Fontana A, Nuzzo T et al (2022) Machine learning algorithm unveils glutamatergic alterations in the post-mortem schizophrenia brain. Schizophrenia (Heidelb) 8:8. https://doi.org/10.1038/s41537-022-00231-1
18. Yu H, Ni P, Zhao L et al (2023) Decreased plasma neuropeptides in first-episode schizophrenia, bipolar disorder, major depressive disorder: associations with clinical symptoms and cognitive function. Front Psych 14:1180720. https://doi.org/10.1080/15622975.2023.2271965

19. Wang X, Xie J, Ma H et al (2023) The relationship between alterations in plasma metabolites and treatment responses in antipsychotic-naïve female patients with schizophrenia. World J Biol Psychiatry:1–10. https://doi.org/10.1080/15622975.2023.2271965
20. Fernandes BS, Karmakar C, Tamouza R et al (2020) Precision psychiatry with immunological and cognitive biomarkers: a multi-domain prediction for the diagnosis of bipolar disorder or schizophrenia using machine learning. Transl Psychiatry 10:162. https://doi.org/10.1038/s41398-020-0836-4

Chapter 9
Conclusions

9.1 Final Considerations

The primary objective of this review was to examine the molecular biomarkers associated with the hypotheses of the pathophysiology of schizophrenia spectrum disorders (SSD) that underwent ROC curve analysis in the past 12 years (2011–2023). The aim was to identify prevalent biomarkers within developed models, potentially forming a distinctive biomarker signature for SSD. The conclusions drawn from this investigation reveal intriguing and promising insights.

Despite China leading in more than 50% of the studies, this research spans over 20 countries and involves a significant cohort of over 10,000 patients diagnosed with SSD. Biomarkers related to neurotrophic, neuroendocrine, oxidative stress, and immunoinflammatory processes demonstrated moderate performance. Generally, models relying on a single biomarker yielded area under the curve (AUC) values less than 70%. The most promising scenarios emerged when models incorporated 3–10 metabolic biomarkers (human or associated with the gut microbiota) from diverse chemical classes and pathways. This trend was further enhanced when cognitive or symptom scores were included.

From the analysis of models incorporating human metabolic biomarkers, a list of 19 peripheral metabolites surfaced, demonstrating the ability to characterize SSD. In disease-directed enrichment analysis, this metabolite panel secured the second position for SSD with a notable significance level (FDR 1.51E-30). Similarly, considering metabolites associated with the gut microbiota, a list of 8 peripheral amino acids (L-aspartic acid, L-tyrosine, L-arginine, glycine, glutamine, leucine, phenylalanine, and serine) yielded a comparable outcome. Notably, six of these amino acids overlapped with the previous list, underscoring the potential of this eight-amino acid panel for SSD diagnosis.

The viability of amino acids as biomarkers for SSD is substantiated by other studies [1–5]. However, meticulous attention is required in future targeted metabolomic analyses, especially in the collection of patient samples, which should be rigorously selected based on specific stages of SSD. Some amino acids may exhibit altered levels depending on the phase of the disease, such as serine [6, 7]. Confounding factors, including the timing of blood collection, dietary intake, and genetic variations in the studied populations, should be considered in multivariate correction.

Moreover, the Bacillota phylum and Lachnospiraceae family emerged as among the most prevalent in all studies, with genera such as *Coprococcus*, *Blautia*, *Roseburia*, and *Dorea* particularly noteworthy. These genera exhibited reduced abundance in patients with SSD. The Lachnospiraceae family, integral to the core of the gut microbiota, establishes colonization in the intestinal lumen from birth and undergoes dynamic changes in species richness and relative abundances throughout the host's life [8]. Renowned for producing anti-inflammatory short-chain fatty acids, Lachnospiraceae maintains a complex relationship with human health, disease, and dietary composition. Furthermore, this family stands out as one of the most prevalent microbial families globally [9], underscoring its potential as a biomarker that transcends geographical and ethnic boundaries.

The quest for biomarkers for SSD remains a formidable challenge, given the complex and interconnected changes coursing throughout the body. These transformations, not necessarily in a specific sequence, are instigated by dysfunctional neurotransmission and altered neuronal signaling. They set off significant oxidative stress, which ripples through the body, impacting the HPA-axis, hormonal regulation, and neurotrophic processes. This cascade of events disrupts glycolysis in the brain, redirects fatty acid metabolism toward ketosis, triggers immune-inflammatory responses, and leads to alterations in enzyme function and production. In addition to these challenges, the impact of this multifaceted process extends to the intestinal flora, leading to dysbiosis.

These dynamic changes manifest in genetic, epigenetic, protein, and metabolic profiles. Importantly, these profiles lack constant patterns, underscoring the uniqueness of individual minds and their distinct ways of feeling and perceiving the world. Unraveling this intricate puzzle demands a profound understanding of the origin of thought and its pervasive influence on various physiological processes. In this endeavor, we are only at the nascent stages of comprehension, with much exploration and discovery lying ahead.

> It has also become clear that cognition may be the royal road to understanding this brain disease—Matcheri Keshavan

9.2 Cautions in Pursuit of Robust Biomarkers

Considerations arose during the composition of this review, emphasizing the need for methodological rigor in conducting a sound study. To ensure the reliability of the data obtained and the accuracy of the reported results, it is imperative to adhere to a comprehensive checklist at each stage of the metabolomic workflow. Given the intricate nature of this study, meticulous attention to detail is paramount. The following checklist was based on insights from the authors cited in this review.

9.2.1 Study Design and Planning

Implementing a more stringent study design is essential for ensuring the robustness and reliability of research outcomes [10–15]. Key considerations for enhancing the study design encompass:

9.2.1.1 Primary Outcome Variables

- Clearly define the primary outcome variables that align with the research objectives.
- Ensure these variables are quantifiable and directly address the study hypotheses.

9.2.1.2 Sample Type

- Carefully select appropriate sample types relevant to the research question.
- Consider different sample sources (e.g., tissues, biofluids) for comprehensive insights.

9.2.1.3 Control Groups

- Establish both positive and negative control groups to validate experimental results.
- Include controls that mimic the experimental conditions while omitting the factor of interest.

9.2.1.4 Confounding Factors and Covariates

- Identify potential confounders and covariates and incorporate strategies to control for them.
- Clearly document how these factors will be addressed in the study design and analysis.

9.2.1.5 Sample Biomass

- Standardize/normalize sample biomass to ensure consistency across samples.
- Validate the adequacy of sample size for the proposed analyses.

9.2.1.6 Sample Collection Methods

- Detail standardized protocols for sample collection to minimize variability.
- Consider factors such as timing, environment, and participant instructions.

9.2.1.7 Equipment and Materials

- Specify the equipment and materials required, ensuring their reliability and calibration.
- Conduct pilot studies to validate the suitability of chosen tools for the study.

9.2.1.8 Processing and Storage of Samples

- Establish rigorous protocols for sample processing and storage to preserve sample integrity.
- Monitor and document any variations introduced during processing.

9.2.1.9 Biological and Technical Variations

- Account for both biological and technical variations in the study design.
- Implement quality control measures to distinguish between true biological changes and technical artifacts.

9.2.1.10 Adequately Powered Studies

- Conduct power analyses to determine the appropriate sample size for desired effect sizes.
- Ensure the study is adequately powered to detect meaningful differences.

9.2.1.11 Metadata

- Collect and document comprehensive metadata associated with each sample.
- Include relevant participant information, experimental conditions, and any other contextual details.

9.2 Cautions in Pursuit of Robust Biomarkers

- Implement pre-registration of study protocols to enhance reproducibility, significance, and representativeness of results [12, 15–21].

9.2.1.12 Longitudinal Designs

- Consider implementing longitudinal designs with multiple collection points to capture dynamic changes.
- Plan for repeated measures to analyze the progression of the disorder over time.

9.2.1.13 Statistical Methods

- Select and justify statistical methods based on the nature of the data.
- Consider advanced statistical approaches for handling complex data structures.

9.2.1.14 Omics Data Integration

- If applicable, outline strategies for integrating omics data to provide a holistic view [22].
- Ensure compatibility between different omics platforms and data types.

9.2.1.15 Standard Measures for Drugs Treatment Response

- Define standardized measures for assessing responses to drug treatments.
- Include quantitative criteria for determining treatment efficacy.

9.2.2 Sample Collection

9.2.2.1 Provide Comprehensive Details on Sample Collection and Preparation

- Standardize sample collection procedures.
- Document sample handling protocols to maintain sample integrity.
- Implement randomization techniques to minimize bias.
- Opt for the sample type that aligns most effectively with the research question [22, 23].
- Elaborate on the methodology, ensuring detailed information on sample collection and preparation procedures.

9.2.2.2 Enable Confounder Adjustment Through Extensive Data Collection

- Gather demographic, biological, and clinical details from patients.
- Include parameters such as ethnic origin, geographic region, all drugs used (including doses), age (mean/standard deviation), sex distribution, cognitive function, symptom severity, genetic and molecular biology data, physiology, duration of treatment and illness, PANSS score, disease stability, time and interval of blood collection, smoking habits, alcohol (and other drugs) use, education, and disease course [13, 22–25].

9.2.2.3 Standardize Patient Inclusion Criteria and Socio-Demographic Conditions

- Clearly define inclusion criteria to minimize heterogeneity in cohort characteristics.
- Specify diagnostic criteria used, reported medical and psychiatric comorbidities, gender balance, and the use of psychoactive substances [10].
- Use reporting guidelines with appropriate reference standards for psychiatric disorders.
- Consider participating in multicenter consortia to increase cohort numbers.

9.2.2.4 Define Exclusion Criteria

- Exclude participants with inflammatory, gastrointestinal, autoimmune, liver, or cardiovascular diseases.
- Consider exclusion for other current or past psychiatric disorders, pregnancy, breastfeeding, endocrine and metabolic disorders, substance abuse, family history of psychiatric disorders, and hormonal variables like contraceptive use, menopausal status, and menstrual cycle phases [10, 11, 17].

9.2.2.5 Consider Confounding Factors Influencing Microbial Diversity Determination

- For extrinsic factors, include non-breastfeeding, environmental factors, lifestyle, diet, bowel habits, infection exposure, antibiotic use, age, medication use, and diagnostic criteria.
- For intrinsic factors, consider metabolite levels, immunity, hormones, and genetic background.
- Include other variables like gender, body mass index (BMI), geographical region (east/west), hypervariable region sequenced (V3-V5 or V4), platform, and instrument [13, 16].

9.2 Cautions in Pursuit of Robust Biomarkers

9.2.2.6 Recruit Patients with Specific Criteria

- Define specific criteria for patient recruitment, such as disease stage, symptom severity, inflammatory parameters, or classic disease categories [23, 26].

9.2.3 Sample Preparation

- Optimize extraction methods for target metabolites.
- Validate sample stability during preparation.
- Implement quality control measures for reproducibility.

9.2.4 Analytical Techniques

- Choose appropriate analytical platforms (e.g., mass spectrometry, nuclear magnetic resonance).
- Validate and calibrate instruments regularly.
- Monitor and control experimental conditions for consistency.
- Provide comprehensive details about the equipment and conditions used.
- Employ various techniques to conduct multiple assays and identify a diverse set of biomarkers [22].

9.2.5 Data Acquisition and Processing

- Employ standardized protocols for data acquisition.
- Implement quality checks to identify and address outliers.
- Utilize robust statistical methods for data preprocessing.

9.2.6 Statistical Analysis

- Conduct power analysis to determine sample size.
- Apply appropriate statistical tests based on data distribution.
- Implement corrections for multiple comparisons, such as False Discovery Rate, ROC curves, or other tools.
- Measure the diagnostic efficiency of biomarkers to avoid bias in selective reporting, false positive findings, and non-reporting of negative findings.

9.2.7 Data Interpretation and Validation

- Employ pathway analysis tools for biological interpretation.
- Validate findings using independent techniques or replicate experiments.
- Consider biological relevance in addition to statistical significance.
- Be cautious in selecting biomarkers. Consider factors like substance abuse, nutrition status, BMI, and psychotherapy interventions.
- Understand that acute phase proteins and cytokine levels can be influenced by various confounding factors [11, 27].
- Prioritize biomarkers that are specific and impactful in clinical decision-making processes [12].

9.2.8 Reporting and Documentation

- Transparently report methods, results, and statistical analyses.
- Share raw data or provide access to it for reproducibility. Open data format facilitates community reanalysis and larger studies [10].
- Clearly articulate limitations and potential sources of bias.
- Refrain from reporting results as increased or decreased levels; instead, provide specific quantitative values, including concentration, fold change, and the respective p-value [22].
- Integrate symptoms with the most differentiated biomarkers in the reporting of results [22].
- Improve technical language and rigor when reporting results [28].

By incorporating these considerations into the metabolomic workflow, researchers can enhance the robustness and reliability of their studies, ultimately contributing to the advancement of scientific knowledge in this complex field.

References

1. Leppik L, Kriisa K, Koido K, Koch K, Kajalaid K, Haring L, Vasar E, Zilmer M (2018) Profiling of amino acids and their derivatives biogenic amines before and after antipsychotic treatment in first-episode psychosis. Front Psych 9:155. https://doi.org/10.3389/fpsyt.2018.00155
2. Taniguchi K, Sawamura H, Ikeda Y, Tsuji A, Kitagishi Y, Matsuda S (2022) D-amino acids as a biomarker in schizophrenia. Diseases 10:9. https://doi.org/10.3390/diseases10010009
3. Parksepp M, Leppik L, Koch K, Uppin K, Kangro R, Haring L, Vasar E, Zilmer M (2020) Metabolomics approach revealed robust changes in amino acid and biogenic amine signatures in patients with schizophrenia in the early course of the disease. Sci Rep 10:13983. https://doi.org/10.1038/s41598-020-71014-w
4. He Y, Yu Z, Giegling I et al (2012) Schizophrenia shows a unique metabolomics signature in plasma. Transl Psychiatry 2:e149. https://doi.org/10.1038/tp.2012.76

5. Avigdor BE, Yang K, Shinder I, Orsburn BC, Rais R, Kano S-I, Sawa A, Pevsner J (2021) Characterization of antipsychotic medications, amino acid signatures, and platelet-activating factor in first-episode psychosis. Biomark Neuropsychiatry 5:100045. https://doi.org/10.1016/j.bionps.2021.100045
6. Zahid U, Onwordi EC, Hedges EP, Wall MB, Modinos G, Murray RM, Egerton A (2023) Neurofunctional correlates of glutamate and GABA imbalance in psychosis: a systematic review. Neurosci Biobehav Rev 144:105010. https://doi.org/10.1016/j.neubiorev.2022.105010
7. De Luca V, Viggiano E, Messina G, Viggiano A, Borlido C, Viggiano A, Monda M (2008) Peripheral amino acid levels in schizophrenia and antipsychotic treatment. Psychiatry Investig 5:203–208. https://doi.org/10.4306/pi.2008.5.4.203
8. Vacca M, Celano G, Calabrese FM, Portincasa P, Gobbetti M, De Angelis M (2020) The controversial role of human gut Lachnospiraceae. Microorganisms. https://doi.org/10.3390/microorganisms8040573
9. Dhakan DB, Maji A, Sharma AK, Saxena R, Pulikkan J, Grace T, Gomez A, Scaria J, Amato KR, Sharma VK (2019) The unique composition of Indian gut microbiome, gene catalogue, and associated fecal metabolome deciphered using multi-omics approaches. Gigascience 8:giz004. https://doi.org/10.1093/gigascience/giz004
10. Rodrigues JE, Martinho A, Santa C, Madeira N, Coroa M, Santos V, Martins MJ, Pato CN, Macedo A, Manadas B (2022) Systematic review and meta-analysis of mass spectrometry proteomics applied to human peripheral fluids to assess potential biomarkers of schizophrenia. Int J Mol Sci 23:4917. https://doi.org/10.3390/ijms23094917
11. Khoury R, Nasrallah HA (2018) Inflammatory biomarkers in individuals at clinical high risk for psychosis (CHR-P): State or trait? Schizophr Res 199:31–38. https://doi.org/10.1016/j.schres.2018.04.017
12. Pinto JV, Moulin TC, Amaral OB (2017) On the transdiagnostic nature of peripheral biomarkers in major psychiatric disorders: a systematic review. Neurosci Biobehav Rev 83:97–108. https://doi.org/10.1016/j.neubiorev.2017.10.001
13. Fan Y, Gao Y, Ma Q et al (2022) Multi-omics analysis reveals aberrant gut-metabolome-immune network in schizophrenia. Front Immunol 13:812293. https://doi.org/10.3389/fimmu.2022.812293
14. Fernandes BS, Steiner J, Berk M, Molendijk ML, Gonzalez-Pinto A, Turck CW, Nardin P, Gonçalves C-A (2015) Peripheral brain-derived neurotrophic factor in schizophrenia and the role of antipsychotics: meta-analysis and implications. Mol Psychiatry 20:1108–1119. https://doi.org/10.1038/mp.2014.117
15. Kumar PS (2021) Microbiomics: were we all wrong before? Periodontol 2000 85:8–11. https://doi.org/10.1111/prd.12373
16. Li Z, Zhou J, Liang H et al (2022) Differences in alpha diversity of gut Microbiota in neurological diseases. Front Neurosci 16:879318. https://doi.org/10.3389/fnins.2022.879318
17. Perkovic MN, Erjavec GN, Strac DS, Uzun S, Kozumplik O, Pivac N (2017) Theranostic biomarkers for schizophrenia. Int J Mol Sci. https://doi.org/10.3390/ijms18040733
18. Navarrete F, García-Gutiérrez MS, Jurado-Barba R, Rubio G, Gasparyan A, Austrich-Olivares A, Manzanares J (2020) Endocannabinoid system components as potential biomarkers in psychiatry. Front Psych 11:315. https://doi.org/10.3389/fpsyt.2020.00315
19. Evans AM, O'Donovan C, Playdon M et al (2020) Dissemination and analysis of the quality assurance (QA) and quality control (QC) practices of LC-MS based untargeted metabolomics practitioners. Metabolomics 16:113. https://doi.org/10.1007/s11306-020-01728-5
20. Beger RD, Dunn WB, Bandukwala A et al (2019) Towards quality assurance and quality control in untargeted metabolomics studies. Metabolomics 15:4. https://doi.org/10.1007/s11306-018-1460-7
21. Begou O, Gika HG, Theodoridis GA, Wilson ID (2018) Quality control and validation issues in LC-MS metabolomics. Methods Mol Biol 1738:15–26. https://doi.org/10.1007/978-1-4939-7643-0_2

22. Fernandes BS, Dai Y, Jia P, Zhao Z (2022) Charting the proteome landscape in major psychiatric disorders: from biomarkers to biological pathways towards drug discovery. Eur Neuropsychopharmacol 61:43–59. https://doi.org/10.1016/j.euroneuro.2022.06.001
23. Lema YY, Gamo NJ, Yang K, Ishizuka K (2018) Trait and state biomarkers for psychiatric disorders: importance of infrastructure to bridge the gap between basic and clinical research and industry. Psychiatry Clin Neurosci 72:482–489. https://doi.org/10.1111/pcn.12669
24. Rodrigues-Amorim D, Rivera-Baltanás T, Bessa J et al (2018) The neurobiological hypothesis of neurotrophins in the pathophysiology of schizophrenia: a meta-analysis. J Psychiatry Res 106:43–53. https://doi.org/10.1016/j.jpsychires.2018.09.007
25. Carvalho AF, Solmi M, Sanches M et al (2020) Evidence-based umbrella review of 162 peripheral biomarkers for major mental disorders. Transl Psychiatry 10:152. https://doi.org/10.1038/s41398-020-0835-5
26. Kroken RA, Sommer IE, Steen VM, Dieset I, Johnsen E (2018) Constructing the immune signature of schizophrenia for clinical use and research; an integrative review translating descriptives into diagnostics. Front Psych 9:753. https://doi.org/10.3389/fpsyt.2018.00753
27. Xiong P, Zeng Y, Wan J, Xiaohan DH, Tan D, Lu J, Xu F, Li HY, Zhu Z, Ma M (2011) The role of NGF and IL-2 serum level in assisting the diagnosis in first episode schizophrenia. Psychiatry Res 189:72–76. https://doi.org/10.1016/j.psychres.2010.12.017
28. Xia J, Broadhurst DI, Wilson M, Wishart DS (2013) Translational biomarker discovery in clinical metabolomics: an introductory tutorial. Metabolomics 9:280–299. https://doi.org/10.1007/s11306-012-0482-9

Chapter 10
Step-by-Step Guide to Building a Diagnostic Model Using MetaboAnalyst

10.1 Introduction

The metabolomic study follows a characteristic workflow with several steps. Initially, one must formulate the biological question, which guides the hypothesis formulation. From this point onward, the journey involves progressing toward the identification of metabolites and the disturbed metabolic pathways. Subsequently, the study includes the development and validation of the discriminatory model used for diagnosing the studied phenotype (Fig. 10.1).

The experimental design in metabolomics is determined by the chosen approach for identifying and quantifying metabolites. The targeted approach focuses on identifying and quantifying a limited set of known metabolites, typically fewer than a hundred. In contrast, the untargeted approach is discovery-based, involving the identification and relative quantification of hundreds of analytes in each sample [1, 2].

According to Xia [3], three prominent workflows are employed in metabolomics studies:

- **Chemometrics (or cheminformatics):** This approach utilizes multivariate mathematical-statistical methods to assess the spectral features obtained in the study. Data is analyzed using methods such as PCA or PLS-DA to recognize patterns and classes associated with experimental conditions. Subsequently, metabolites are identified, tabulated, and subjected to enrichment analysis to determine significant biological pathways for interpreting the biological system. The most significant/important metabolites identified are utilized, among other applications, to construct discriminatory models (biomarker analysis) for predicting/diagnosing diseases or assessing the toxicity and adverse effects of drugs.
- **Metabolic profiling:** This involves the identification and quantification of a selected number of pre-defined metabolites or all metabolites and derived products, whether identified or not, in a biological sample using a specific analytical

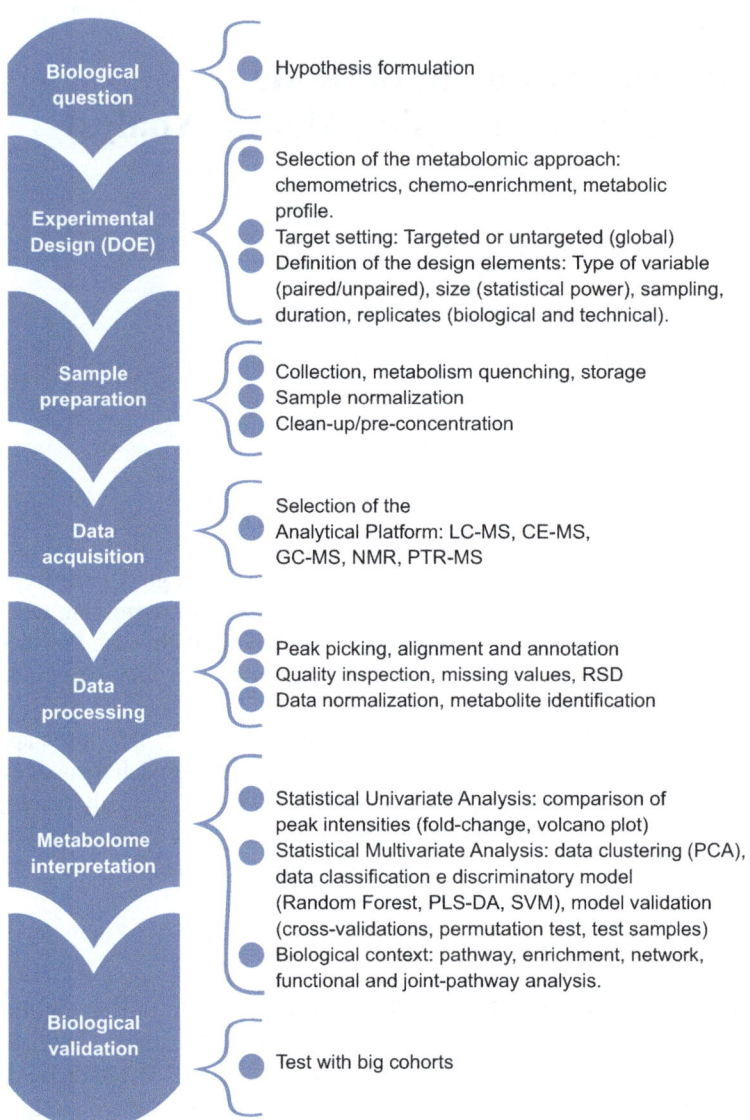

Fig. 10.1 Metabolomics pipeline. (Based on Sussulini [4])

technique. This process can be employed, for example, in disease diagnosis or to elucidate the function of entire pathways or intersecting pathways [5].

- **Chemo-enrichment analysis:** This method connects spectral features with biological interpretations without explicitly identifying the compounds. Metabolite identification is typically performed post hoc, meaning it is done after the initial analysis that identifies enriched biological processes. The Mummichog algorithm is a commonly used tool for chemical enrichment analysis in untargeted

10.1 Introduction

metabolomics. In this approach, the list of significant features (spectral peaks or other chemical signals) is compared against a database to find potential matches for pathways and metabolic networks. The result is then compared with those obtained based on peak lists drawn randomly from reference peaks to calculate statistical significance.

Study groups, cohort sizes, sample types, and preparations are determined based on the biological question and the chosen analytical platform for data acquisition. Among the most widely used platforms, LC-MS represents a type of two-dimensional analysis. The first separation occurs in the chromatograph due to the interaction between metabolites and the mobile and stationary phases. The time at which a metabolite elutes from the chromatographic column is known as retention time (RT). The second separation occurs in the mass spectrometer, where metabolites are ionized and separated based on their mass-charge ratio (m/z). The entire analysis path is termed a run, and the result (raw data) is obtained in a 3D graphic format. Here, the first axis represents RT, the second axis displays m/z values, and the third axis shows the intensity of each metabolite separated in the analysis [6]. The 3D raw data needs to be transformed or translated into data sheets, allowing for the comparison of columns and rows in various subsequent analyses. This process culminates in the correlation between metabolites and the biological question, followed by subsequent validation.

Data obtained in profile format (continuum) and in raw extension can be analyzed directly by the vendor's software or by some open-source software pipelines such as MZmine 3.0 [7] and MS-DIAL [8], which allow processing data acquired in either Data-Dependent (DDA) or Data-Independent (DIA) mode. However, the MetaboAnalyst platform requires centroid format and specific data extensions, such as mzML, mzXML, mzData, or CDF. To convert raw data to mzML or mzXML, users can utilize Proteowizard [9]. A brief tutorial on this conversion process is available on the website [10] or in the step-by-step protocol by Holman et al. [11]. It's worth noting that Proteowizard has limitations in handling data acquired by DIA mode, requiring a previous procedure before conversion [12].

The original data obtained in profile format are more comprehensive, with each column of points representing the signal measured at a specific time and stored in a spectrum. However, the file size is often excessively large. Centroid transformation addresses this by condensing all points in a column to a single point, located in the row of greatest intensity. This forms a representative signal for an ion in each spectrum, as described in [13]. While the reduction in data size accelerates processing, it comes at the cost of losing some data. After the data transformation, the Savitzky-Golay smoothing filter is applied. This filter considers the weighted average of intensities along a quadratic curve, enhancing the shape of peaks and valleys (noise) while preserving peak height [14, 15].

After converting the files to the appropriate extension, the user should open one of them with Notepad (*.txt extension) to verify the presence of "MS1 spectrum, level 1" mass data and "MSn spectrum, level 2" fragmentation data. The latter should be in centroid format, as illustrated in the text below:

```
<dataProcessing id="pwiz_Reader_Thermo_conversion">
<processingMethod order="0" softwareRef="pwiz">
<cvParam cvRef="MS" accession="MS:1000544" name="Conversion to mzML" value=""/>
</processingMethod>
<processingMethod order="1" softwareRef="pwiz">
............. . . .
<cvParam cvRef="MS" accession="MS:1000579" name="MS1 spectrum" value=""/>
<cvParam cvRef="MS" accession="MS:1000511" name="ms level" value="1"/>
<cvParam cvRef="MS" accession="MS:1000130" name="positive scan" value=""/>
<cvParam cvRef="MS" accession="MS:1000127" name="centroid spectrum" value=""/>
................ . . . .
<cvParam cvRef="MS" accession="MS:1000285" name="total ion current" value="4.9244048e07"/>
................ . . . .
................ . . . .
<cvParam cvRef="MS" accession="MS:1000580" name="MSn spectrum" value=""/>
<cvParam cvRef="MS" accession="MS:1000511" name="ms level" value="2"/>
```

MetaboAnalyst 6.0 includes 18 modules subdivided into the following categories (for more details, refer to [16, 17]):

- **Data Processing:** LC-MS Spectra Processing (with or without MS2, Peak annotation)
- **Exploratory Statistical Analyses:** Statistical Analysis (one factor)
- **System Biological Analysis**: Functional Analysis, Enrichment Analysis, Pathway Analysis, and Network Analysis
- **Translational Analysis:** Biomarker Analysis, Dose Response Analysis, and Power Analysis
- **Meta-Analysis**: Functional Meta-Analysis, Statistical Meta-Analysis
- **Other utilities (internal and external resources):** Compound ID conversion, batch effect correction, lipidomics profiling, ontology enrichment analysis for lipidomic data (LION), etc.

From now on, we will be using some terms related to mass spectrometry and metabolomics, each defined below:

Annotated metabolite: A metabolite putatively identified based on physicochemical and/or spectral similarities to compounds in a spectral library [18].

10.1 Introduction

BPI (Base peak intensity chromatogram): This corresponds to the set of the most intense *m/z* signals from a spectrum at each point in the analysis [13].

Classifier: Is an algorithm that automatically orders or categorizes data into one or more of a set of classes [19].

Collinearity: Refers to a situation where two or more predictor variables (features) in a model are highly correlated. In other words, there is a linear relationship between two or more independent variables. This correlation can cause issues in certain types of models, particularly linear regression [20].

EIC (extracted ion chromatogram): Encompasses a series of BPI within a mass tolerance window around the *m/z* of a given metabolite at each point in the analysis [13].

Feature: Corresponds to each *m/z*-RT pair resulting from the LC/MS raw data processing [21].

Feature annotation: Refers to the process of assigning different metabolic features (adducts, isotopologous, fragments, and neutral losses) into a single value [22].

Feature extraction: Involves creating new and more relevant features from the original features [23].

Feature selection: Refers to the process of selecting important features while filtering out irrelevant ones [23].

Isotopic trace: Refers to a set of signals coming from the same metabolite state of charge combination but differing in weight due to the presence of certain isotopes, such as ^{16}O, ^{17}O, and ^{18}O.

Isotopic envelope trace: This is produced by a series of isotope traces.

Metabolite annotation: Refers to the process of assigning tentative identifications or candidates to signals or features detected in metabolomic analyses based on certain criteria such as mass, retention time, and spectral characteristics [22].

Metadata: Consists of "data about the data." It describes the relevant information needed to perform a given analysis, for example, in the MetaboAnalyst platform [23].

Peak: This term is used for one-dimensional signals, both for *m/z* peaks in the centroid spectrum and for RT peaks in the chromatogram.

Predictor: This is a variable (features or independent variables) that will be used to predict the value of the target variable [19].

Spectrum: This is formed by all the *m/z* signals detected in each retention time.

TIC (total ion chromatogram): Represents the sum of all signals from a spectrum at each point in the analysis [13].

Well-behaved peak group: Corresponds to the peaks of a given *m/z*-RT window that are present in most samples [6].

For additional information about isotope trace, isotopic envelope trace, total ion spectrum (TIS), and total ion chromatogram (TIC), refer to [24–27].

10.2 LC-MS Spectra Processing

10.2.1 Data Upload

On the MetaboAnalyst 6.0 web page, users can navigate to the "LC-MS Spectra Processing" module (Fig. 10.2). If the number of files exceeds the platform loading limit (a maximum of 200 samples simultaneously), users should download the MetaboAnalystR program to perform processing offline.

Create your project and click on "Start," then "Select," and upload the individually zipped files, which should include Quality Controls (QCs) and blanks. This enables the program to perform subsequent extraction of contaminants and subtraction of signals present in the blanks.

Prepare a spreadsheet (metadata) with two columns (filename and class/group), as exemplified in Fig. 10.3. Save this spreadsheet in the txt extension before loading it, along with the converted files.

In Version 6.0, users are provided with the flexibility to select between entering MS1 only, MS1+DDA, or MS1+SWATH-DIA data, based on the analytical platform utilized for acquiring the data.

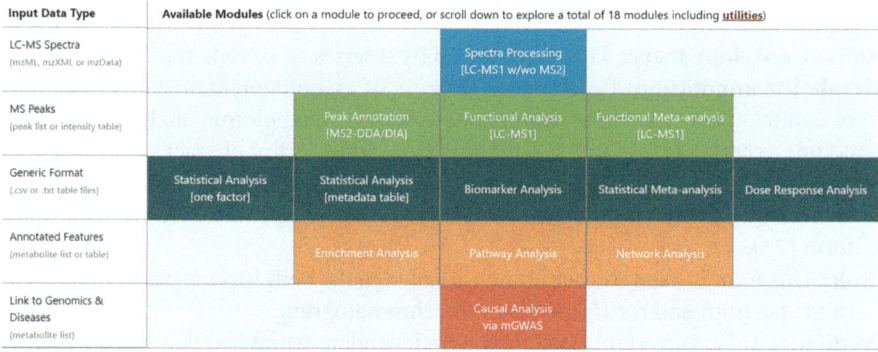

Fig. 10.2 Access to the data processing module on the MetaboAnalyst 6.0 platform

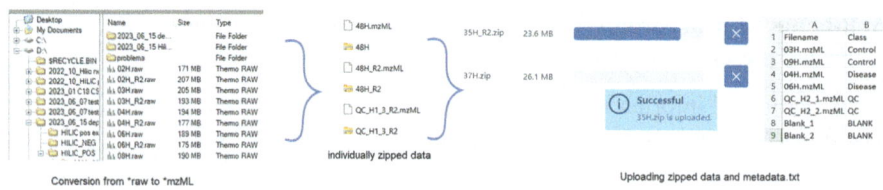

Fig. 10.3 Uploading data to the MetaboAnalyst platform

10.2 LC-MS Spectra Processing

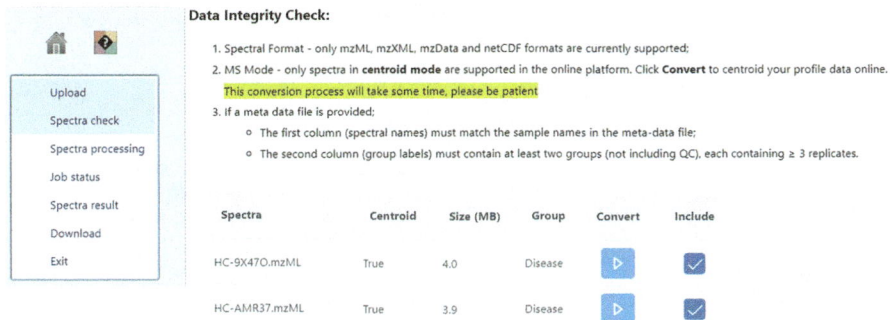

Fig. 10.4 Checking the integrity of uploaded data

10.2.2 Data Integrity Check

After a successful upload, in the "Data Integrity Check" window, users can verify the format, size, and grouping of the data. If the data is not in centroid format, it can be converted by clicking on the "Convert" button (Fig. 10.4). However, it is recommended that this conversion be performed previously. Once verification is complete, click on "Proceed."

10.2.3 Data Processing

Most processing methods share common functions, including deconvoluting analytical signals, filtering noise, detecting and aligning chromatographic peaks, baseline correction, and gap filling. Selecting the optimal method for each step is no simple feat. A good review, which merits reading, compares the most utilized processing methods. It emphasizes the challenges associated with each approach and highlights potential enhancements that can be implemented [28].

To proceed, click on the "LC-MS Platform" tab to select the equipment that was used to obtain the data (Fig. 10.5).

The objective of processing is to convert the 3D data into two-dimensional table/matrix, with rows representing features (m/z-RT pairs) and columns representing the abundance (intensity) of ions measured in each sample within an experiment. Data processing encompasses several steps such as peak picking, peak alignment, and peak annotation, which also involves MS2 processing [4].

10.2.3.1 Peak Picking

To detect/identify a peak, some algorithms initially divide the LC/MS data into slices, typically 0.1 m/z wide. They then operate within these slices in the chromatographic time domain. Within each slice, the signal is determined by taking the

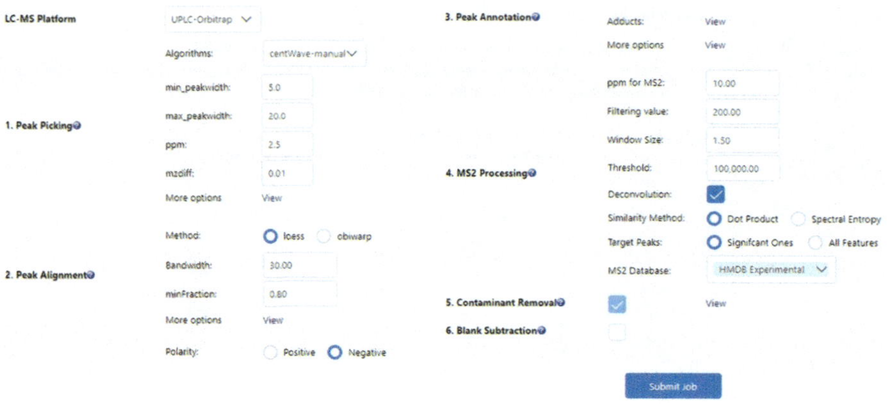

Fig. 10.5 Configuring parameters for the processing step

maximum intensity at each time point in the slice, resulting in the extracted ion chromatogram. The algorithm then creates overlapping combined chromatograms, applies filters, integrates signals, and selects peaks based on an ideal cutoff value, which also serves as the basis for noise elimination. For more details refer to [29].

MetaboAnalyst 6.0 offers four algorithm options for peak picking. Among them, Centwave can be utilized in either "Auto" or "Manual" mode.

10.2.3.1.1 Asari

This algorithm operates by taking centroid mzML files as input and constructing chromatograms for each as mass tracks. To prioritize contemporary mass resolution, *m/z* alignment is initially conducted to create a MassGrid, assisted by isotopic landmarks. Aligned mass tracks across samples undergo RT correction using LOWESS regression on a subset of high-quality elution peaks, and then they are aggregated into composite mass tracks. All composite mass tracks are stored in a "Composite Map." Elution peak detection is subsequently performed on the composite mass tracks, and a feature table is generated by referencing the corresponding peak areas in each individual sample. Annotation groups further simplify features into empirical compounds, and reference databases are employed to match the *m/z* values in empirical compounds. The "composite mass track" represents data from all samples by summing the signals in the corresponding mass tracks after RT alignment. The authors of this algorithm assert that this design enhances reproducibility. In contrast, conventional algorithms typically construct chromatograms, detect elution peaks, align retention times, and match the peaks [30].

10.2.3.1.2 Massifquant

This algorithm is used to enhance sensitivity for low intensity features. For a comparison between Massifquant, Matched Filter, and Centwave methods, refer to [31].

10.2 LC-MS Spectra Processing

10.2.3.1.3 Matched Filter

This algorithm is used to process data obtained in low resolution.

10.2.3.1.4 Centwave-Auto

Centwave is an algorithm tailored for processing data acquired at high resolution. It achieves chromatographic peak separation by integrating the density technique to identify the most abundant areas (Regions of Interest—ROIs) and the Wavelet approach to detect nearby and partially overlapping peaks with varying widths. The continuous wavelet transform (CWT) serves as the mathematical function in this process, chosen for its capability to distinguish weak signals from noise, thereby enhancing sensitivity [21]. The ROIs are extracted from the spectrum, followed by an optimization of the Design of Experiment (DoE). This optimization aims to identify the parameter combination that best suits the groups of stable and well-behaved peaks, facilitating their subsequent detection in all spectra [32].

This method can be utilized by users who are not familiar with the specific parameters required for the "Peak Picking" and "Peak Alignment" steps. However, in the "Peak Annotation" tab, it is essential to select the polarity of the experiment and the expected adducts. If both QCs and blanks were loaded, the "Contaminant Removal" and "Blank Subtraction" steps will be automatically selected.

10.2.3.1.5 Centwave-Manual

In this option, all parameters are manually specified by the user.

- **Parameters to be configured:**

Xia's lab protocol provides detailed information about MetaboAnalyst parameters [17]. Details on all arguments to be set are found at [33]. Some examples from the literature are presented in Table 10.1.

min_peakwidth: This specifies the minimum expected width of the chromatographic peak in the RT dimension. Appropriate values can be estimated based on EICs from internal standards, for example, or known compounds that are present in the sample. Preset value is 5 s [21].

max_peakwidth: This specifies the maximum expected width of the chromatographic peak. This value can be obtained from the EIC. Preset value is 20 s.

ppm (tolerated mass deviation in part per million): This specifies the maximum expected deviation of the m/z values from the centroids corresponding to a chromatographic peak. Typically, it is set to a generous multiple of the mass accuracy of the mass spectrometer. For instance, users may choose to set it at 30 ppm for a spectrometer with an accuracy ranging from 3 to 5 ppm [21]. The value assigned to this parameter is used in the initial identification of ROIs. In contrast to random

Table 10.1 Examples of configured parameters for peak picking (using Centwave Algorithm) and peak alignment (using ObiWarp Algorithm) according to literature data

Parameters	Mehl et al. [34]	Zhang et al. [35]	Gowda et al. [36]	Quintas et al. [37]	Rafiei and Sleno [38]	Kuhl et al. [39]
LC-MS platform	UPLC-QTOF	UPLC-QTOF	LC-QTOF	UPLC-QTOF	UPLC-QTOF	UPLC-QTOF
min_peakwidth[a]	5	5	10	5	5	5
max_peakwidth[a]	30	20	60	25	30	12
ppm[a]	20	NI	15	20	10	30
mzdiff[a]	0.01	NI	NI	NI	NI	NI
snthresh[a]	10	NI	NI	12	NI	6
noise[a]	5000	NI	NI	NI	NI	NI
prefilter peaks[a]	3	NI	NI	5	NI	2
prefilter intensity[a]	10,000	NI	NI	5000	NI	200
bw[b]	5	5	5	NI	5	3
mzwid[b]	0.015	0.015	0.015	NI	0.015	0.015
minfrac[b]	0.5	0.5	0.5	NI	0.5	NI
minsamples[b]	NI	NI	NI	NI	NI	NI
ProfStep[b]	1	1	1	NI	NI	NI
maxfeature[b]	NI	NI	NI	NI	NI	NI
Integration[b]	1	NI	NI	NI	NI	NI
span[b]	NI	NI	NI	NI	NI	NI
extra[b]	NI	NI	NI	NI	NI	NI

NI not informed
[a]Peak picking parameters
[b]Peak alignment parameters

noise, the actual signal from an ion is expected to produce stable m/z values on consecutive scans. The dispersion of the m/z values around the actual m/z value of the ion should be inversely related to its intensity. In the CentWave method, all data points that differ by the specified ppm value in consecutive spectra are combined into an ROI, which is then subjected to peak detection based on CWT. When the m/z deviation is large, peak determination becomes more ambiguous. This parameter is related to the equipment resolution and increases for low-intensity signals.

mzdiff: This specifies the minimum m/z difference between peaks that have the same retention time (overlapping RT).

More Options:

> **snthresh (signal/noise threshold)**: This sets the baseline cutoff value for peak selection. The usual value is 10. Peaks below this value will be discarded [29].

10.2 LC-MS Spectra Processing

noise: This corresponds to the baseline noise filter. A value can be determined by inspecting chromatograms.

prefilter: This determines the number of peaks per slice. In other words, this parameter indicates the minimum number of points (minimum scan number) a peak must possess to qualify as a peak.

value of prefilter: This specifies the minimum intensity (abundance) considered in peak determination. Peaks below this value will be discarded. A value can be determined by inspecting chromatograms.

10.2.3.2 Peak Alignment

The peaks identified in individual samples need to be correlated with the peaks in all other samples to perform calculations of RT offsets and compare relative ion intensity. However, during chromatographic separation, the RT of a given metabolite may undergo slight non-linear shifts in different samples due to various factors: (i) Changes that may occur in the mobile and stationary phases; (ii) Variations in temperature, flow, and pressure of the system; (iii) Interference caused by the matrix; (iv) Column wear, especially after long periods of analysis with very small flows (nano or microliters) [6].

Therefore, it is necessary to align the RTs. For this, the algorithm must map the well-behaved peaks. As these peaks represent signals from the same species of ions, they are used to estimate RT shifts between samples. The union of identities discovered in this way approaches completeness over time, and the quantities could also be compared for features without identity in each run [40]. Using the peak density matching method, the algorithm groups all chromatographic peaks under the same peak density in overlapping *m/z* "boxes" and fits the RTs using non-linear fit curves and smoothed distribution of the peaks in the determined chromatographic time to form a feature. Very few samples will have no peaks or will have more than one peak marked (Fig. 10.6).

There are two methods available for peak alignment:

10.2.3.2.1 LOESS (or Savitzky-Goley Filter)

This method performs the alignment of the RTs by local polynomial regression, a non-parametric analysis that uses LOESS (locally estimated scatterplot smoothing) or LOWESS (locally weighted scatterplot smoothing). These two algorithms use locally weighted regression to smooth the curve, making the relationship between the variables visible and predicting trends [41]. This method is recommended to remove RT deviation when internal standards are not added to samples [17].

At the beginning and end of LC/MS runs, where there are no well-behaved peak groups available for fitting, the deviation function is flattened to a constant value (Fig. 10.7). The resulting RT profile is then compared to the original RT profile to assess the quality of the alignment, a technique known as pairwise comparison. For

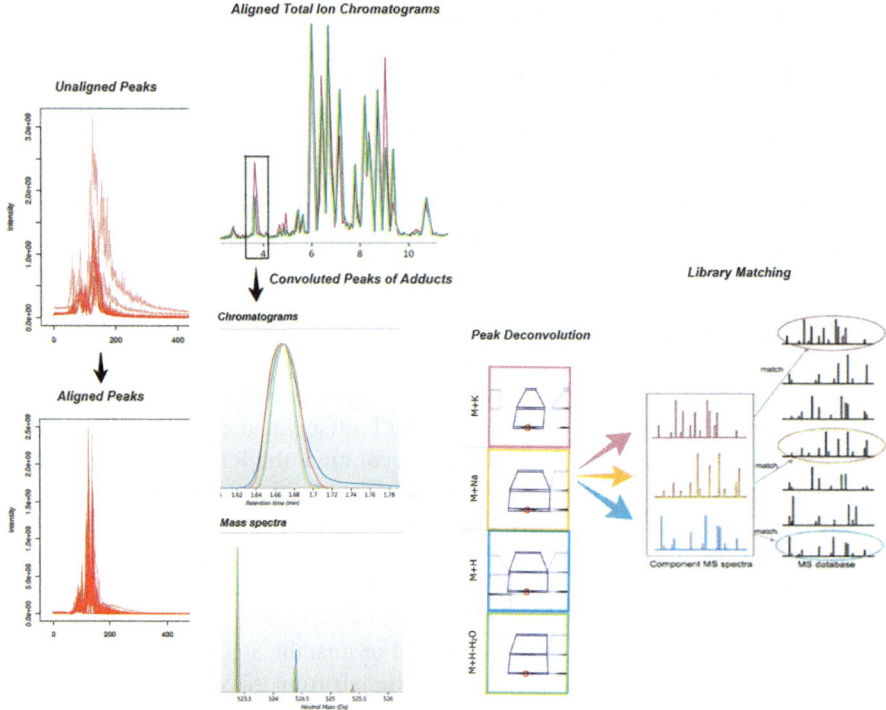

Fig. 10.6 Alignment, deconvolution, and library matching for peak identification

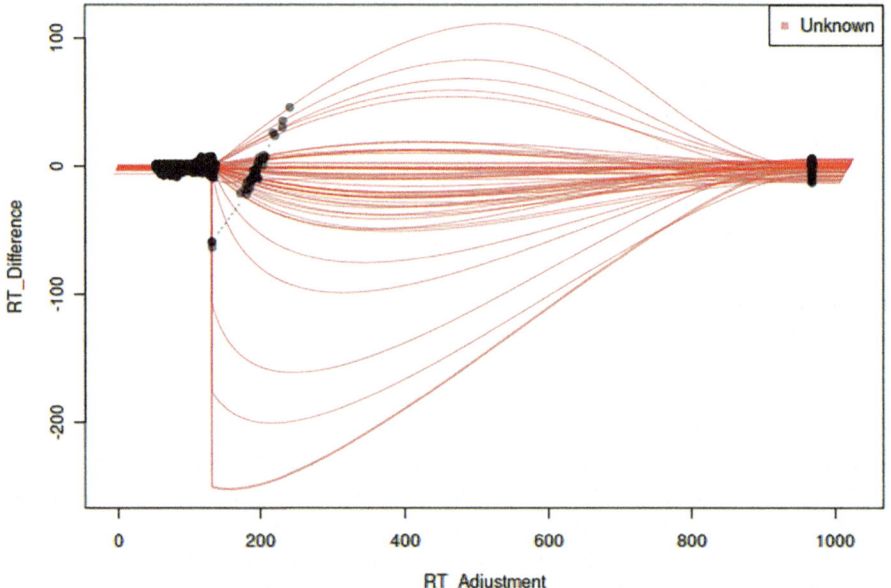

Fig. 10.7 Retention time alignment map of 52 plasma samples analyzed by HILIC LC-MS

10.2 LC-MS Spectra Processing

each pair of runs, all peaks present in both runs are used to calculate the "Median Retention Time Difference (MedRD)". The values from all possible pairs of runs can be combined into a MedRD matrix. The closer to zero the individual values in this matrix are, the better the alignment, indicating that no remaining systematic change in RT has occurred [6]. A positive RT difference (deviation) indicates that the sample was eluting after the MedRD, while a negative RT deviation indicates that the sample was eluting before the MedRD. The matching/alignment procedure can be repeated iteratively, successively discerning more and more groups of well-behaved peaks for increasingly precise alignment [29].

10.2.3.2.2 Obiwarp (Ordered Bijective Interpolated Warping)

In this method, a bijective distortion map is used as a basis to build a smooth deformation function that will be applied to each run through interpolation. It is based on dynamic time warping (DTW), which is an approach to align single or multivariate signals over time while preserving the internal ordering of the signals [40].

This method supports the alignment of multiple samples by aligning each of them against a central sample, which can be QC samples. It forms a median chromatogram from the deformation of the data, which serves as a mold where all the others are deformed to fit. Alignment is done directly on the profile matrix and can be performed independently of peak detection or peak clustering. The subset of parameters allows for defining the samples that must be aligned. Samples that do not fit the parameters are left out of the alignment models' estimation, but their retention times are later adjusted based on the alignment results of the closest sample in the subset.

- **Parameters to be configured** (see examples from the literature in Table 10.1):

    bandwidth: This determines the degree of smoothing for the peak density function. Xia's lab recommends a range of 2–5 for Q-exactive Orbitrap and 5–15 for TOF [17].
    minfraction (possible values between zero and 1): This defines the proportion of samples in which a group of well-behaved candidates should be detected. For example, a value of 0.8 means that a well-behaved peak group must be present in 80% of all samples in the experiment.

- **More Options:**

    span: This parameter sets the degree of smoothing for the LOESS function used in regression curve fitting. A small value, considering only a few points at each retention time (RT), or zero, will result in a high quality of fit but is also prone to overfitting. On the other hand, a large interval, such as 1, may not accurately represent the local trend and could introduce a constant linear offset, which is not suitable as the RT deviation is not linear. Values between 0.4 and 0.6 appear to be reasonable for most experiments [6].

maxfeature: This specifies the maximum number of features in an *m/z* slice for peak grouping. Default is 100 [17].

minsamples: This specifies the minimum number of samples in which a peak must be present. Default is 1 [17].

mzwid: This defines the width of the data slice in *m/z* into which peaks will be clustered.

extra: This sets the maximum number of additional peaks for all samples to be assigned to a well-behaved peak group for RT correction. For example, for a dataset containing 6 samples, "extra = 1" means that the program will use all peak groups with a total peak count ≤6 + 1.

ProfStep (profile step): This determines the step size (in *m/z*) to be used for profiling of the raw data files. For example, "ProfStep = 0" is used when one wants data to be loaded without generating the profile matrix (data spreadsheet).

integrate: There are two methods for integration: (i) uses Mexican hat type wavelets to correctly determine peak bounds, while (ii) determines the peak bounds on the baseline. The second method is more accurate but is subjected to the influence of noise.

The matrix (spreadsheet) obtained after the selection and alignment of peaks may contain gaps (values = 0, known as missing values) in some samples. This can occur due to instrumental limitations, such as the metabolite concentration being below the detection limit of the method, the absence of the metabolite in the sample, or ineffective processing of the data (Table 10.2) [42, 43]. In each case, no chromatographic peaks will be assigned to a feature. Some algorithms estimate this loss based on the nearest neighbor/sample approach. For more details, refer to [44].

The quality of the features must be checked in the matrix obtained by performing:

- Verification of the concomitant presence of a certain feature in a study sample and in the QCs, as these represent the set of all samples.
- Verification of the analytical coefficient of variation (CV = RSD, relative standard deviation) attributed to each feature ((Fig. 10.16; Table 10.2). When QC samples are available, Vinaixa et al. recommend removing any feature with a CV higher than 20% from the dataset [45].
- Verification of the percentage of missing values in the samples (Fig. 10.15). Some authors recommend that missing values exceeding 20% be removed from the data [44].

Note: The inclusion of missing values may reflect the consistency of feature detection in the CV result, while the exclusion of missing values may reflect the consistency of feature quantification [42].

10.2.3.3 Peak Annotation

After ionization, a metabolite can give rise to one or more ionic species, including isotopologues, fragments, adducts, etc. However, for the interpretation of systemic biology, what matters is the metabolite that gave rise to this redundancy. Therefore,

10.2 LC-MS Spectra Processing

Table 10.2 Example of spreadsheet containing missing values and coefficients of variation (CV)

Sample Label	CV%	2 Baseline-SSD	20 Baseline-SSD	6 Treated-SSD	12 Treated-SSD	QC-1 QC	QC-2 QC	Bl-1 Blank	Bl-2 Blank
5.06_387.0342	25.90	994.1852	789.0567	0	0	712.3444	772.4356	0	0
4.70_459.3085	21.73	122.9344	0	31.8341	27.1014	159.1118	141.9569	4.0251	3.7374
3.74_239.2375	25.81	679.2496	612.8146	171.5759	127.0298	765.3327	743.2987	0	0
4.74_1150.8084	24.24	235.5464	202.5588	22.4147	83.8533	322.5621	226.5462	0	7.1096
4.68_277.1046	22.31	169.9580	121.5583	75.7667	36.0297	112.4922	118.0439	0.81416	0
5.42_335.0191	29.26	9155.699	7383.0744	5572.3540	5983.7280	6953.8230	8510.6000	0	0
4.77_989.9171	23.73	207.9573	258.1679	97.0496	70.2345	468.6246	334.7645	0	0

Note: The number of samples has been reduced to fit the window. MetaboAnalyst requires a minimum of three replicates for each sample

two steps are necessary for peak annotation to be carried out by the algorithm CAMERA: the grouping of all features that arise from the same metabolite and the annotation of the types of ionic species [39].

There are four levels of identification for features and the metabolites they represent [46]:

- Positively identified: by correspondence with patterns.
- Putatively identified: by matching MS + RT or MS/MS + RT. Most of the time, this will be our case.
- Putatively identified in a class of compounds.
- Unknown compounds.

- **Parameter settings**:

10.2.3.3.1 Polarity

It sets the detection mode of the experiment, whether positive or negative.

10.2.3.3.2 Adducts

It determines which redundant peaks are expected, such as isotopologues, sodium and potassium adduct, neutral losses, oligomers, etc. Click on "View" to add the expected adducts, as shown in Fig. 10.8.

10.2.3.3.3 More Options

These options are in accordance with [47]:

Perc_fwhm: This determines the percentage of the full width at half maximum (FWHM) around the centroid to establish a specific RT window where the most intense features will be grouped and annotated [39]. The default is 60% or 0.6.

Mz_abs_iso e Mz_abs_add: These settings determine the allowed variance for the search for isotopologues and adducts annotation. The default values are 0.005 and 0.001, respectively.

Max_charge: This setting determines the maximum number of isotope charges. The default value is 2.

Max_iso: This setting determines the maximum number of isotopic peaks. The default is 2 peaks per isotope.

Corr_eic_th: This determines the threshold for intensity correlations across samples (refer to graphs in the supplementary material by [39]. The default value is 0.85.

10.2 LC-MS Spectra Processing

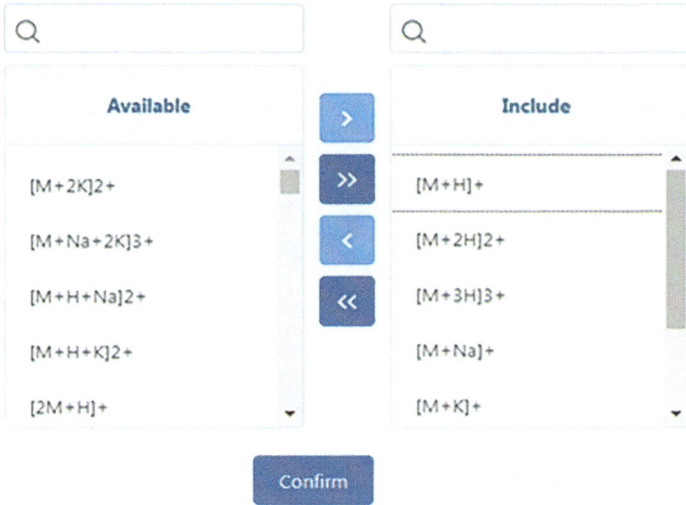

Fig. 10.8 Selecting expected adducts for metabolites

10.2.3.4 Contaminant Removal

Peaks arising from contaminants are automatically excluded if auto-optimized mode is selected. In this case, removal is based on QC samples.

If the default/manual mode is chosen, it is important to inspect some chromatograms before deciding which compounds will be removed. To do this, click on "View." A new window will open, showing, by default, 3D and 2D graphs of a random QC sample, if provided. Users can change the sample by selecting the file tab and switch to other spectral regions based on m/z and RT ranges, as demonstrated in Fig. 10.9:

File: This allows the selection of the sample on which the contaminant removal will be based.
mz range: This specifies the mass/charge range where the algorithm should look for the contaminant.
RT range: This defines the RT range (in seconds) where the algorithm should look for the contaminant.
Image style: 3D (default) (m/z x RT x intensity) or 2D (m/z x RT).
Resolution: This can be increased up to 1000 to show more details in the graphs.

After filling in all the parameters, click "Update." A window will open, displaying the new 3D map after removing contaminants from the selected sample.

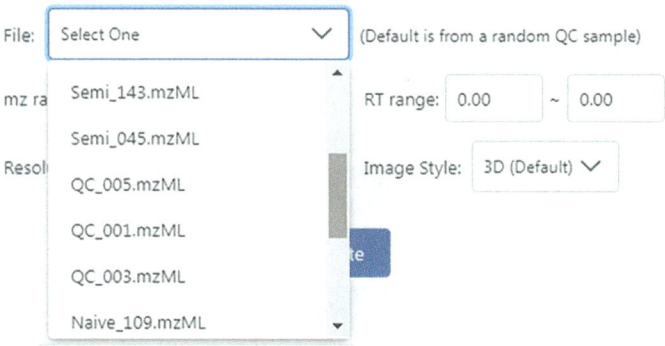

Fig. 10.9 Removing contaminants manually

10.2.3.5 Blank Subtraction

Select this option if you have loaded a BLANK file among the samples, and this file is specified in the metadata (*.txt list). Click "Confirm" in the dialog box, wait for processing, and receive the message that the work has been saved. It's crucial to recall that both quality controls (QCs) and blank samples need to have a minimum of three replicates to be accepted into MetaboAnalyst.

10.2.3.6 MS2 Processing

In metabolomics, MS2 processing refers to the analysis and interpretation of the tandem mass spectrometry (MS/MS or MS2) data obtained during a metabolomics experiment.

In MS/MS analysis, precursor ions of interest are selected and fragmented to produce a spectrum of fragment ions. These fragment ions provide information about the chemical structure of the original precursor ion. The resulting MS/MS spectra are then analyzed to identify the fragment ions present. This process, known as spectral annotation, involves matching the observed fragment ions with those expected based on reference spectra in databases or using computational algorithms to predict the chemical structures of the metabolites.

Reference MS/MS spectra are typically derived from pure chemical standards. However, in metabolomics research, MS/MS spectra are obtained by analyzing complex sample matrices containing numerous compounds. When multiple compounds are fragmented simultaneously in the same MS/MS experiment, the resulting spectra are termed "chimeric." In metabolomics workflows using data-independent acquisition (DIA), the instrument is configured to apply predefined rules to determine which precursor ions are fragmented and MS/MS spectra acquired. The precursor ion selection window used in DIA is typically wide (e.g.,

10.2 LC-MS Spectra Processing

>20 *m/z* in SWATH), or in some cases, all ions are fragmented (named as all ion fragmentation). Consequently, most fragmentation data obtained are chimeric, posing a computational challenge to deconvolute the MS/MS data and accurately assign precursor ions. Conversely, metabolomics workflows using data-dependent acquisition (DDA) typically employ a narrow MS/MS isolation window (e.g., 1–3 *m/z*). While DDA workflows aim to fragment a single precursor in each MS/MS experiment, this approach often results in numerous chimeric spectra originating from background peaks or contaminants. Consequently, deconvolution is also required to increase the number of identified metabolites in DDA-based metabolomics analyses [48].

Metaboanalyst provides two similarity methods to match tandem mass spectrometry (MS/MS) spectra against experimental or in silico mass spectral libraries:

10.2.3.6.1 Dot Product (or Inner Product)

Dot product is the mathematical operation commonly used to assess the similarity between two vectors by measuring the cosine of the angle between them. The dot product essentially measures the extent to which the two vectors point in the same direction. If the vectors are perfectly aligned (i.e., pointing in the same direction), the dot product will be maximized. Conversely, if the vectors are orthogonal (i.e., perpendicular to each other), the dot product will be zero, indicating no similarity. To ensure that the cosine similarity lies within the range of −1 to 1, the dot product is normalized by dividing by the product of the magnitudes of the vectors. This normalization accounts for differences in the magnitudes of the vectors, ensuring that the cosine similarity focuses solely on the direction of the vectors. The closer to its value of 1, the more similar the vectors (MS/MS spectra) are.

10.2.3.6.2 Spectral Entropy

To assess the similarity between two MS/MS spectra, this algorithm initially applies Shannon entropy to the power spectral density (PSD) of the signal. This step effectively captures the distribution of intensity values across each individual spectrum, providing a comprehensive understanding of the spectral complexity. As the algorithm progresses, it takes into consideration the inherent increase in entropy (chaos), when mixing two sets of molecule ions. To address this, the entropy of a mixed spectrum, formed by combining the input spectra, such as an experimental spectrum and a library spectrum, is calculated. Subsequently, the entropy of this mixed spectrum is subtracted from the entropy of each individual input spectrum to yield the entropy distance, a measure of the dissimilarity between the spectra. To provide a clear and interpretable measure of similarity, the algorithm further computes the unweighted entropy similarity. This is achieved by dividing the entropy distance by the normalized maximum entropy difference. In essence, this methodology enables the algorithm to discern subtle differences between highly similar structures, even amidst the presence of low-abundance ions. Through rigorous evaluation, this approach has demonstrated superior performance compared to alternative similarity methods, boasting an impressive AUC of 95.8%. For more comprehensive insights into this methodology, refer to the work by Li et al. [49].

- **Parameter settings:**

    ppm: allowed signal deviation in *m/z* dimension.
    Filtering value: the minimum intensity (abundance) considered in peak determination. A value can be determined by inspecting chromatograms.
    Window size: the width in *m/z* used to select precursor ions.
    Threshold: the maximum intensity (abundance) considered in peak determination.
    Deconvolution: select this option for both DIA and DDA mode.
    Target peaks: users can choose between selecting only the significant peaks or including all of them.
    MS2 database: users can select the most appropriate database according to the sample or purpose of the analysis: HMDB (the "experimental" option is for learning purposes), GNPS, MINEs, LIPIDBLAST, MoNa, Mass Bank, Riken, ReSpect, MS-DIAL, and BMDMS. For more information about each database, refer to [8, 50–54].

10.2.4 Job Submission

After all parameters are configured, click "Submit Job" (Fig. 10.5). The "Job Status View" window opens, displaying the percentage of work processed. When the auto-optimized mode is selected, data processing takes longer. By defining the parameters, processing is faster.

If registration and login were not performed before the data upload, click on "Create Job URL" to generate a link that makes data processing available for 2 weeks. A dialog box will open (Fig. 10.10). Click on "Copy" and paste the link in a specific place to retrieve the processing data later. When "Job Progress" reaches 100% completion, click "Proceed" or "Spectra Result", in the navigation tree, to open the window containing the results.

10.2.5 Results of Data Processing

After clicking on the "Proceed" button, the data is subjected to principal component analysis (PCA). As already mentioned in Chap. 2, this is an unsupervised learning technique. In the results window, the "PCA Visualization" tab provides 3D graphics. In Fig. 10.11, the scatterplot on the left shows the clustering pattern of all samples, which corresponds to the three types of samples analyzed distinguished by

10.2 LC-MS Spectra Processing

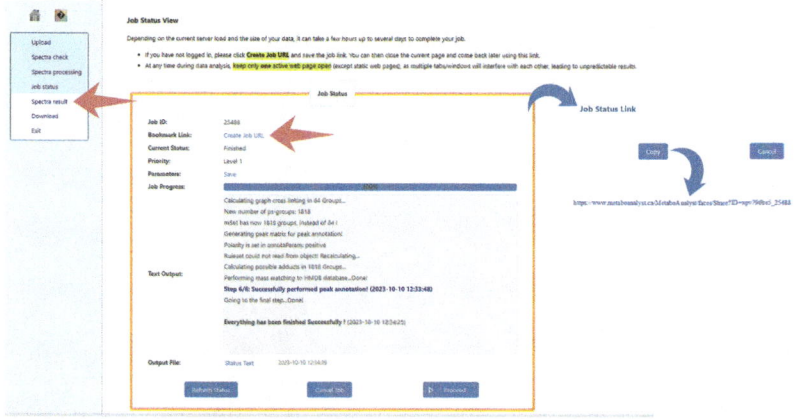

Fig. 10.10 Creating data processing link

different colors. The combined percentage contributions of PC1 and PC2 explained about 81% of the total variation in the data. Double-clicking on a node shows the total ion chromatogram (TIC) of the selected sample. Clicking the "TIC Plot" tab displays the overlapping TICs of all samples.

The loading plot showcases all detected features (Fig. 10.11, on the right). By double-clicking on a node corresponding to a feature (*m/z*-RT), boxplots of the concentrations of that feature in the analyzed samples are displayed (Fig. 10.12, left). Further, double-clicking on a node within one of the boxplots reveals the respective extracted ion chromatogram (EIC). When consecutively clicking on a node from another sample, the EICs overlay (Fig. 10.12, right). Users can set the resolution and extension of the figure by clicking on the palette icon. Additionally, clicking on the icon allows users to delete the overlapped EICs [17].

The "Intensity Stats" tab (Fig. 10.13) displays boxplots of sample concentrations ($\log_2$intensity), with each boxplot differentiated by colors based on PCA clustering.

The "RT Correction" tab displays a graph corresponding to RT adjustment based on the median RT differences, as explained in Sect. 10.2.3.2.1. In addition, the "BPI plot" and "Aligned BPI" tabs showcase chromatograms before and after peak alignment, respectively, as demonstrated in Fig. 10.6.

The bottom half of the result page contains three tables, with the "Result Summary" tab presenting information about the raw spectra processing (Fig. 10.14).

In the "Spectra/Sample Table" tab (Fig. 10.15), users can check whether the number of peaks detected in the replicates of a sample is reproducible (Peaks No.), and whether the percentage of missing values is within the expected range (Missing %). If a list containing sample labels (metadata) was loaded along with the raw data, the sample will be labeled accordingly (Group), otherwise, it will be labeled as "unknown."

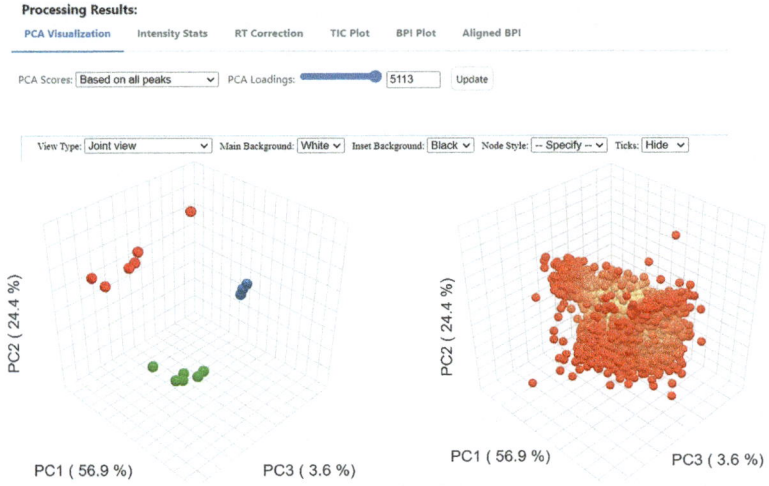

Fig. 10.11 Principal component analysis 3D plots: scatter plot (left) and loading plot (right)

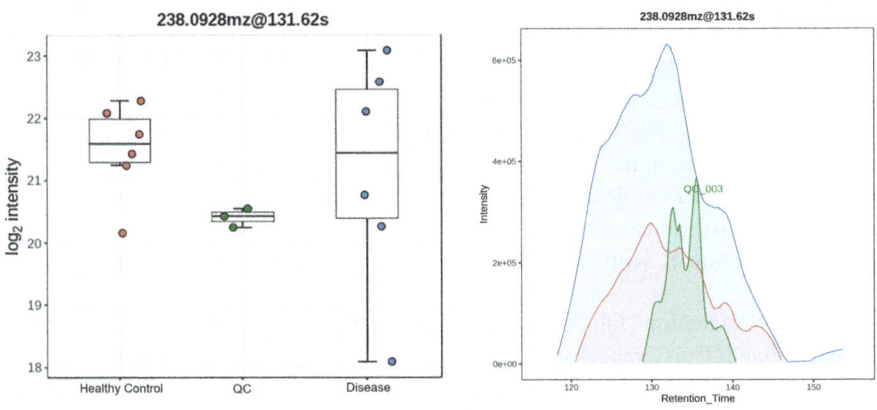

Fig. 10.12 Boxplots of concentrations of a feature in samples (left) and superimposed extracted ion chromatograms (right)

In the "Feature/Peak Table" tab (Fig. 10.16), users can check the coefficient of variation (CV), the significance (p-value), and the false discovery rate (FDR) of a given feature. In the "Annotations" column, the number of isotopes is in brackets, the precursor ion is represented by "M", and the different types of adducts detected are in braces, followed by the actual mass. Empty cell means the corresponding feature is a precursor ion or no annotations are available. Clicking an icon in the

10.2 LC-MS Spectra Processing

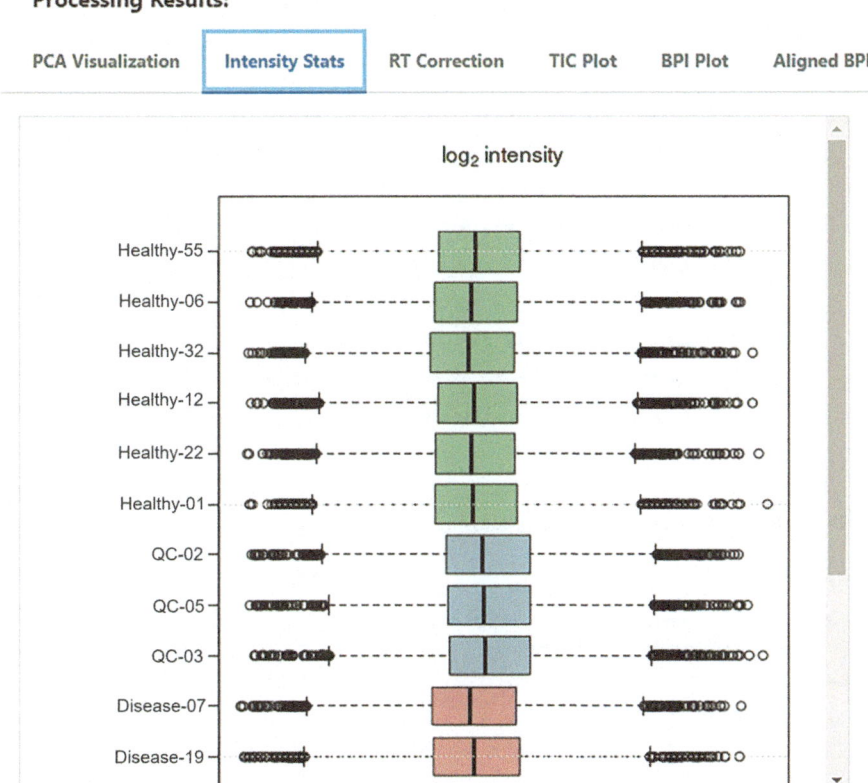

Fig. 10.13 Intensity status tab: boxplots with different colors according to PCA clustering

"View" column shows boxplots, as illustrated in Fig. 10.12. Clicking an icon in the "Putative IDs" column provides links to the feature's possible identities in the Human Metabolome DataBase (HMDB).

After viewing the results and saving the figures with the desired resolution, click on "Download" in the tree navigation (Fig. 10.17, red arrow on the left). Then, click the "Generate Report" button to obtain a pdf file containing information about all processing steps. Finally, click on "Download.zip" to get all the files compressed and downloaded together. The files "annotated_peaklist.csv" or "metaboanalyst_input_clean.csv" contain the information needed to perform subsequent statistical analyses in MetaboAnalyst.

Users can test various processing methods until finding the one that yields the best results, such as the highest number of detected peaks and annotated adducts and the greatest total variance explained by principal components 1 and 2, among other criteria. To do this, return to the navigation tree and click on "Spectra Processing" to initiate the entire process with the new mode, which can be switched between "default/manual" and "auto optimized" [17].

Result Summary | Spectra / Sample Table | Feature / Peak Table

Raw Spectra Processing Result Summary:

MetaboAnalyst has finished raw spectra processing with OptiLCMS (1.0.5):

There are 15 samples of 3 groups included for processing!

Total of 3673 features have been detected and aligned across the whole sample list.

The mass deviation of this study was estimated/set as 4.3 ppm.

951 features (25.88%) have been annotated as isotopes.

872 features (23.73%) have been annotated as adducts.

526 unique formulas have been matched to HMDB database.

1354 potential compounds have been matched to HMDB database.

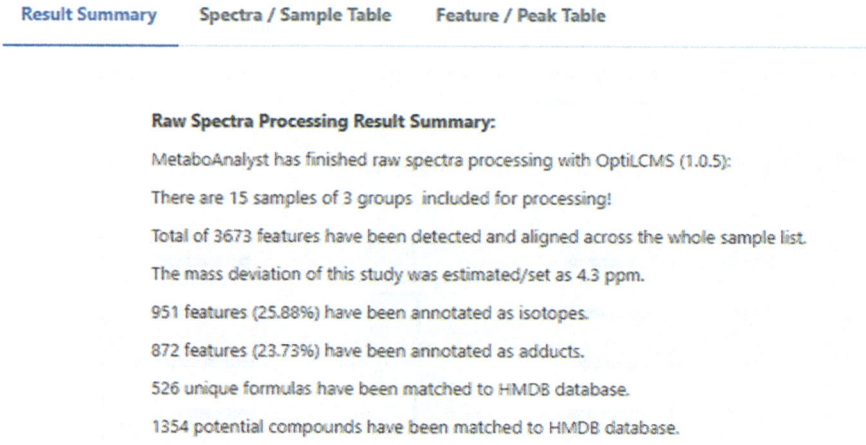

Fig. 10.14 Result summary of spectra processing

Result Summary | Spectra / Sample Table | Feature / Peak Table

Spectra ⇅	Group ⇅	Peaks No. ⇅	Missing (%) ⇅	RT Range	m/z Range	View
Healthy-55	Healthy	3102	15.55	20.67~298.43	85.084~1266.511	
Healthy-09	Healthy	3100	15.6	20.67~298.43	85.084~1266.511	
Healthy-11	Healthy	3011	18.02	20.67~298.43	85.084~1266.511	

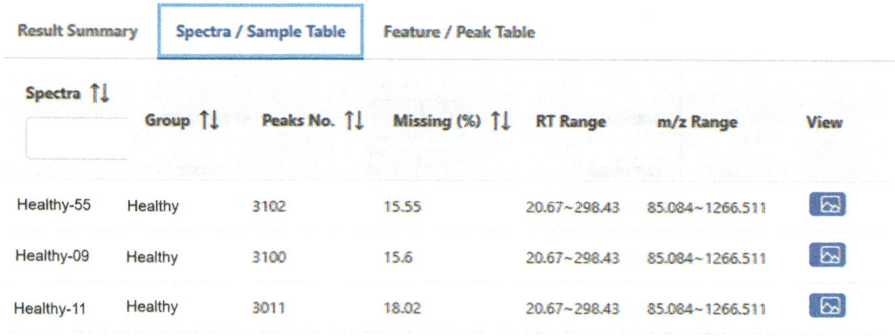

Fig. 10.15 Contents of the Spectra/Sample Table

Users can proceed from here to a new module by clicking the "Start New Journey" tab, selecting the desired analysis, and clicking the "Go" button, as demonstrated in Fig. 10.18. Alternatively, spreadsheets generated after data processing can be modified to suit further statistical analysis.

10.3 Statistical Analyzes (One Factor)

To carry out this task, the user must upload the most appropriate data sheet. Next, the most informative features are extracted and normalized/scaled so that they are suitable for the various statistical treatments to which they will be subjected.

10.3 Statistical Analyzes (One Factor)

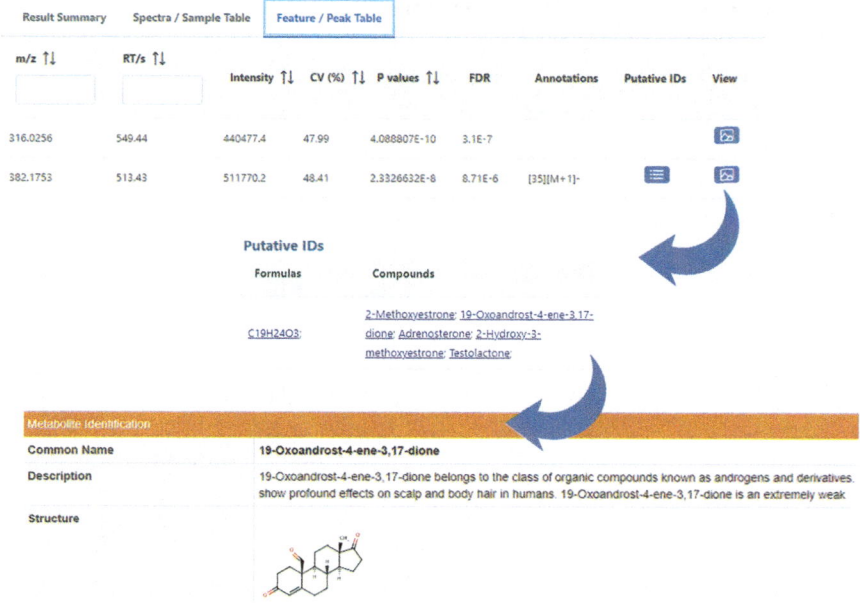

Fig. 10.16 Contents of the Feature/Peak Table: mass/charge ratio (*m/z*), retention time (RT), coefficient of variation (CV), false discovery rate (FDR), and putative identities (IDs) from the selected database

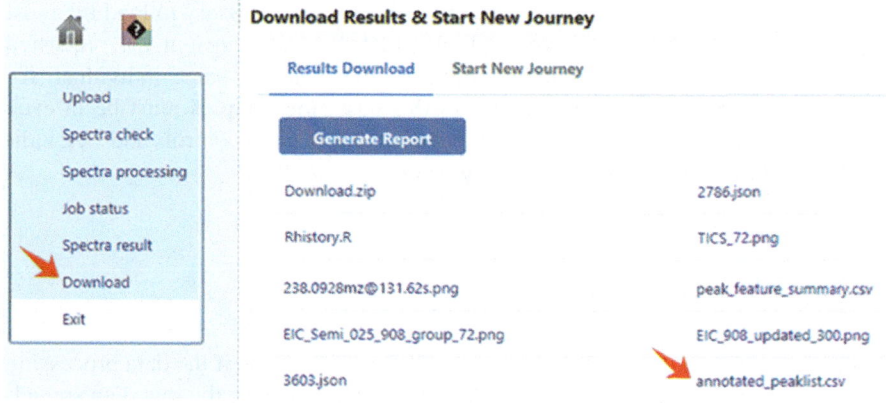

Fig. 10.17 Downloading the results of data processing

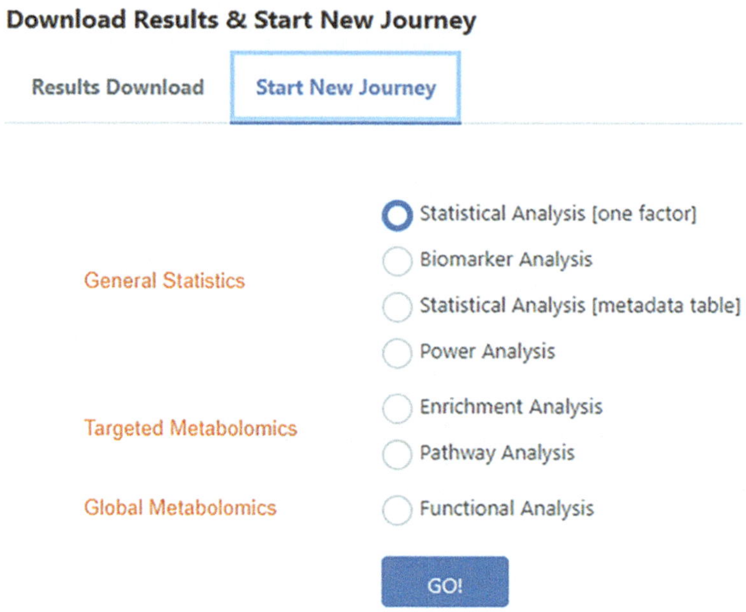

Fig. 10.18 Statistical analyses available in MetaboAnalyst 6.0

10.3.1 Data Uploading

Users can start a new statistical analysis by choosing the desired module on the MetaboAnalyst home page (Fig. 10.2). In this case, it is necessary to load the most appropriate spreadsheet as shown in Fig. 10.19. When filling out the requested information, remember that paired samples come from the same individual, for example, before and after a treatment. In this case, the samples must be in even numbers. Unpaired samples are independent, such as healthy controls and sick individuals, and the number of samples may be odd.

10.3.2 Data Processing

At this stage, to minimize excessive manipulation of the data, if the data processing was conducted in software other than MetaboAnalyst, submit the raw data spreadsheet without statistical treatment (i.e., not normalized), as normalization and scaling may be performed later.

10.3 Statistical Analyzes (One Factor)

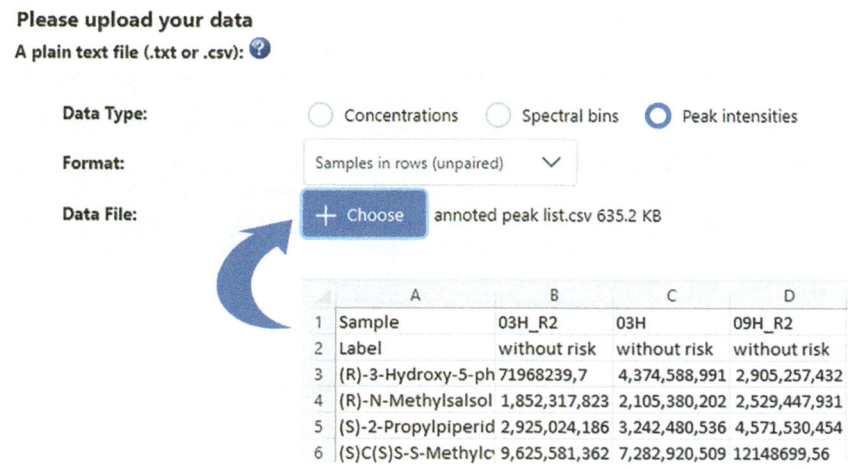

Fig. 10.19 Loading the data sheet for statistical analysis

10.3.2.1 Data Check

After submitting the data sheet, the "Data Integrity Check" window will open (Fig. 10.20). If the data content is successful, click "Proceed." If the samples have not been labeled, a dialog box will open, warning that they need to be edited. Click "Edit Groups" and name the samples appropriately, for example, blank, QC, control, and disease.

10.3.2.2 Missing Value

In the navigation tree, select "Missing value" to configure this parameter if the total of missing values exceeds 20% [44], which can be checked in the "Data Integrity Check" window. By default, the remaining values will be replaced with small values, such as the feature detection limit (LoDs: 1/5 of the minimum positive value of each variable), or with estimates derived from the other algorithms provided below the LoD method, as shown in Fig. 10.21 [55].

It is worth reading the evaluation on the performance of different methods for imputation of missing values, normalization, transformation, and scaling in the processing of data from UPLC and GC/MS platforms in [43, 56, 57].

10.3.2.3 Data Filter

Many features detected by untargeted metabolomics may not be biologically relevant, as they could represent background signals from sample processing or multiple signals arising from the same analyte (such as adducts, isotopes, or in-source

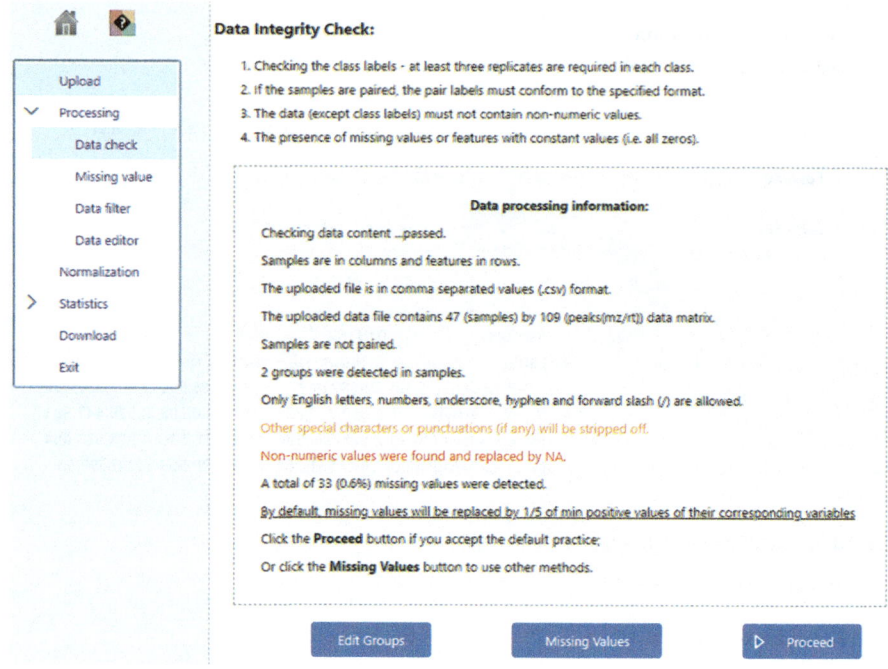

Fig. 10.20 Data integrity check window

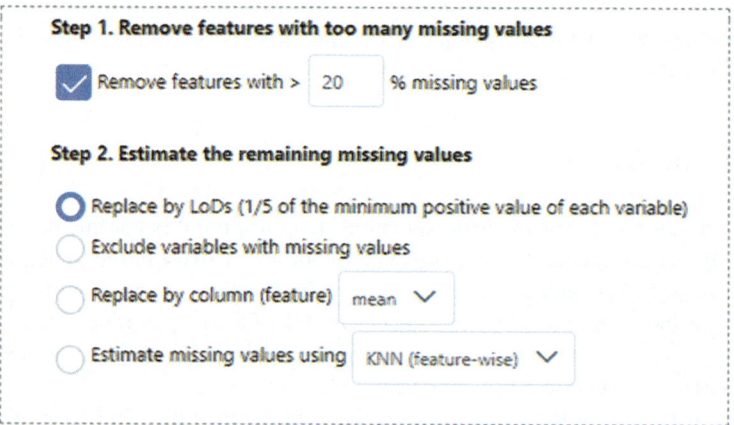

Fig. 10.21 Configuring missing values

fragmentation). Inadequate feature filtering can significantly impact subsequent statistical analyses. For example, if a high-quality feature is erroneously filtered out, it will not be considered as a candidate biomarker in univariate tests of significance for association with biological factors of interest or in metabolic pathway analysis.

10.3 Statistical Analyzes (One Factor)

Users should carefully evaluate whether the options presented here are suitable for their data. Filtering methods should be data-adaptive, meaning they tailor filtering to the specific characteristics of the dataset rather than relying on predefined methods. A data-adaptive pipeline adjusts filtering based on the attributes of a given dataset. To establish data-dependent thresholds, such as for blank-based contaminants, coefficients of variation (CVs), and handling missing values, users must visualize differences in characteristics between the extracted ion chromatograms of known high- and low-quality features. This approach enables noise reduction without compromising the underlying biological signal. For more detailed guidance on implementing this methodology, refer to [58].

To define the technique that will filter noise or uninformative features, select "Data Filter." Correction is carried out according to some noise/features characteristics provided by the platform as follows (Fig. 10.22):

- **Reliability filter:** This option can be selected if there are quality controls (QC) among the uploaded samples. Features with a large standard deviation (RSD) in relation to the QC samples can be considered to have low repeatability and should be excluded. The suggested limit is 20–30% for LC-MS and 30% for GC-MS.
- **Variance filter:** Features with near-constant (low variance) values across all samples are often indicative of contaminants. These features, representing noise, can be identified using statistical metrics such as the standard deviation (SD), which quantifies the dispersion of data points around the mean, or the interquartile range (IQR), offering a more robust measure. Other methods, such as median absolute deviation (MAD) and relative standard deviation (RSD), are also available for this purpose.
- **Abundance filter:** Signals with very small values, often appearing close to the baseline or detection limit, can be considered noise. These signals may be effectively identified and filtered out using statistical measures such as the mean or median.

Fig. 10.22 Filtering noise and uninformative features

10.3.3 Normalization

10.3.3.1 Sample Normalization

As we know, features can present very large ranges of concentrations/intensities in the samples. To prevent the algorithm from being biased toward variables with a higher order of magnitude, normalization transforms the variables to the same order of magnitude, within the range [0, 1] or [−1, 1], making them comparable.

MetaboAnalyst offers some algorithms to perform normalization depending on the target, whether the feature or the sample [59], as demonstrated in Fig. 10.23.

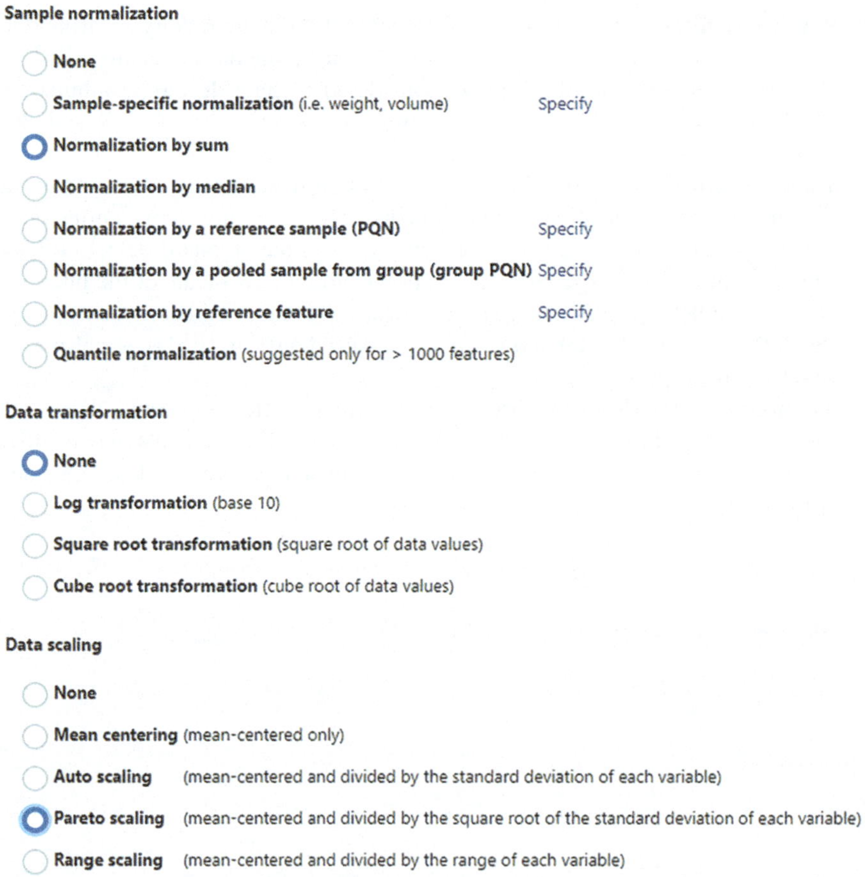

Sample normalization
- None
- Sample-specific normalization (i.e. weight, volume)   Specify
- ● Normalization by sum
- Normalization by median
- Normalization by a reference sample (PQN)   Specify
- Normalization by a pooled sample from group (group PQN) Specify
- Normalization by reference feature   Specify
- Quantile normalization (suggested only for > 1000 features)

Data transformation
- ● None
- Log transformation (base 10)
- Square root transformation (square root of data values)
- Cube root transformation (cube root of data values)

Data scaling
- None
- Mean centering (mean-centered only)
- Auto scaling   (mean-centered and divided by the standard deviation of each variable)
- ● Pareto scaling   (mean-centered and divided by the square root of the standard deviation of each variable)
- Range scaling   (mean-centered and divided by the range of each variable)

Fig. 10.23 Methods available in MetaboAnalyst for data normalization, transformation, and scaling

10.3 Statistical Analyzes (One Factor)

No method works for all cases. Select the most appropriate one, or none, for your data based on the descriptions below:

- **Normalization by sum (or total spectral area normalization):** Forces the sum of the data values (intensity/peak area) for each sample to be equal to 1 or to remain constant, resulting in closed data. This means that an increase in the concentration of a given metabolite imposes a decrease in the concentration of all others. Noise from high-intensity peaks is transformed into systemic variation, which can lead to the emergence of spurious components when the data is subjected to PCA. In some cases, performing a logarithmic transformation before normalization can help mitigate these effects and prevent erroneous conclusions, especially in the study of biomarkers [57, 60–62].
- **Probabilistic quotient normalization (PQN):** This method assumes the intensity of most signals is an exclusive function of dilution (the so-called size effect). The PQN algorithm performs an integral normalization of the different samples to equalize their magnitudes. Then, it calculates the quotients between all the features of the studied sample and a reference sample, or a group of them. Next, it calculates the median of all quotients. Subsequently, the quotient between all features of the test sample and the median of all quotients produces the normalization factor (the most probable quotient), which replaces the total integral as a marker of the sample concentration and is representative of its dilution factor in relation to the sample reference. PQN maintains robustness against high-intensity peaks but is sensitive to missing values, which must be removed from the sample, and to features that co-vary with factors other than dilution, such as urinary proteins. For a good visual explanation, refer to [63, 64].
- **Normalization by median:** The normalization performed by this method results from dividing the intensity of each feature by the median of the intensities of each sample. It is worth reading the discussion by Välikangas et al. [65] on the effect that different types of normalization can have on intra group variation between technical replicates and on the AUCs of differential expression analysis.
- **Quantile normalization:** This is a multi-sample normalization technique focused on removing technical variations, specifically uncontrolled changes that occur in experimental conditions. Instead of estimating a normalization factor for each sample, it creates a target profile based on the average peak intensity/area, forcing sample distributions to be identical. To achieve this, the feature values in each column of the data table are sorted in ascending order. The average of the values in each row formed after sorting is then calculated. The resulting average values replace the original feature values in each row of the new data table. Then the average values are reordered to their original order. As a result, the columns of the new data table consist of the normalized samples, while the original order of the features is preserved. It is important to check whether the variability between groups is greater than the variability within groups (variation between technical replicates). Otherwise, this method is not the most suitable [66].

- **Normalization by reference feature:** In this type of normalization, a standard compound with a known concentration or a reference gene can be added to each sample. The area/intensity of the standard will then be used to calculate a normalization factor for each sample. However, this method is not convenient for untargeted metabolomics, as the source of unwanted variation is not only related to sample introduction into the instrument, but also variations in dilution factors [67].

10.3.3.2 Data Transformation

Select whether and how the data needs to be transformed. Sometimes, the quantities to be compared have very different values and require transformation, such as taking logarithms, to exhibit homoscedasticity (constant residual variance) and have residuals that are near-normal or near-Gaussian (Fig. 10.24). For more information, refer to [68].

10.3.3.3 Data Scaling

Many users get confused with the terms scaling and normalization as both are used in machine learning to bring data into a fixed range. Scaling is a more general term that can refer to various methods of changing the range of the data. Normalization is a specific type of scaling that aims to bring all the features into a similar range, often using the minmax scaling method.

The choice between scaling and/or normalization and which method to use depends on the specific characteristics of the data and the requirements of the machine learning algorithm or statistical analysis one wants to use.

If the variables in a dataset are all on the same scale (e.g., intensity) but exhibit extreme differences in mean values, centering is applied before scaling. This process brings the means to the origin (zero) in the coordinate system, adjusting for

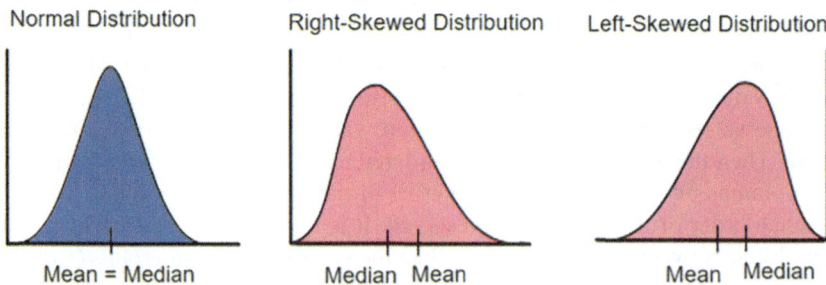

Fig. 10.24 Curves of data distribution. Left: symmetric (Gaussian). Middle and right: asymmetric (skewed)

differences in the balance between more abundant and less abundant metabolites. Scaling factors can be calculated based on a measure of dispersion (e.g., standard deviation or median absolute deviation) or on a measure of size (e.g., mean or median), as described in Table 10.3. For more information, refer to [57, 69–71].

After defining the normalization or scaling method, click on the "Normalize" button, followed by the "View Result" button. A new window will open, displaying the feature or sample distribution curve before and after applying the chosen method. Following normalization, the optimal method should align the sample distribution with a near-Gaussian curve, as illustrated in Fig. 10.25.

Equally, the most effective scaling method should produce a narrow peak, indicative of low feature dispersion (Fig. 10.26). If these desired outcomes are not achieved, data transformation may be necessary.

It is worth recalling that the term normal just applies to the entire population and not to the sample data [45].

Normalizing or scaling the data is not a strict requirement. In many cases, the model will still perform well when trained using the original features. It may be beneficial to experiment and determine which procedure yields better results for the specific task [19].

Close the "View Result" window. Click the "Proceed" button to continue the analysis.

After completing the data processing step, in the navigation tree, select "Statistics" and choose the next analysis, as shown in Fig. 10.27.

10.4 Chemometrics Analysis

Chemometrics is a chemical discipline that involves the application of statistical and mathematical methods to analyze chemical data. It integrates principles from chemistry, mathematics, and statistics to extract meaningful information and patterns from complex chemical datasets [72]. Given the multidimensional nature of chemical datasets, chemometrics often employs multivariate statistical techniques to analyze and interpret data with multiple variables. Principal Component Analysis (PCA), Partial Least Squares (PLS), and cluster analysis are common methods in this context [73].

10.4.1 Principal Component Analysis

As previously mentioned in Chap. 2, PCA is an unsupervised technique, which means working with samples without prior knowledge of their labels. For more details, refer to [74, 75].

Under **Chemometrics Analysis**, select the "PCA" link. In the window that opens, several tabs are available, each offering different graphic options. The

Table 10.3 Methods for data scaling

Method	Approach	Advantages	Disadvantages
Autoscaling (unit variance scaling, standardization)	After mean centering, each variable is divided by its SD, resulting in a variance equal to 1 and equalizing the weights of all variables to the same importance. The comparison between variables occurs through correlations instead of covariances	A robust method for identifying metabolites with significant changes in relative quantity (percentage), especially in PCA cluster analysis when the data exhibit technical errors such as poor alignment signal	Inflates measurement errors, mainly if SD is small, and introduce noise. Outliers have the potential to skew the scaling factors
Range scaling	After mean centering, each variable is divided by the difference between the minimum and maximum fold-change over the samples (the range concentration). Places all variables on an equal footing and compares them based on the biological response range	Removes instrumental response factors from the data, generating relative concentrations for each variable	Very sensitive to outliers as extreme values can make the range (denominator) larger. A robust interval estimator is necessary to minimize this limitation. Inflates measurement errors
Pareto scaling	After mean centering, each feature intensity is divided by the square root of its SD, up-weighing larger and intermediate features while keeping baseline noise down	Identifies metabolites with changes in their absolute quantities and increases the data contribution of signals with small SD. Useful when the distribution of feature values is skewed	Sensitive to very large fold changes and outliers
Mean centering (level scaling)	After mean centering, the variable is divided by the average of the intensities over the samples. Converts changes in intensities to changes relative (percentage) to average intensities	Appropriate for cases where large relative changes are of specific biological interest such as the identification of relatively abundant metabolites or biomarkers. Robust to outliers	Sensitive to technical errors and may generates inconsistencies between the correlation coefficient and VIP values of discriminatory metabolites. Inflates measurement errors

SD standard deviation, *VIP* variable importance in the projection

"Overview" tab (Fig. 10.28) displays graphs for each pairwise combination of the 5 principal components (PC) considered in the analysis, along with the respective percentage contributions of the PCs to the cluster. One can observe a moderate separation between the healthy and diseased populations, particularly noticeable between PC1 and PC2.

10.4 Chemometrics Analysis

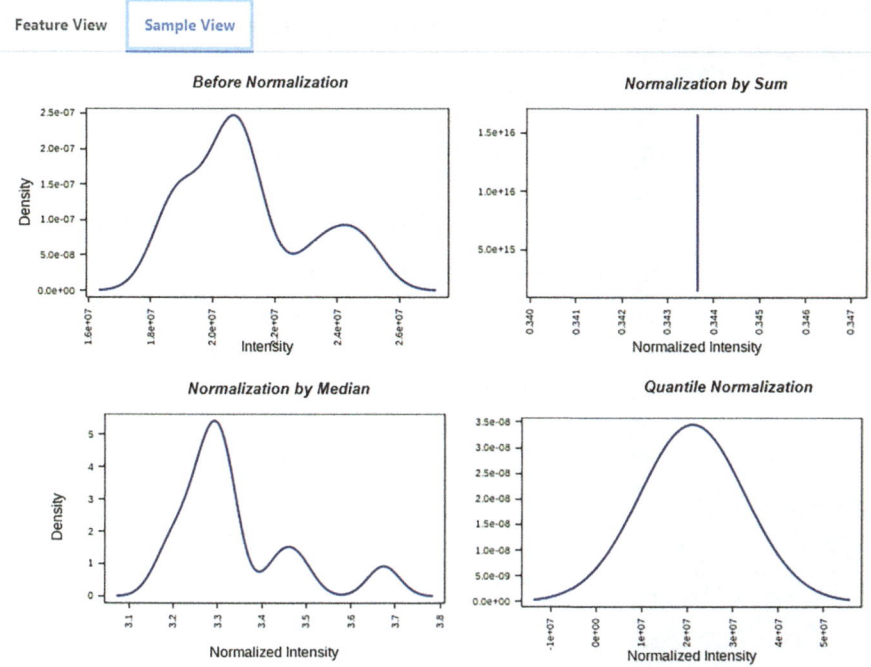

Fig. 10.25 Effect of different normalization methods on sample distribution

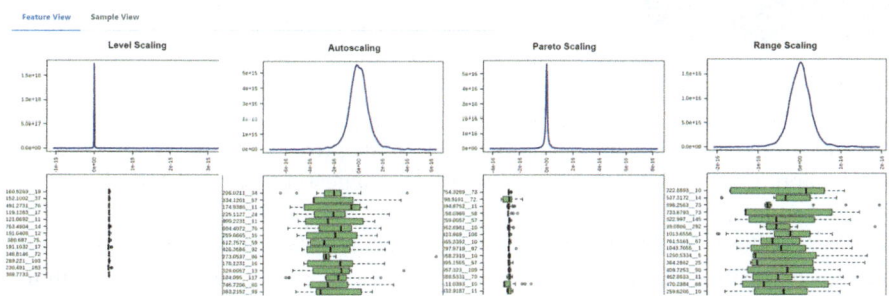

Fig. 10.26 Effect of different scaling methods on the range of feature intensities. Kernel density plots (above) are based on all data while the boxplots (underneath) show at most 50 features due to space limitations

The 2D and 3D "Scores Plot," as previously mentioned in Chap. 2 (Sect. 2.4.1, Fig. 2.3), demonstrates the projection of samples onto the span of the principal components.

The "Scree Plot" (Fig. 10.29) illustrates how much variation each PC captures from the data individually and cumulatively, indicating how many PCs should be retained in the analysis. This corresponds to the minimum number of PCs that

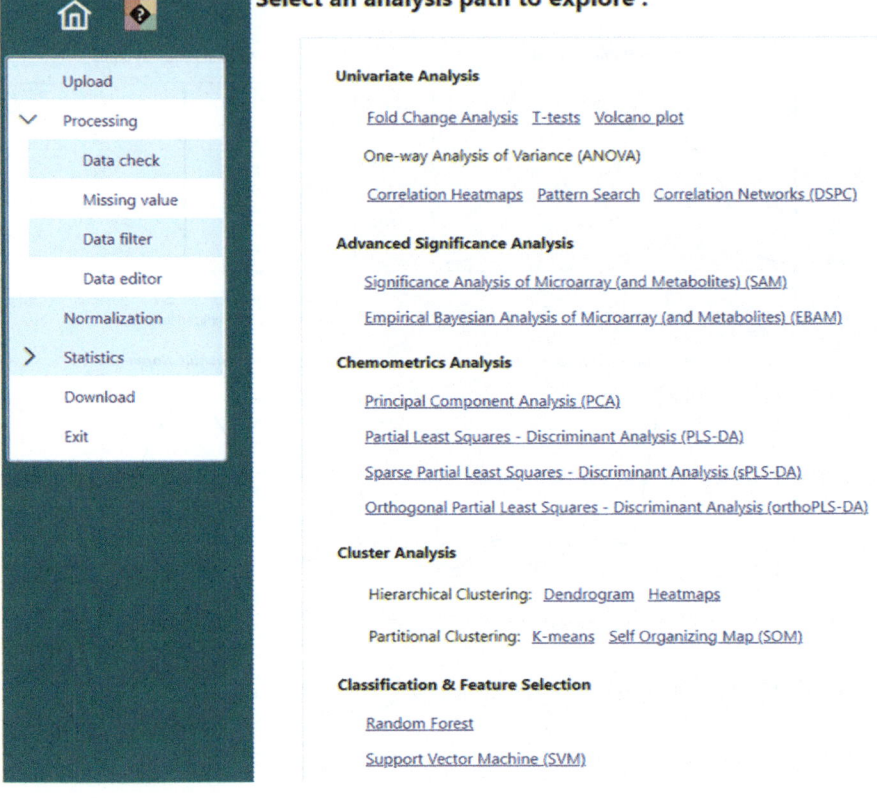

Fig. 10.27 Selecting the statistical analysis

account for the greatest variation in the data. In the provided graph, it is shown that 5 PCs captured only 45% of the total variance, suggesting the need for more informative features.

The "Loadings Plot" (Fig. 10.30) illustrates the weight (influence) of the original features in defining the principal component. The distance of each feature from the origin (zero) of the coordinate system shows its influence on the PC. The closer to zero, the weaker its influence. However, if a feature is very far away, it may even be considered an outlier.

Click on the spreadsheet icon to open the "Feature Details Table" (Fig. 10.31) and locate the coefficients of each feature associated with PC1 and PC2 on the X- and Y-axes, respectively.

Click 'View' to access the box plot illustrating the concentration of a given feature in each sample, both before and after normalization.

The "Biplot" tab (Fig. 10.32) displays variables (features) and observations (samples) in the same space. Features are represented by arrows (vectors) emanating from the origin, indicating their weight (loadings) and direction of variation.

10.4 Chemometrics Analysis

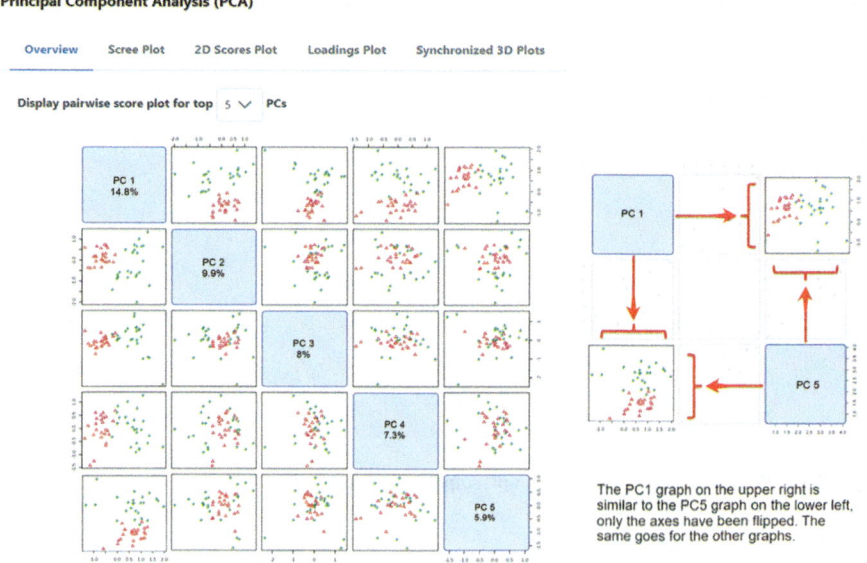

Fig. 10.28 Pairwise scores plot between 5 principal components. The explained variance of each component is shown in the corresponding diagonal blue cell. Red node: health; green node: disease

Fig. 10.29 Scree plot displaying 5 principal components (x-axis), the accumulated variance explained (green line on top), and the variance explained by each PC (blue line underneath)

Fig. 10.30 PCA Loadings Plot between principal component #1 and #2

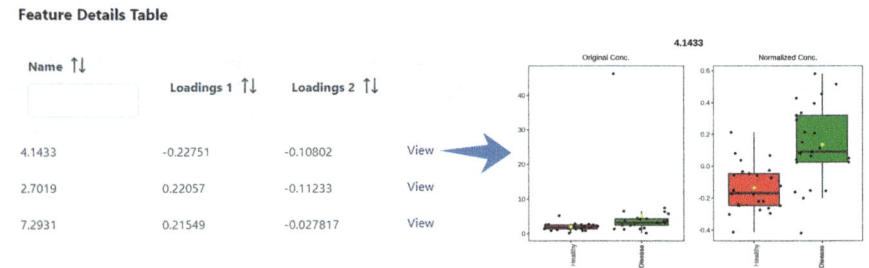

Fig. 10.31 Feature Details including loadings and concentrations. The "Loadings 1" column shows the coefficients of the linear combination that defines principal component #1 (X-axis), and the "Loadings 2" column shows the coefficients for principal component #2 (Y-axis)

Samples are denoted by their input codes, and the axes depict the principal components.

This type of graph reveals tendencies and correlations. For instance, a variable vector points toward increasing values of variance, and its length shows its influence on the PC. When we encircle the arrows with a bordering circle, the vectors that extend all the way to the edge are those whose information (variance) has been fully captured by the principal component. Therefore, the longer the vector, the stronger the relationship, as exemplified by the vector at 4.1433. Vectors in exact opposite directions indicate a negative correlation, as is the relation between the vector at 7.5802 and the vector at 4.1433, separated from each other by an angle of 180°. Positively correlated vectors point in the same direction and are very close to each other, as seen in the case of features at 7.4292 and 7.3659. Since Loadings Plot is part of Biplot, they must match up for the top hits.

10.4 Chemometrics Analysis

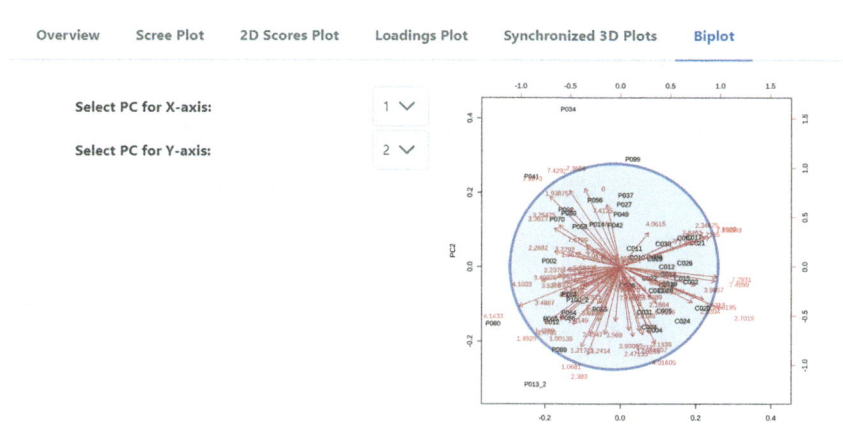

Fig. 10.32 PCA Biplot between principal components #1 and #2. The highlighted circle borders the most informative vectors (features) captured by the principal components

10.4.2 Partial Least Squares Discriminant Analysis

Partial Least Squares Discriminant Analysis (PLS-DA) is a supervised technique whose objective is to find a set of latent variables (linear combinations of the original variables) that maximize the separation between predefined groups or classes in multivariate data [76, 77].

In the process of building a classification model using PLS-DA, it is necessary to define the complexity of the model, i.e., the optimal number of latent variables or components, and subsequently evaluate the model's performance [78].

Under **Chemometrics Analysis**, click the "PLS-DA" link to perform the analysis and obtain the results. "Overview," "2D scores plot," and "Loadings plot" show graphs like the PCA graphs explained previously in Sects. 10.4.1 and 2.4.1.

To obtain the optimal number of components for classification and estimate the error of the model, click on the "Cross Validation" tab (Fig. 10.33). Choose the number of principal components to search and one of the three cross-validation (CV) methods available, as defined below:

- **k-Fold CV** (where k can be five or ten): The original dataset is randomly divided into five or ten roughly equal-sized subsets, or "folds." The modeling process is then repeated five or ten times. In each iteration (repetition of training and validation), nine (or four) folds are used as the training set, while one of the folds is used as the validation set. Performance metrics (accuracy, Q2 and R2) are calculated based on the model's predictions on the validation set. Subsequently, the permutation test quantifies the statistical significance of the diagnostic statistics [78, 79]. This method reduces overfitting.

Fig. 10.33 PLS-DA model validation by fivefold CV. The red asterisk indicates the optimal number of components based on the highest value obtained for Q2

- **LOOCV** (**L**eave-**O**ne-**O**ut **C**ross-**V**alidation): This is a specific form of CV where the number of folds is equal to the number of observations in the dataset. For each data point in the dataset, the model is trained on all data except one data point. The model is then validated on the data point that was left out. This process is repeated for all data points so that each data point serves as both training and validation data, ensuring that the model is evaluated on almost every possible combination of training and validation sets. Subsequently, the above-mentioned metrics for model performance and overall performance are applied. LOOCV can provide a less biased estimate of a model's performance compared to other forms of CV, especially when the dataset is small [79].

Q2 is considered the standard diagnostic statistic to validate PLS-DA models in metabolomics and represents the estimate of the model's predictive capacity [78]. However, as mentioned in Sect. 2.4.2, the goodness of prediction (Q2) should not be considered alone, but in conjunction with the goodness of fit (R2) and the accuracy of the model. R2 and Q2 should have values above 0.7. When the difference between R2 and Q2 is too large (R2 >>> Q2), it indicates potential overfitting of the model [80–82].

Click on the spreadsheet icon to obtain the accuracy, R2 and Q2 values on the five components initially defined to search. Click on the palette icon to obtain the graphic with the desired extension and resolution.

Click on the "Permutation" tab (Fig. 10.34) to configure the test conditions.

A permutation test, also known as a randomization test, is a non-parametric method used to evaluate the significance of a statistic by shuffling or permuting the labels of a dataset several times (e.g., 100, 500, or 1000 times). It tests the null

10.4 Chemometrics Analysis

hypothesis that two random variables are exchangeable, assuming there is no difference between the classes. The algorithm simulates the null hypothesis and determines whether the observed result is unlikely to occur by chance [83]. There are two test statistics available:

- **Separation Distance (B/W)**: The difference or effect for each permutation is calculated by taking the ratio of the sum of the squares of the errors (variability) between groups (B) and within groups (W). In Fig. 10.34, the histogram on the left shows the distribution derived from the permuted data (called null distribution). The bar on the right, indicated by the red arrow, represents the B/W ratio of the original data (the observed test statistic). If the B/W ratio of the original class assignment falls within the null distribution, there is no dissimilarity between the two class assignments. Conversely, if the B/W ratio of the original class is far enough from the permuted class distribution, the model is efficient in separating the classes [84]. A p-value below the established significance level (e.g., $\alpha = 0.05$) indicates that the difference between classes is significant, so the null hypothesis is rejected [85].

The greater the variance between groups compared to the variance within groups, the better the latent variables originated from these groups and, consequently, the better the model for discriminating between the groups.

- **Prediction accuracy:** This procedure occurs similarly to the one mentioned above, but it involves using the prediction accuracy (precision) obtained on the original (non-permuted) data (Fig. 10.35). The idea is to assess how extreme or unlikely the observed performance is under the assumption of no true relationship.

The number of permutations needs to be "large enough" to sample the tails of the distribution. The lower limit on the number of permutations is dictated by the

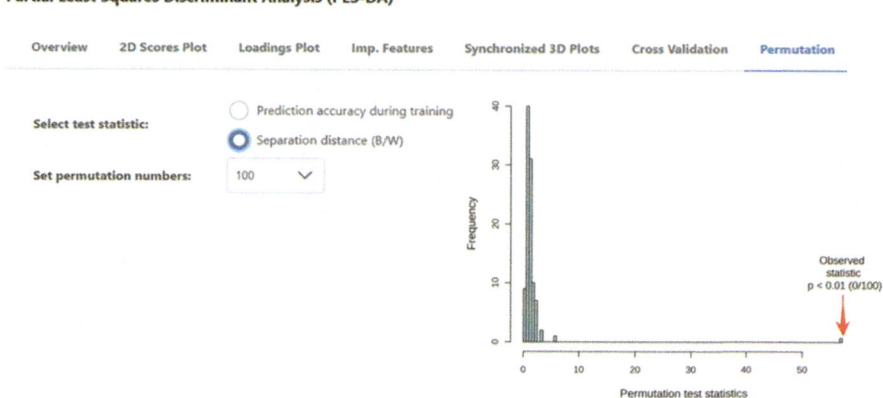

Fig. 10.34 Permutation test carried out using "separation distance (B/W)" test statistic and 100 permutations

required statistical significance. For example, to achieve a *p*-value of 0.01, at least 100 permutations are needed. However, this number may not be sufficient to adequately sample the tails of the distributions. An optimal number may be difficult to infer [78].

Following the completion of cross-validation and permutation procedures, users can navigate to the "Imp. Features" tab, as illustrated in Fig. 10.36. Within this tab, users have the flexibility to define the number of features showcased in the graph, such as selecting the top 15. Furthermore, users are presented with two distinct methods for assessing the importance of these features, as described below:

- **Variable importance in projection (VIP):** In this approach, features are filtered/selected based on their weighted sum of squares of the PLS loadings, considering the amount of explained Y-variation in each dimension (i.e., VIP scores are calculated in relation to each component). When a feature is permuted, the increase in the model's prediction error is calculated. A feature is deemed "important" if the permutation results in an increase in error in the model, indicating that the model relied on the feature for prediction. Conversely, a feature is considered "unimportant" if the permutation leaves the model error unchanged, suggesting that the model ignored the feature for prediction. Features with high VIP scores (>1) are considered more influential in the model [86], as illustrated in Fig. 10.36. However, if two or more features are correlated, the importance of the permuted feature may be influenced by unrealistic data instances, such as a person measuring 2 m and weighing 30 kg, potentially reducing the importance of the correlated feature. The choice of threshold may involve a trade-off between sensitivity and specificity, depending on whether the emphasis is on including all potentially relevant features or focusing on the most impactful ones [81].

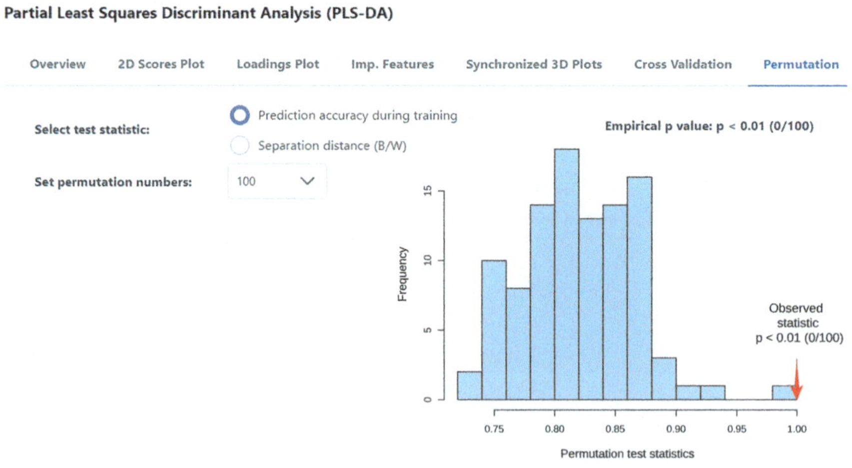

Fig. 10.35 Permutation test carried out using the "prediction accuracy during training" test statistic and 100 permutations

10.4 Chemometrics Analysis

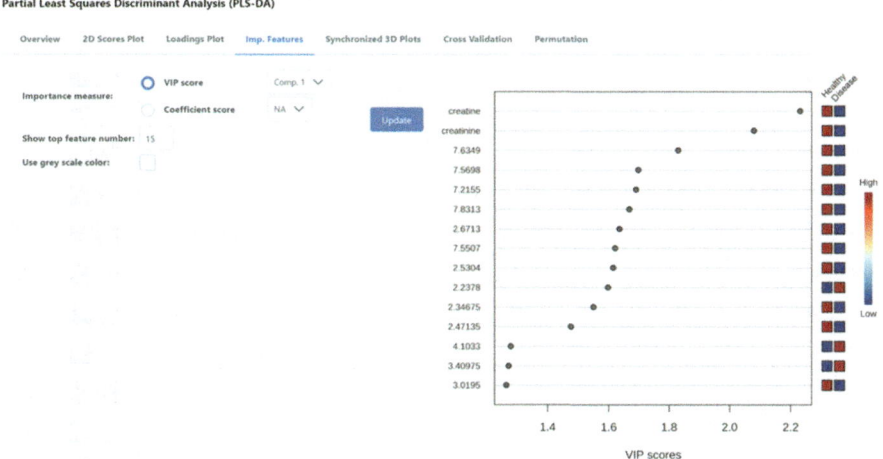

Fig. 10.36 Selection of the 15 most influential features on PC #1 using the VIP method. The colored boxes on the right indicate the relative concentrations of the corresponding feature in each class under study. The color bar represents the concentration gradient

- **Coefficient score:** This is calculated by taking the absolute values of the regression coefficients and summing them up, with each coefficient being multiplied (or not) by a weight. The weights can be used to give certain predictors more influence in the sum, reflecting their perceived importance. A large coefficient indicates a stronger overall impact of the predictor on the dependent variable or model, as illustrated in Fig. 10.37. However, like the VIP method, this method is susceptible to highly correlated predictors, and interpretation should be approached with caution. Collinearity can inflate coefficient estimates, making it challenging to attribute changes in the response variable to specific predictors [87].

10.4.3 Cluster Analysis

In machine learning, clustering is an unsupervised learning technique aimed at grouping similar data points together. There are various clustering algorithms, categorized based on different criteria, such as partitioning-based algorithms (K-Means) and connectivity-based algorithms (hierarchical clustering). MetaboAnalyst provides three types of algorithms, as summarized in Table 10.4. The choice depends on the specific characteristics of the data and the analysis goals. For more information, refer to [23, 88–90].

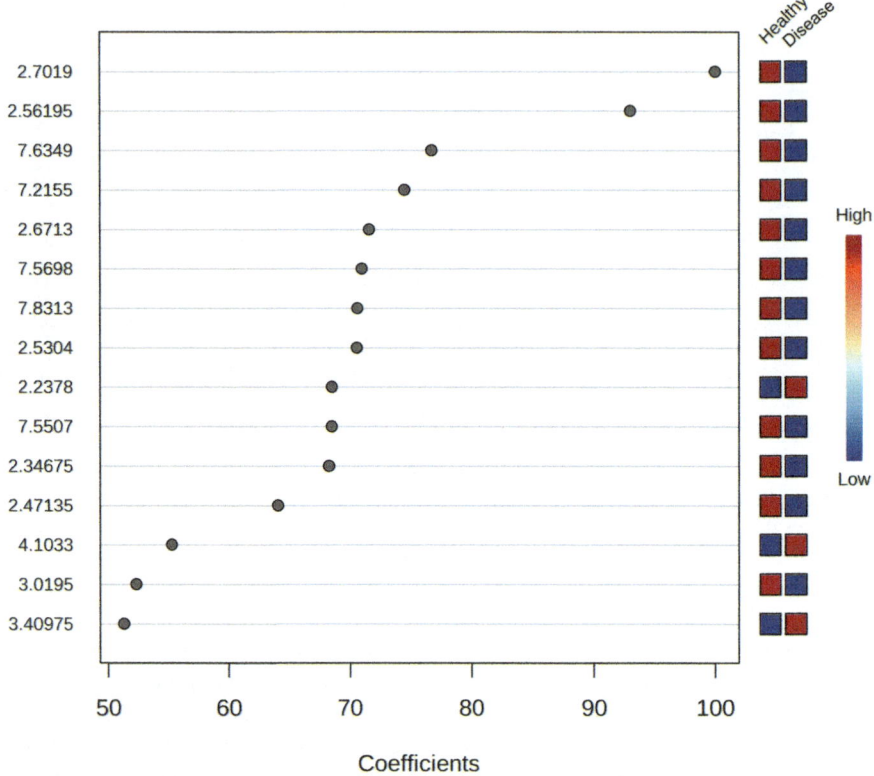

Fig. 10.37 Selection of the 15 most influential features on PC #1 using the Coefficient Score method

10.4.3.1 Hierarchical Clustering Analysis (HCA)

Hierarchical Cluster Analysis (HCA) summarizes the data in a hierarchical manner, initially forming clusters of more similar samples and nesting them through short branches at the base of the dendrogram. Subsequently, additional samples are incorporated into clusters already created or into new clusters, which have higher dissimilarity values and, therefore, are connected by wider branches [91].

The results can be visualized using dendrograms. Alternatively, a heatmap can be added between two dendrograms (e.g., samples vs top 25 features) to offer an overview of the data table [92]. The color intensity on the map is proportional to the concentration value of the features in each sample, with higher concentrations represented by warmer colors and lower concentrations by colder colors. Regions with equal color intensity suggest similarity (indicating a small distance) among the samples concerning the variable values. This allows for the identification of features responsible for the grouping or formation of classes.

10.4 Chemometrics Analysis

Table 10.4 Clustering algorithms available on MetaboAnalyst

Algorithm	Characteristics	Approach	Advantages	Disadvantages
K-Means	Requires specifying the number of clusters (k) in advance Minimizes the sum of squared distances between data points and their assigned centroids	Iteratively assigns data points to the nearest centroid Updates centroids as the mean of the assigned data points.	Efficient and easy to implement Scales well to large datasets Often provides well-separated, spherical clusters	Sensitivity to initial cluster centers can lead to different results Assumes clusters are spherical and equally sized, which may not be the case in all datasets
Hierarchical	Forms a dendrogram representing the merging or splitting of clusters Provides flexibility in exploring clusters at different levels of granularity	**Agglomerative (bottom-up):** Starts with individual data points as clusters and merges them **Divisive (top-down):** Starts with one cluster and divides it recursively	Provides a hierarchical structure, allowing the exploration of clusters at different levels Can handle clusters of arbitrary shapes	May be sensitive to noise and outliers The output can be hard to interpret with a large number of data points
Self-Organizing Maps (SOM)	Topological relationships between data points are preserved Clusters are formed based on the activation of neurons in the grid	Uses a neural network to map high-dimensional data onto a lower-dimensional grid Neurons compete to represent different regions of the input space	Capable of preserving the topology of the input space Handles non-linear relationships and complex data distributions	Requires tuning of hyperparameters, such as the size of the map and learning rates Interpretability can be challenging for large maps

In the navigation tree, click on Statistics > HCA > Heatmap, to set the parameters (Fig. 10.38).

The way the clustering algorithm defines dissimilarity and the method used to measure distance (dissimilarity) between the clusters will influence how the map is created. Subsequently, the choice of feature/sample scaling and/or normalization depends on the selected method for measuring distance. In the "Other view options," users can highlight a few top features (e.g., select "use top 25") in the analysis by choosing among available variable ranking methods (e.g., *t*-test/ANOVA, PLS-DA VIP, etc.).

If one or both axis variables in the map are categorical in nature, it might be worth considering a change in the order in which those axis variable levels are plotted to better grasp patterns in the data. In such cases, users can customize the display by swapping the selection of the "do not cluster" button. As illustrated in Fig. 10.39, the heatmap with associated dendrograms on the right allows users to promptly

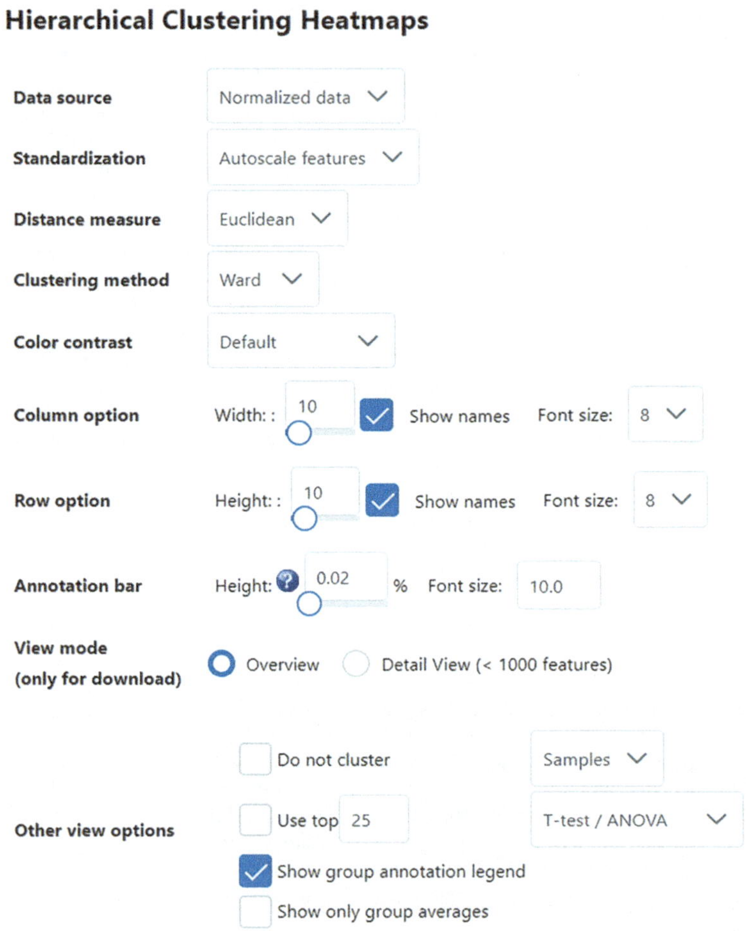

Hierarchical Clustering Heatmaps

Data source	Normalized data
Standardization	Autoscale features
Distance measure	Euclidean
Clustering method	Ward
Color contrast	Default
Column option	Width: : 10 ☑ Show names Font size: 8
Row option	Height: : 10 ☑ Show names Font size: 8
Annotation bar	Height: ❓ 0.02 % Font size: 10.0
View mode (only for download)	○ Overview ○ Detail View (< 1000 features)
Other view options	☐ Do not cluster — Samples ☐ Use top 25 — T-test / ANOVA ☑ Show group annotation legend ☐ Show only group averages

Fig. 10.38 Parameter setting for heatmap

identify clusters of samples with similar patterns, as well as discern groups of discriminating metabolites that drive sample clustering.

The visualization of a large number of features of interest (the default is the top 1000 features ranked by p-value) can be obtained by leaving the "use top features" button blank, as depicted in Fig. 10.40.

Clustering algorithms available [93, 94]:

- **Single linkage (the nearest neighbor method)**: Dissimilarity is defined as the shortest distance between any two points in the two clusters. This method is sensitive to outliers and noise.
- **Complete linkage (the furthest neighbor method)**: Dissimilarity is defined as the longest distance found between any two points in the two clusters.
- **Average linkage (centroid)**: Dissimilarity is defined as the average distance between all pairs of points in the two clusters.

10.4 Chemometrics Analysis

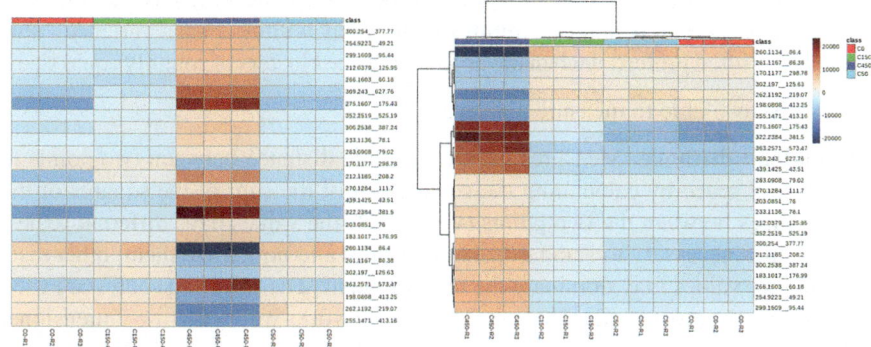

Fig. 10.39 Heatmaps created with the top 25 features selected by *t*-test, normalized data, Euclidean distance, Ward's method, and no standardization. On the left, "do not cluster" was selected. On the right, both samples and features were clustered, highlighting the differences among samples more clearly

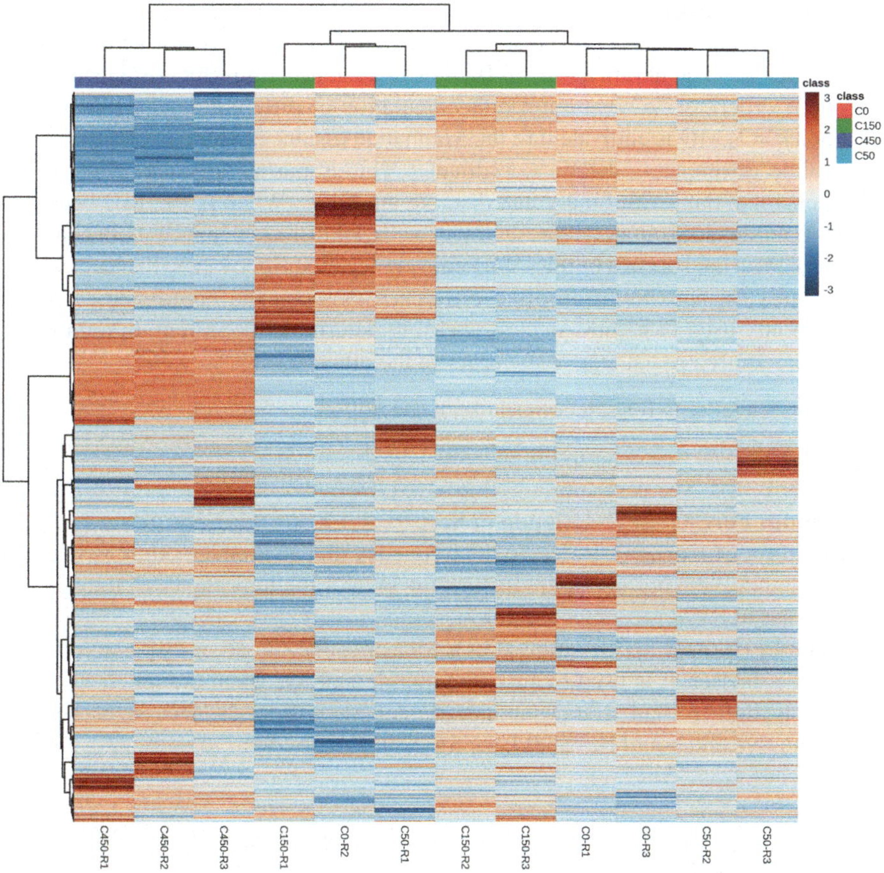

Fig. 10.40 Heatmap created with all features ("use top features" was not selected)

- **Ward's linkage**: Two clusters are joined when the smallest increase in heterogeneity is observed. It aims to minimize the total within-cluster variance. This method is often preferred when the goal is to create clusters of roughly equal sizes.

Methods available to measure distance:

- **Euclidean distance:** Measures the straight-line distance between two points in Euclidean space by calculating the square root of the sum of the squares of the differences between the two points (vectors). If the values have different scales, it becomes necessary to standardize (auto scale) and normalize them; otherwise, the higher values may dominate the measurement. The resulting scores are used to identify the most similar clusters. This metric works well with low-dimensional data, especially when measuring the magnitude of vectors is crucial. It is sensitive to the scale and can be affected by outliers.
- **Pearson correlation:** Measures the linear relationship between two variables by calculating the ratio between their covariance and the product of their standard deviations. This method is applicable when data have a normal distribution, and linearity is anticipated. However, it is sensitive to outliers.
- **Minkowski distance (MD):** Is a measure used in normed vector space (real n-dimensional space), where distances are represented as a vector with a length. It serves as a generalization of Euclidean distance, introducing a parameter called "order" or "p" that allows the calculation of different distance measures. When $p = 1$, MD is equivalent to the Manhattan distance, which is the sum of the absolute differences of their coordinates. This is particularly useful for high-dimensional data with binary and discrete attributes. When $p = 2$, MD becomes the Euclidean distance, representing the straight-line distance between two points. As p approaches infinity, MD converges to the Chebyshev distance, which measures the maximum absolute difference between the coordinates. This is commonly used in games, such as chess, to determine the minimum number of moves required to get from one corner to the other.

Methods available for selecting/ranking the important features:

- ***t*-test ANOVA**: See description in Sect. 10.5.2.
- **PLS-DA VIP**: See description in Sect. 10.4.2.
- **Random Forest:** See description in Sect. 10.4.4.1.
- **Abundance (average):** Refers to the frequency that a variable/feature is selected in the model; the higher the frequency, the greater the importance of the feature. It is used in SVM (see Sect. 10.4.4.2).
- **Variance (interquartile range-IQR):** Measures the spread or dispersion of values in a dataset. It is used mainly in network analysis when the number of features is higher than a thousand. Variables with low variance may be less informative and could be candidates for exclusion.

10.4.3.2 Partitional Clustering (PCA K-Means)

The PCA K-means algorithm (KM) combines PCA to reduce multidimensional data and the k-means method to minimize the intra-cluster variance for a chosen number of k clusters. After partitioning n observations into k centroids (the center of the cluster, calculated by averaging the data) in a dataset, every data point is allocated to the nearest cluster while keeping the centroids as small as possible [95, 96].

In biomarker analysis, the K-means algorithm is integrated with the Least Absolute Shrinkage and Selection Operator (LASSO) approach to simultaneously identify clusters and select a subset of relevant features, thereby reducing redundancy (refer to Sect. 10.8.3). LASSO regression aims to find the subset of predictors that minimizes the prediction error for a quantitative response variable [23]. The combination of both algorithms is referred to as sparse K-means [95].

In the navigation tree, click on "Statistics" > "Partitional Clustering" > "K-means" to set the parameters (Fig. 10.41). The optimal cluster number can be defined according to cross-validation carried out in PLS-DA (Sect. 10.4.2). After clicking the "Submit" button, a PCA 2D plot appears in "Sample Overview." In the given example, sample C0-R1 (red node) can be considered an outlier as it is significantly separated from the others. However, outlier determination depends on the clustering algorithm used and one should be cautious about its removal. Click on the spreadsheet icon to display the table containing the cluster to which each sample belongs.

The "Layout Separate" tab depicts a separated graph of feature concentration for each cluster of samples formed, while the "Layout Overlay" tab shows the graphs overlapped (Fig. 10.42). In the separated graphs, one can observe that there are a few features with concentrations that stand out either downwards or upwards. However, it is in the summation of the graphs that the differences become more evident.

10.4.4 Classification and Feature Selection

There are two supervised learning algorithms available to classify and select the most important features: Random Forest and Recursive Support Vector Machine (R-SVM).

10.4.4.1 Random Forest

Random Forest is an algorithm suitable for high-dimensional datasets, working in four main steps as follows [97–99]:

- **Bootstrapped sampling:** During the training of each individual tree in the forest, a random subset of the original dataset is selected as a replacement. This means that some instances may be repeated in the sample, while others, the out-of-bag (OOB) instances, may be left out.

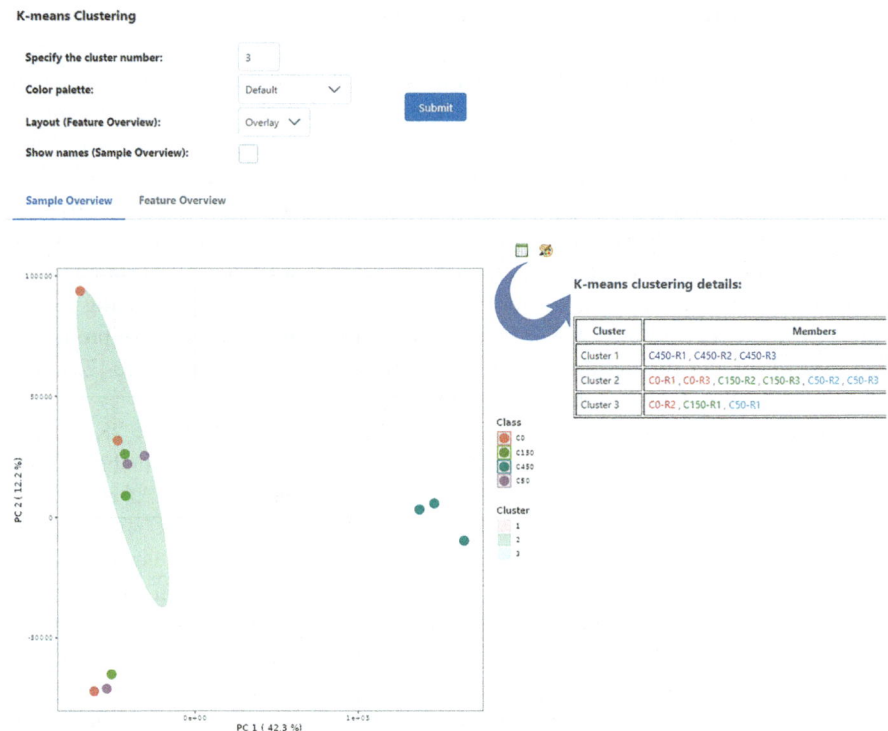

Fig. 10.41 2D scores plot from PCA K-Means cluster analysis using 3 clusters. Ellipsoidal confidence intervals were calculated only for clusters containing more than four samples. The percentage variance explained in principal component #1 and #2 is in brackets

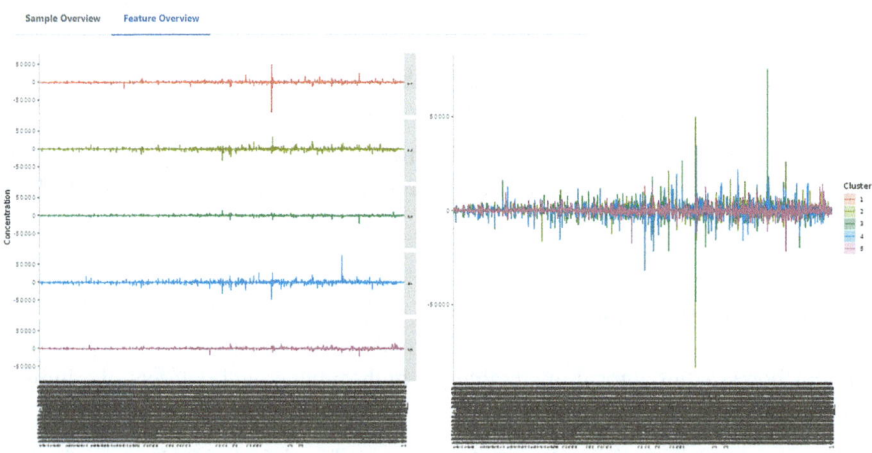

Fig. 10.42 PCA K-Means cluster analysis performed with 5 clusters (left: "Layout Separate"; right: "Layout Overlay"). The x-axis represents variable indices, and the y-axis represents relative intensities (concentration). The lines depict the median intensities of features in their respective clusters

10.4 Chemometrics Analysis

Fig. 10.43 Cumulative error rates are based on the number of trees used in Random Forest models containing seven predictors. The overall error rate is depicted by the red line, and the OOB error for 500 trees is displayed in the confusion matrix. "Class. error" column shows the proportion of misclassified samples

- **Validation:** The one-third of the instances left out of the bootstrap sample serve as a natural validation set, where the performance (accuracy) of each tree is assessed.
- **Estimating OOB error:** For classification tasks, this involves comparing the predicted class with the true class labels, and for regression tasks, it entails comparing the predicted values with the true values. The error is computed as the ratio between the number of misclassified cases and the total number of cases.
- **Bagging or bootstrap aggregation:** The OOB errors from all trees in the Random Forest are averaged or otherwise aggregated to obtain an overall estimate of the model's predictive performance.

The outlier measures are based on the proximities during tree construction [92].

In the navigation tree, click on "Statistics" > "Classification and Feature Selection" > "Random Forest" to set the parameters (Fig. 10.43). Determining the ideal number of trees (n-estimators) and predictors (features) for a Random Forest analysis often involves a process of experimentation and validation. For example, one can start with the default settings (500 trees and the square root of the total number of features for classification tasks; or one-third of the features for regression tasks) and monitor the OOB error as the number of trees increases. A stable or minimum OOB error suggests an optimal number of trees. The ideal number of predictors can be defined by Support Vector Machine based on the error rate, as described in Sect. 10.4.4.2.

In machine learning algorithms, randomness is often introduced for tasks such as bootstrapping or random feature selection to create diversity in the model. However, for reproducibility and consistency in results, it is essential to set a fixed (current or constant) random seed. A seed serves as the starting point for a sequence of pseudo-random numbers generated by a computer algorithm. Fixing the random seed ensures that the same random numbers are used during each run of the algorithm, making the entire training process deterministic—meaning predictable and reproducible. This can be performed in R using the command "set.seed ( )" [100]. The actual value of the random seed is often arbitrary. It's common to use values like 42, 10,000 or any other number.

After clicking the "Update" button, the "Classification" tab (Fig. 10.43) displays the cumulative error graph and the confusion matrix containing the proportion (class. error) of misclassified samples. In the given example, all three replicates of samples A15 and A45 were correctly classified (error = 0.0), while the three replicates of A5 and M0 were 100% (error = 1.0) misclassified as M0 and A5, respectively.

The "Outlier Detection" tab (Fig. 10.44) displays the top five potential outlier samples. One can choose to remove these samples from the analysis to reduce errors by navigating back to the "Data Editor" > "Edit Samples" > selecting the sample(s) and clicking the "add" button. However, if the number of samples is limited, it is better to remove outlier features detected in the PCA biplot or loadings plot. To do this, click on the "Edit Features" tab, fill in the feature values (m/z_RT), and click "Submit" (Fig. 10.45, on left). A window will open to redo normalization and scaling.

Then, redo the Random Forest Analysis and observe the new values for OOB error and the proportion of misclassified samples. Repeat this procedure as many times as necessary. In the given example (Fig. 10.45, on right), all errors were reduced, indicating that the removed features were indeed contributing to them.

The "Var. Importance" tab (Fig. 10.46) displays the 15 most important features selected based on the mean decrease in model accuracy produced by each one when permuted. The greater the reduction in accuracy, the more important the feature. Clicking on the spreadsheet icon opens the "Feature Details Table," which shows the decrease in accuracy produced by each feature. Clicking on "View" shows box-plots of original and normalized concentrations of the selected feature in each sample.

10.4.4.2 Recursive Support Vector Machine

SVM (with linear kernel) is a type of supervised machine learning algorithm used for classification or regression tasks. It works by finding a hyperplane that best separates the data into different classes. The objective is to maximize the margin between classes while minimizing the classification error. However, data from metabolomics studies are inherently noisy and multidimensional. To tackle this challenge,

10.4 Chemometrics Analysis

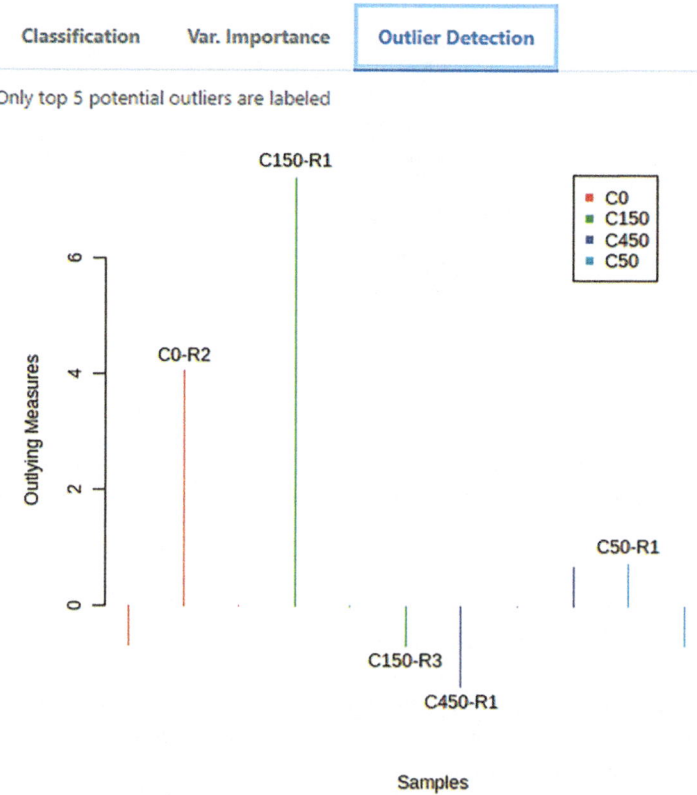

Fig. 10.44 The top five outlier samples determined by Random Forest analysis

Fig. 10.45 Monitoring out-of-bag (OOB) error in Random Forest models after outlier removal. The red rectangle in the confusion matrix highlights around a 70% decrease in the prediction error of the M0 sample in comparison with the model of Fig. 10.44

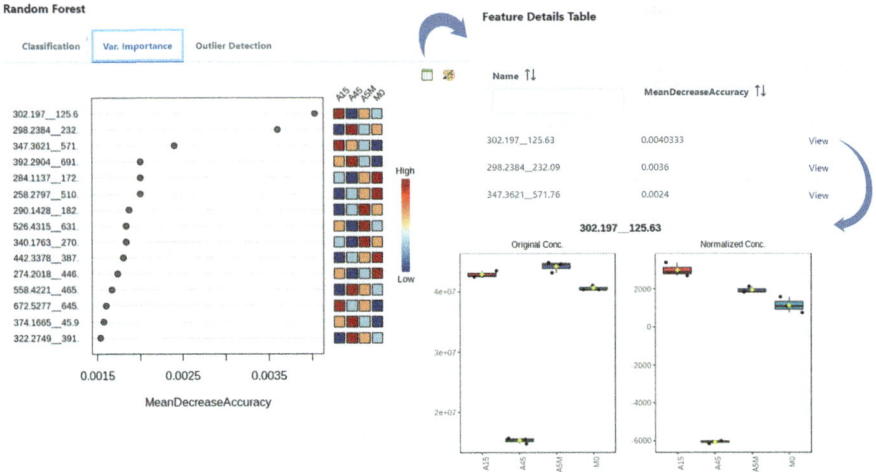

Fig. 10.46 The top 15 features were selected by Random Forest based on the mean decrease in model accuracy when permuted. In the boxplot, the black dots represent the concentrations of the selected feature from all samples. The notch indicates the 95% confidence interval around the median of each group. The mean concentration of each group is indicated with a yellow diamond. If the notches do not overlap, the medians are likely different

recursive SVM (R-SVM) was developed to be more robust to noise and outliers, functioning similarly to recursive feature elimination (RFE) algorithms. These algorithms iteratively eliminate less important features (reducing dimensionality) and train an SVM at each iteration on different subsets of features. Features are selected based on their relative contribution (weight) in the classification using cross validation error rates, and they are ranked by the frequency with which they were selected in the best models. The process is repeated until an ideal number of features is reached, i.e., the number of features that results in the minimal error rate. Finally, the features used by the best model are plotted [101–103].

In the navigation tree, click on "Statistics" > "Classification and Feature Selection" > "SVM." As a rule of thumb, the input data should contain only two classes. There is a single parameter to be set: the cross-validation method (Fig. 10.47). Out of the three methods available, K-fold CV and LOOCV have already been discussed in PLS-DA (Sect. 10.4.2), bootstrapping is presented below:

- **Bootstrapping cross-validation:** Involves sampling the dataset with replacement. It is intended to be run multiple times so that the average validation error across the repetitions can be used to estimate error rates more accurately [100, 104].

According to some authors, tenfold CV is considered better than bootstrap and LOOCV in model selection and evaluation. However, as the sample size grows, the differences in performance among these resampling methods decrease [105–107].

After selecting the CV method and submitting the task, the "Classification" tab displays a graph depicting the error rate estimate according to the number of

10.4 Chemometrics Analysis

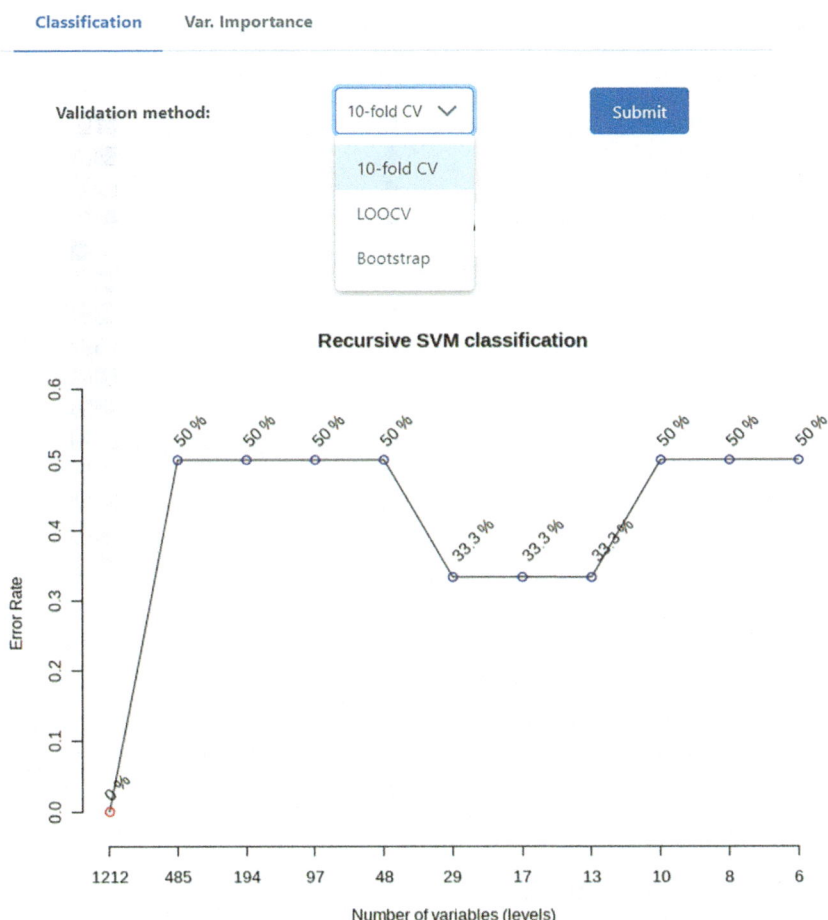

Fig. 10.47 Selection of the ideal number of predictors for building an SVM model based on the error rate estimate. The red circle denotes the best model, identified as the one with the lowest error rate, utilizing 1212 predictors

predictors used to build the model (Fig. 10.47). In the given example, the minimal error rate (0%) was achieved using 1212 predictors. A more cost-effective model was attained with either 29, 17, or 13 predictors but exhibited a 33.3% error rate, indicating the need for more informative features or an increase in the sample number.

The "Var. Importance" tab (Fig. 10.48) shows the top 15 features ranked by the number of frequencies they were selected during SVM model building.

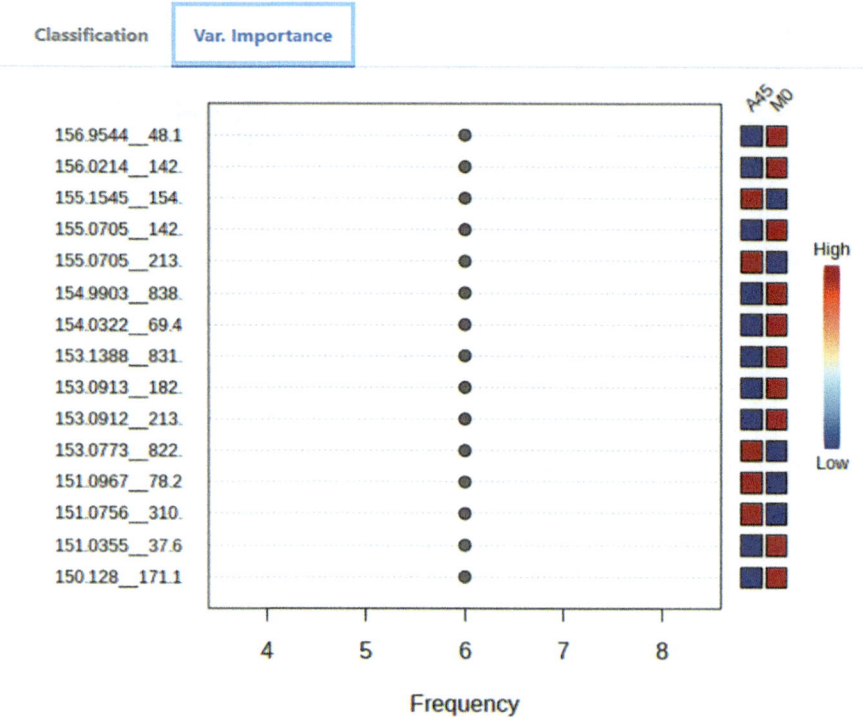

Fig. 10.48 Top 15 features ranked by frequency of selection in the best SVM models

10.5 Univariate Analysis

Univariate analysis methods are commonly used in metabolomics for exploratory data analysis. For two-group data, MetaboAnalyst offers Fold Change (FC) analysis, *t*-tests, and a volcano plot, which is a combination of the first two methods. All three methods support both paired and unpaired analyses. In multi-group analysis, MetaboAnalyst provides two types of analysis: one-way analysis of variance (ANOVA) with associated *post-hoc* analyses, and correlation analysis to identify significant compounds following a given pattern. Univariate analyses offer a preliminary overview of features that may be significant in discriminating the studied conditions.

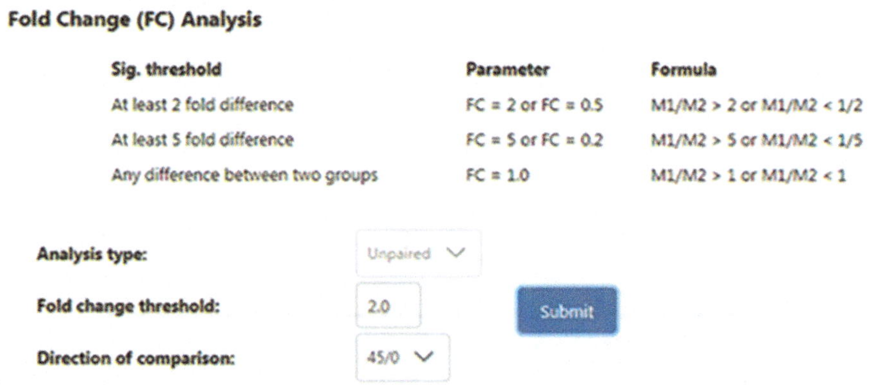

10.5 Univariate Analysis

Fold Change (FC) Analysis

Sig. threshold	Parameter	Formula
At least 2 fold difference	FC = 2 or FC = 0.5	M1/M2 > 2 or M1/M2 < 1/2
At least 5 fold difference	FC = 5 or FC = 0.2	M1/M2 > 5 or M1/M2 < 1/5
Any difference between two groups	FC = 1.0	M1/M2 > 1 or M1/M2 < 1

Analysis type: Unpaired

Fold change threshold: 2.0

Direction of comparison: 45/0

Fig. 10.49 Setting parameters for fold change analysis

10.5.1 Fold Change Analysis

Fold change (FC) is a measure that describes the proportional increase or decrease in the abundance of a biological feature under two different conditions or experimental treatments. It provides insights into biological processes and responses to various stimuli or treatments [45].

The interpretation of the result depends on the specified threshold. However, in statistical differential analysis within non-targeted metabolomics, the central inquiry is whether an FC value significantly deviates from an FC of 1, indicating no change. It is essential to account for uncertainty, as the minimum relevant fold changes escalate with increasing observed standard uncertainty of group-averaged metabolite intensities. Consequently, the commonly employed fold change threshold of 2 remains valid only when the within-group variation is below 20% [108].

As a rule of thumb, the input data should consist of only two classes **with original data not scaled or transformed**. In the navigation panel, click on "Statistics" > "Univariate Analysis" > "Fold Change Analysis." The "FC analysis" window opens (Fig. 10.49) explaining the analysis goal, the way FC is calculated based on the type of sample, and the available thresholds for comparison. The analysis type, fold change threshold and direction of the comparison between the classes must be indicated.

After submitting the task, a graph appears (Fig. 10.50) showing the number of features ranked according to their significance in fold change, which is depicted as log base 2 transformed. Clicking on the spreadsheet icon displays a table containing the fold change and respective log transformation for each feature.

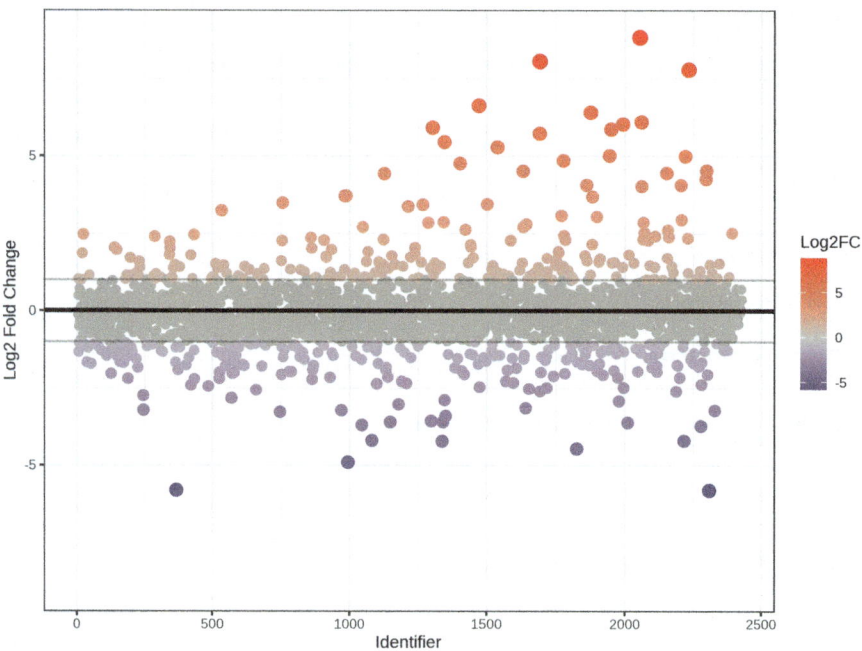

Fig. 10.50 Important features selected by Fold Change Analysis using a threshold of 2. The red nodes represent significantly upregulated features ($n = 219$), the purple nodes represent significantly downregulated features ($n = 248$), and gray nodes are not significant ($n = 1961$). The values are on a log scale, enabling the symmetrical plotting of both upregulated and downregulated features from the zero baseline. The color bar on the right represents the gradient of the log2-transformed fold change values

10.5.2 t-Tests

In an untargeted metabolomics study, *t*-tests are used to select features with potential to distinguish different groups in a data set. Users must consider the nature of the data to decide the type of test to be used. For instance, the choice of tests depends primarily on factors such as the population distribution (parametric or non-parametric), as well as the size and independence of the samples (paired or unpaired), the homogeneity of variances (homoscedasticity or heteroscedasticity), and the approach to significance (raw *p*-value or adjusted *p*-value/FDR). Users should follow the flowchart for univariate analysis suggested by [45].

Parametric tests assume that the data is homoscedastic and follows a normal distribution. They are better suited for continuous (numeric) variables. On the other hand, non-parametric tests are considered distribution-free. They are more robust and better applied to non-normal populations, unequal variances, and unequal small sample sizes.

10.5 Univariate Analysis

To select important features through t-tests, the null hypothesis assumes no difference between the means/medians of features in the studied groups. The probability threshold for rejecting the null hypothesis when, in fact, it should not be (a false positive result or type I error) is generally set at 5% (α). Then, the probability (p-value) of null hypothesis rejection is calculated in the hypothesis testing. If p-values are smaller than the predefined α value, the null hypothesis is rejected, indicating significant differences in means between the groups; otherwise, there is no evidence to reject the null hypothesis. However, after thousands (representing the number of features to be tested) of hypothesis tests, the probability of false positives increases. This accumulation of false positives, termed the multiple testing problem, can be corrected/controlled by the False Discovery Rate (FDR), i.e., the rate of significant features being false. FDR calculates a p-adjusted value (termed q-value) for each tested metabolic feature. The q-value represents the expected proportion of false positives considered when a feature is declared to be significantly varied. In short, a p-value of 0.05 implies that 5% of all tests will result in false positives, while a q-value of 0.05 means that 5% of the significant tests will result in false positives. A significant feature must have both values below 5% [45]. Thus, one cannot assume features are potential predictors based solely on their small p-values. The biological context, including fold change, pathway relationships, and other statistics, must also be evaluated. For a comprehensive discussion on interpreting and using p-value, refer to [109].

In the navigation tree, click on "Statistics". In Univariate Analysis, select "T-test". The "Two-sample t-tests" window opens to set the parameters (Fig. 10.51, on the left). MetaboAnalyst uses the typical t-test, comparing the means of two groups, when the user selects "Group variance equal". If "Non-parametric tests" is chosen, the Wilcoxon rank-sum test, which compares the distribution of ranks between two groups, is used. For a large data set (>1000 variables), both the paired information and the group variance are ignored, and the test is conducted with the default parameters, which are not specified. The analysis can be carried out using the level of significance based on raw p-value or FDR as the threshold.

After clicking on the "Update" button, a graph appears (Fig. 10.52) displaying all features ranked according to their statistical significance. The further a feature is from (0,0), the more significant it is. Clicking on the spreadsheet icon reveals the "Feature Details Table" (Fig. 10.51, on the right), containing all the statistics obtained in the analysis.

10.5.3 Volcano Plot

Volcano plot combines the fold change analysis with t-test into one single graph. The plot often resembles a volcano, where the most significant and biologically relevant features rise to the top, forming the peak of the volcano. In this way, one can quickly identify features that exhibit both a substantial fold change and statistical significance. Typically, significant features are represented by points that are

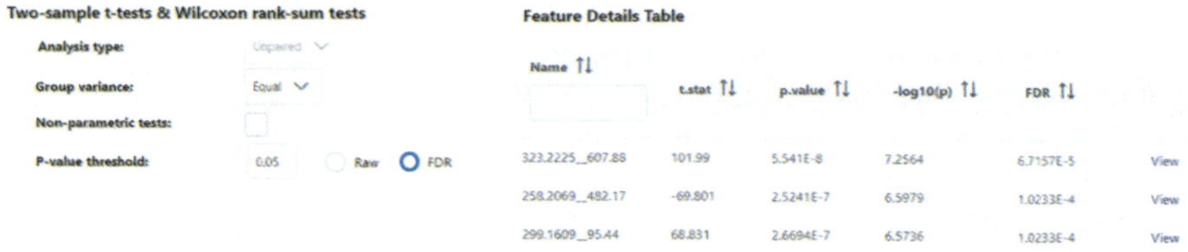

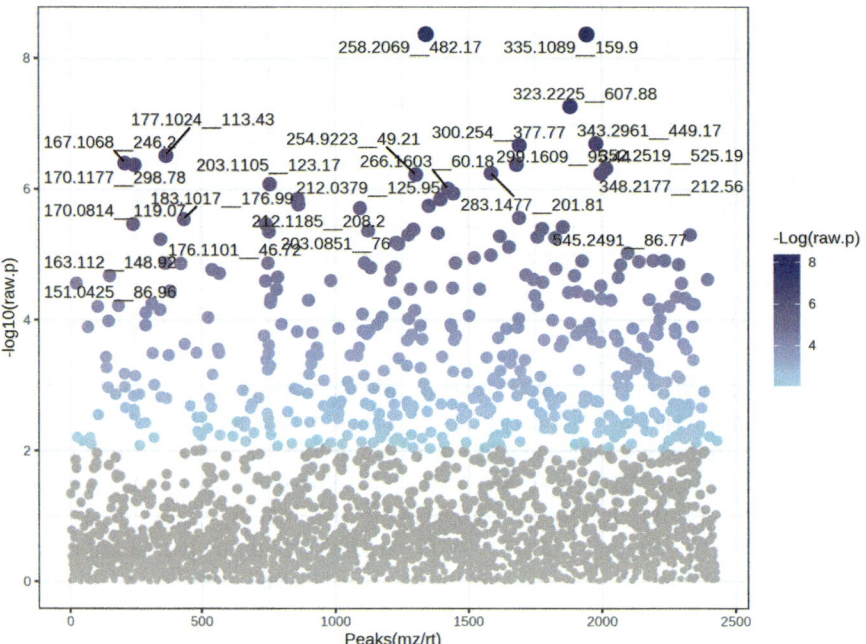

Fig. 10.51 Setting parameters for *t*-tests (left) and observing the results of the analysis (right)

Fig. 10.52 Important features (454) selected by *t*-test with a significance level of 5% (*p*-value and FDR). Raw *p*-values are -log10 transformed to display more significant features (with smaller *p*-values) higher on the graph. The color bar on the right represents the gradient of the log-transformed raw *p*-values

located far to the right or left on the x-axis (indicating a large fold change) and high on the y-axis (indicating a low *p*-value).

In the navigation tree, click on "Statistics." In Univariate Analysis, select "Volcano Plot" to open the "Volcano Plot" window and set the parameters (Fig. 10.53), which are essentially the same as those mentioned for the two analyses separately.

After clicking the "Update" button, the volcano plot appears (Fig. 10.54).

10.5 Univariate Analysis

Fig. 10.53 Setting parameters for Volcano Plot (left) and observing the results of the analysis (right)

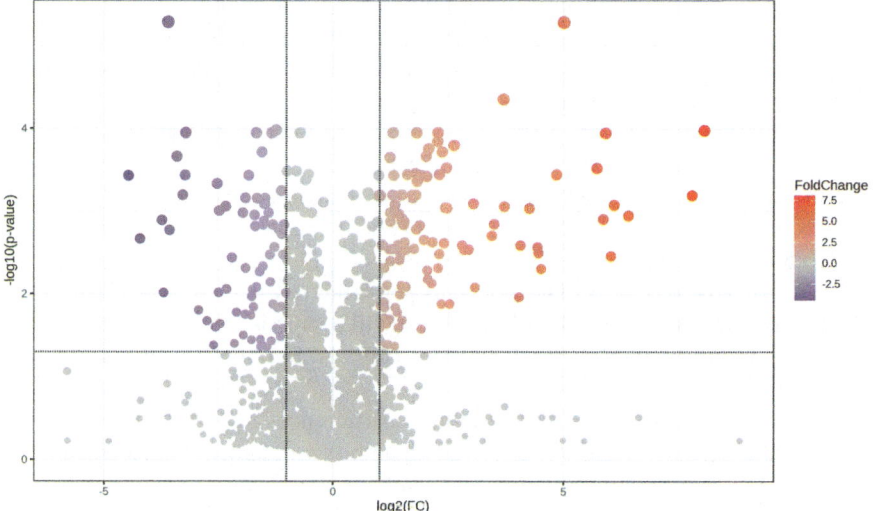

Fig. 10.54 Important features selected by volcano plot with a significance level of 1% (FDR < 0.01). Raw p-values (Y-axis) are -log10 transformed to display more significant features (with smaller p-values) above the zero baseline. Fold change values (X-axis) are log2 transformed to enable symmetrical representation of both upregulated (148 red nodes) and downregulated (129 blue nodes) features from the zero baseline. The color bar on the right represents the gradient of the log2-transformed fold change values

10.5.4 One-Way Analysis of Variance (ANOVA)

ANOVA is an extension of the t-test used to compare means among three or more groups.

The basic idea behind ANOVA is to partition the total variance observed in a set of data into different components, such as the variance within each group and the variance between groups. If the variance between groups is significantly larger than the variance within groups, this suggests that there are differences in the means of the groups.

There are different types of ANOVA, and the choice of which one to use depends on the specific experimental design. The three main types are:

- **One-Way ANOVA:** Used when there is one independent variable with three or more levels or groups.
- **Two-Way ANOVA:** Applied when there are two independent variables. It examines how these variables impact a dependent variable, as well as any interaction between them.
- **Repeated Measures ANOVA:** Employed when measurements are taken on the same subjects at multiple time points or under different conditions.

ANOVA produces an F-statistic by taking the ratio of the variance between groups to the variance within groups to assess whether the means of the groups are different. Then, the *p*-value is calculated to determine whether there are significant differences among the group means. If the p-value is below a chosen significance level, the null hypothesis is rejected, and one can conclude that there are significant differences in at least one pair of group means.

In the navigation tree, click on "Statistics." In Univariate Analysis, select "One-way ANOVA" to open the "One-way ANOVA" window and set the parameters (Fig. 10.55).

Parametric ANOVA with FDR of 0.05 is performed by default; otherwise, select "non-parametric" and the Kruskal-Wallis' test will be applied to the data after clicking the "Submit" button. A graph appears (Fig. 10.56) showing the significant features.

To determine the significance of the mean differences identified by parametric ANOVA, click the spreadsheet icon. A dialogue box will appear to select the *post-hoc* test. Click the "Submit" button and wait for the "Feature Details Table" to appear with the results.

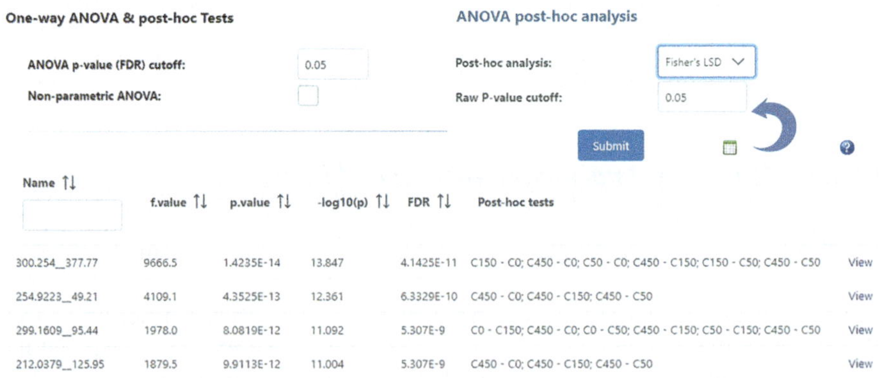
One-way ANOVA & post-hoc Tests
ANOVA p-value (FDR) cutoff:
0.05
Non-parametric ANOVA:
ANOVA post-hoc analysis
Post-hoc analysis:
Fisher's LSD
Raw P-value cutoff:
0.05
Submit
Name
f.value
p.value
-log10(p)
FDR
Post-hoc tests
300.254_377.77 9666.5 1.4235E-14 13.847 4.1425E-11 C150 - C0; C450 - C0; C50 - C0; C450 - C150; C150 - C50; C450 - C50 View
254.9223_49.21 4109.1 4.3525E-13 12.361 6.3329E-10 C450 - C0; C450 - C150; C450 - C50 View
299.1609_95.44 1978.0 8.0819E-12 11.092 5.307E-9 C0 - C150; C450 - C0; C0 - C50; C450 - C150; C50 - C150; C450 - C50 View
212.0379_125.95 1879.5 9.9113E-12 11.004 5.307E-9 C450 - C0; C450 - C150; C450 - C50 View

Fig. 10.55 Setting parameters for ANOVA (above) and observing the results (underneath). *Post-hoc* tests carried out with Fisher's LSD

10.5 Univariate Analysis

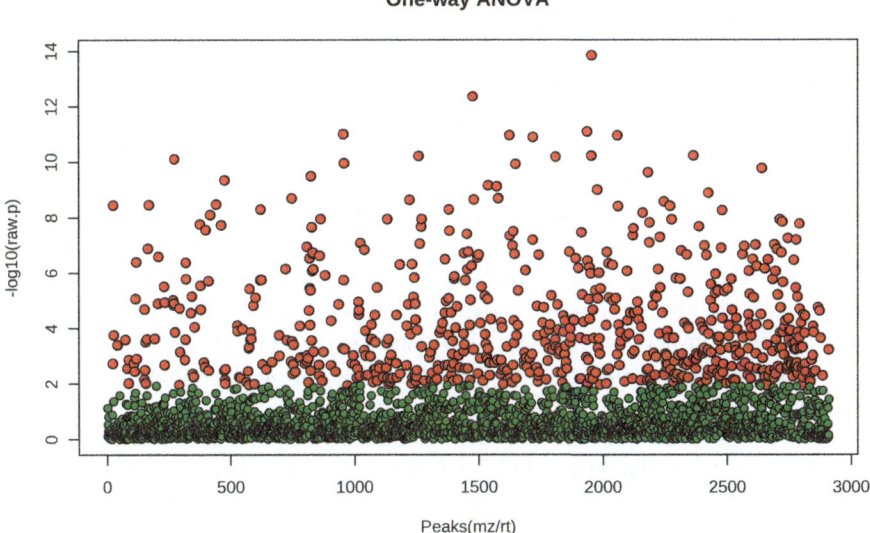

Fig. 10.56 Important features selected by One-way ANOVA and Fisher's LSD *post-hoc* analysis with a significance level of 5% (*p*-value and FDR)

Two *post-hoc* tests are available:

- **Fisher's Least Significant Difference (LSD):** Involves calculating the LSD, which is the smallest difference between two group averages that is statistically significant at a chosen significance level. Subsequently, the difference between the means of each pair of groups is compared to the LSD. If the absolute difference between the means is greater than the LSD, the difference is considered statistically significant. To control for multiple comparison errors, FDR is calculated. This test is sensitive to Type I error and may not be the most appropriate when conducting multiple pairwise comparisons [110, 111].
- **Tukey's Honestly Significant Difference (HSD):** Involves calculating a critical value, called the Honestly Significant Difference (HSD), which represents the minimum difference between two group averages that is required to be considered statistically significant at a chosen significance level. Subsequently, the difference between the means of each pair of groups is compared to the HSD. This test is designed to control the familywise error rate, which is the probability of making one or more Type I errors in a set of comparisons [112].

10.5.5 Pattern Hunter

Like nodes in a network are associated (connected) based on some measure of similarity, in metabolomics, the similarity between metabolites, and therefore their association, is usually expressed using correlation indices [113].

The degree to which a change in one variable is associated with a change in another variable can be described mathematically in terms of the covariance of the variables. Covariance is like variance, but whereas variance describes the variability of a single variable, covariance is a measure of how two variables vary together [114]. The correlations observed in metabolomics data are generally small (<0.6) due to the systemic nature of metabolic control. On the other hand, a high correlation between two metabolites may be dissociated from their proximity in a metabolic pathway or from their participation in the same reaction. It is worth noting that correlation inferences (and data analysis, in general) can be affected by data preprocessing, such as scaling, transformation, and normalization [113]. For these reasons, one should not rely solely on the correlation coefficient but rather graph the data for a visual inspection of the trend of relationships between the variables, such as exemplified in Fig. 10.57, and account for other factors contributing to a given correlation, including its statistical significance [115].

In the navigation tree, click on "Statistics." In Univariate Analysis, select "Pattern Huntter" to open the "Pattern Search" window and set the parameters (Fig. 10.58). Correlation analysis can be performed either against a given feature or against a given pattern, selecting a predefined profile. After clicking the "Submit" button, a graph appears containing the 25 compounds that exhibited the highest correlation with the feature of interest. Clicking the spreadsheet icon displays the "Feature Details Table" with the results.

There are three methods available to measure correlation:

- **Pearson's Correlation Coefficient (r):** This is a dimensionless measure of covariance used for two quantitative continuous variables. It results in values

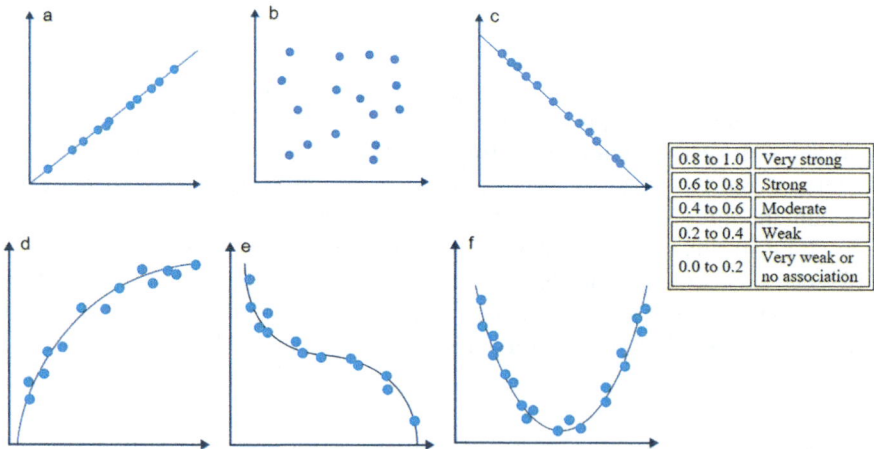

Fig. 10.57 Scatter plots of the different types of correlation. Linear correlation: a perfect positive ($r = +1$), b no correlation ($r =$ zero), c perfect negative ($r = -1$). Non-linear: d monotonic positive, e monotonic negative, f U-shaped. On the right, table containing correlation scores (in absolute value) and respective interpretation

10.5 Univariate Analysis

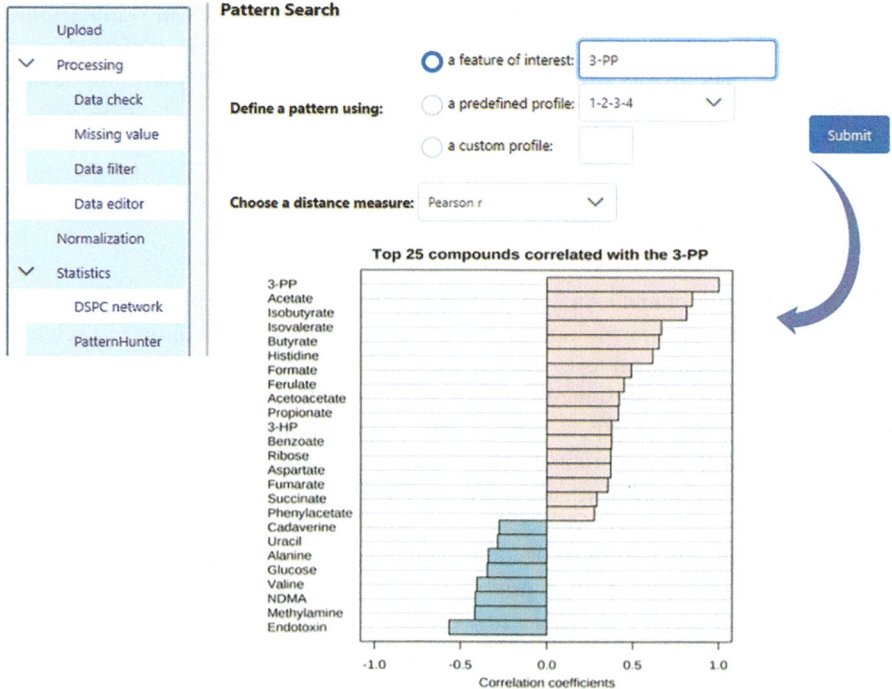

Fig. 10.58 Pattern search—Pearson correlation analysis showing the 25 metabolites most correlated with the metabolite 3-PP. Each row represents the most significant variable identified in the test (p and FDR < 0.05). Positive correlations are depicted above zero (pink bars), and negative correlations below zero (blue bars)

from −1 to +1, where zero indicates no linear association, and −1 or +1 represent perfect negative and perfect positive correlations, respectively. **Requirements:** independent samples, a normal distribution, the absence of relevant outliers, and a linear relationship (variables move in the same direction and at a constant rate; the correlation is a straight line) between the variables. The t-test with FDR is applied to determine the significance of the correlation ($r \neq 0$ and $p < 0.05$) or its absence (null hypothesis: $r = 0$ and $p > 0.05$) [23, 114].

- **Spearman Rank Correlation Coefficient (r_s):** This is calculated using the ranking scores of the values of each of the two variables rather than their actual values. It shares a range and interpretation like Pearson's coefficient but exhibits robustness against outliers. **Requirements:** no assumption of linearity, the use of paired samples, a linear or monotonic relationship (implies that variables move in the same direction but not necessarily at a constant rate, forming either a straight line or a curve). This method is particularly suitable for ordinal random variables.
- **Kendall's Tau Rank Correlation Coefficient** (τ): Like Spearman's coefficient, this method is also based on the ranks of the data and exhibits robustness against

outliers. **Requirements:** no assumption of linearity, variables in ordinal scale values, a linear or monotonic relationships. This method is preferred over Spearman's coefficient when dealing with very few data points that exhibit many rank ties.

10.6 Enrichment Analysis

Metabolite Set Enrichment Analysis (MSEA) is designed to interpret metabolomics data in the context of known biological pathways or sets of metabolites. To achieve this, it performs statistical tests to assess whether the input metabolite list is overrepresented or underrepresented within specific metabolic pathways or sets compared to what would be expected by chance. Subsequently, the analysis identifies pathways or sets that are enriched with the measured metabolites. Enrichment indicates that the input metabolites are more prevalent in those pathways than would be expected by random chance. MSEA helps gain insights into the biological processes and pathways that are perturbed or influenced in each experimental condition based on changes in metabolite levels. It provides a more interpretable and context-specific understanding of metabolomics data by linking it to established biological knowledge [116, 117].

After clicking on the "Enrichment Analysis" button and uploading data, a window opens, offering three different enrichment analyses for metabolomic studies: overrepresentation analysis (ORA), single sample profiling (SSP), and quantitative enrichment analysis (QEA). ORA requires only a list of compound names, while SSP and QEA require both compound names and compound concentrations, which are more commonly used in targeted metabolomics.

10.6.1 Overrepresentation Analysis (ORA)

This analysis typically involves comparing a user-provided list of metabolites (of interest or relevance) against a reference or background set of metabolites. The objective is to identify whether the user's metabolite list shows a statistically significant overrepresentation in certain pathways or sets compared to what would be expected by chance [118]. This technique, widely utilized in the field of metabolomics, plays a crucial role in identifying impacted pathways. However, users should exercise caution when setting parameters, as the background set, metabolite selection method, and pathway database can significantly impact the results [18]. Limitations of ORA include an oversimplification of the biology, a thresholding decision issue when generating the input metabolite list and a lower power for capturing subtle and coordinated changes within a pathway [119].

To execute ORA, users are required to input a list containing either the names of the metabolites of interest or their corresponding database IDs, provided from

10.6 Enrichment Analysis

targeted or untargeted approach. This list of metabolites can be acquired through feature selection methods, as detailed in Sect. 10.5. To streamline this process, the "Input Type" tab should be completed as demonstrated in Fig. 10.59 on the left. In the "Feature Type" tab, users should select either "metabolite" or "lipid" based on the analytical platform and specific assay employed to gather the data. Upon clicking the "Submit" button, the program will take a few moments to perform the matching process between the input metabolite IDs and the database, as illustrated in Fig. 10.59, on the right. Instances where no matches are found will be highlighted in yellow. In such cases, users are required to rectify the discrepancies in the names to align with the database template and then resubmit the corrected list for analysis.

After clicking the "Proceed" button, a new window will appear (Fig. 10.60), allowing users to choose the metabolite set library for one of the five available analyses: (i) Pathway-based analysis, representing the pathway to which the altered metabolite belongs; (ii) Disease signatures analysis, indicating metabolites altered according to specific diseases related to the analyzed sample (biofluid), such as glucose/blood/diabetes; (iii) Chemical structures analysis, representing the chemical class of the altered metabolites; (iv) Other types of analysis, including metabolites related to single nucleotide polymorphisms (SNPs), predicted metabolites altered in the case of dysfunctional enzymes, or altered according to their location; (v) Self-defined analysis, such as all compounds assayed (targeted metabolomics) or all annotatable metabolites (untargeted metabolomics).

In the subsequent options, users should select the "Only use metabolite sets containing at least 2 entries" tab, signifying that these sets will be considered relevant if they consist of at least two distinct metabolites. For enhanced precision, users can choose to increase the number of entries [18].

After clicking the "Submit" button, a new window will appear displaying the results in various visual formats, including an interactive pie plot (Fig. 10.61), a dot plot (Fig. 10.62), a bar chart (Fig. 10.63), and a network view (refer to Fig. 3.2).

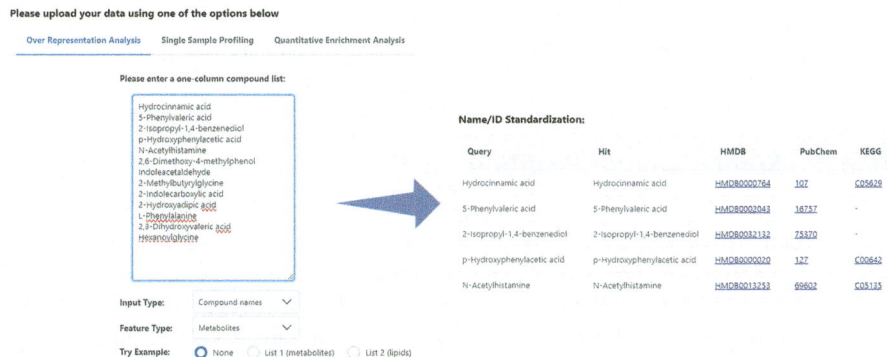

Fig. 10.59 Setting parameters for overrepresentation analysis (left), and checking the matching of the input names with the registered entries in the available databases (right)

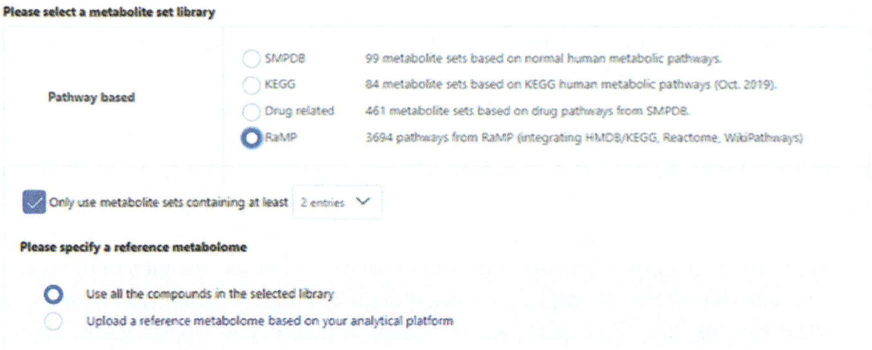

Fig. 10.60 Selecting the metabolite set library for overrepresentation analysis

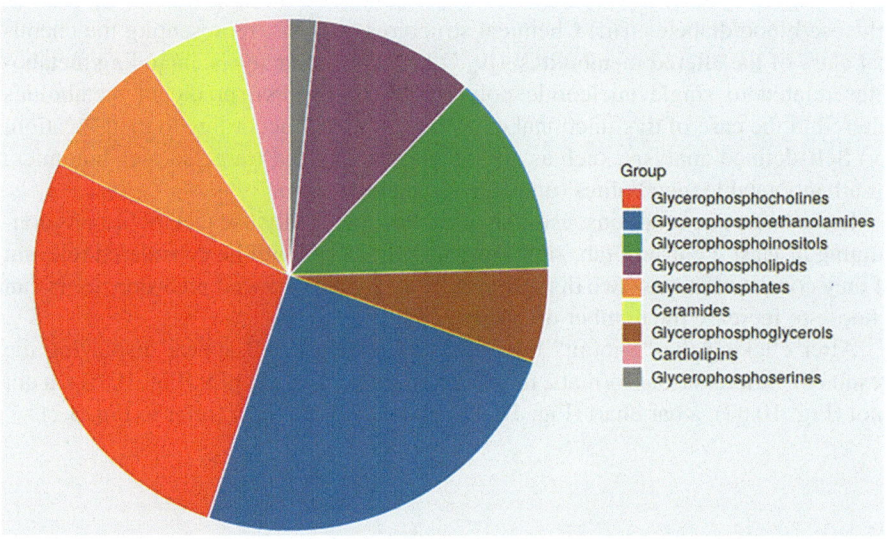

Fig. 10.61 Pie plot of the enriched chemical classes of a list containing 123 significant metabolites (FDR < 0.05)

10.6.2 Single Sample Profiling (SSP)

SSP typically involves quantifying the expression or activity levels of genes, proteins, or metabolites in each sample and subsequently mapping these measurements onto predefined pathways or biological processes. This mapping enables researchers to gain insights into the functional state of the biological system under study. SSP proves to be a valuable tool in personalized medicine, diagnostics, and other applications where understanding the unique characteristics of an individual sample

10.6 Enrichment Analysis

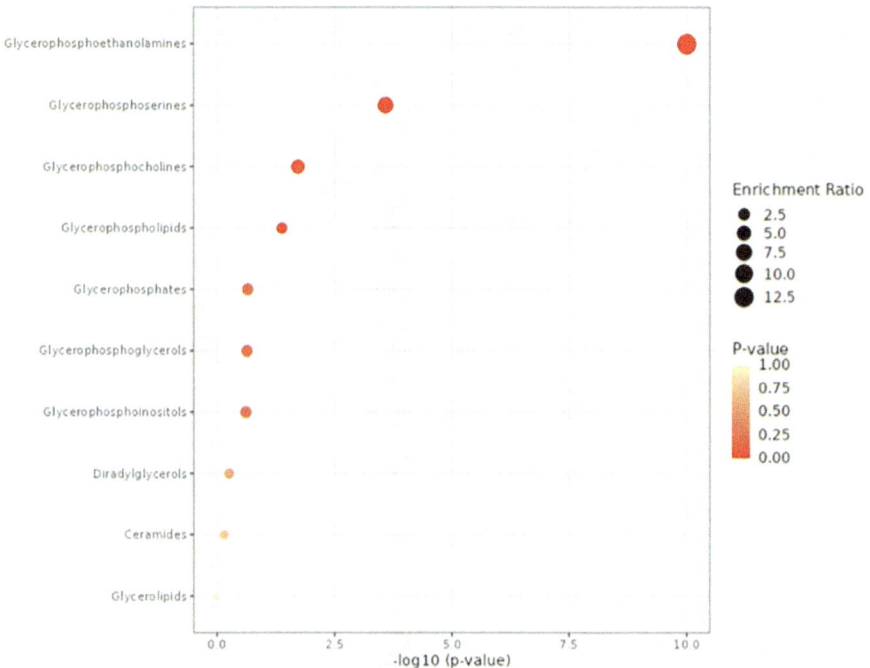

Fig. 10.62 Overview of enriched chemical signatures of a list containing 123 significant metabolites (FDR < 0.05)

is crucial. For instance, a doctor can analyze patients' results by correlating detected metabolites in biofluids (such as blood, serum, urine, etc.) against a reference list containing concentrations observed in normal cases. This approach provides a means to assess the functional implications of molecular data within a specific biological context without relying on comparisons between different samples or groups [120].

The process closely resembles ORA, with the distinction that the list of metabolites is accompanied by their respective concentrations, and users must specify the type of biofluid used in the analysis. After clicking the "Submit" button, verifying the nomenclature of the compounds, and clicking the "Proceed" button, a new window opens (Fig. 10.64), presenting a comparison table between the concentrations of the loaded compounds and the reference list. These concentrations are labeled as high (H), medium (M), or low (L). Metabolites with high concentrations are flagged in the "Include" icon, and upon clicking "View," a window appears, displaying a comparison graph of reference studies with the concentration of the analyzed metabolite.

Toward the end of this table, upon clicking the "next" button, a new window will open. In the "Disease Signatures" section, select the sample used. After

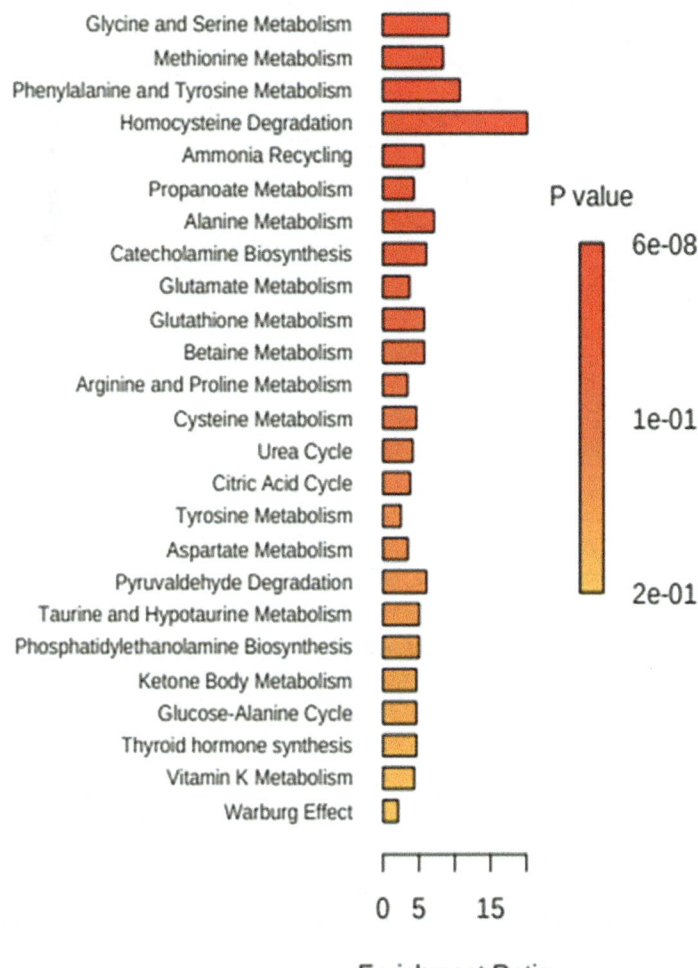

Fig. 10.63 Top 25 enriched pathways obtained by overrepresentation analysis of a list of 123 significant metabolites (FDR < 0.05) using RaMP database

clicking "Submit," a new window will open with the results. The bar graph illustrates potential diseases that the patient may have in relation to the list of metabolites whose levels are increased, as compared to the reference list of normal controls (Fig. 10.65).

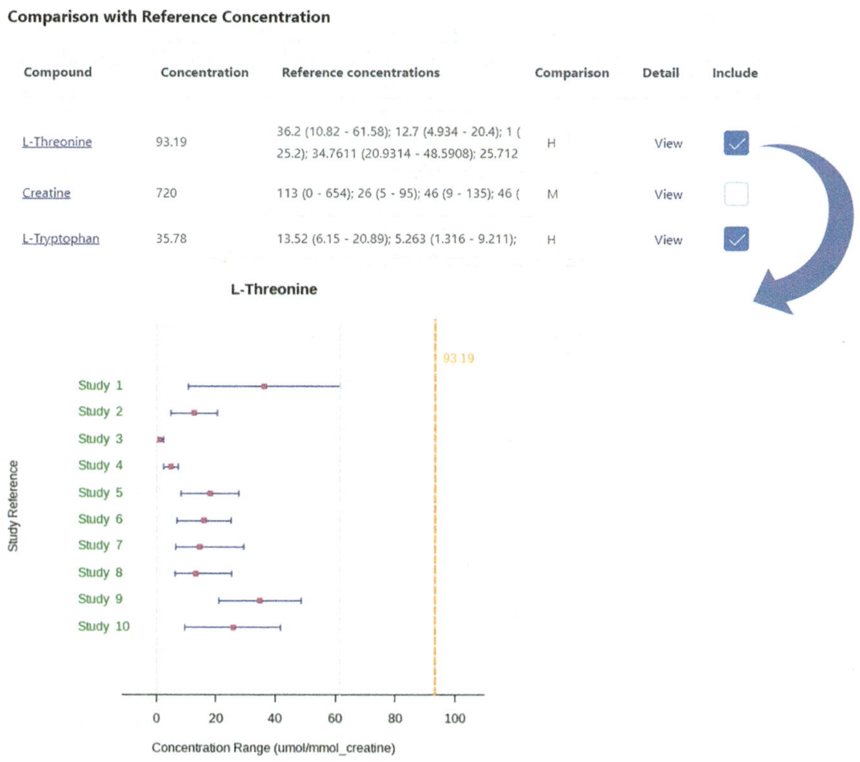
Comparison with Reference Concentration
Compound
Concentration
Reference concentrations
Comparison
Detail
Include
L-Threonine
93.19
36.2 (10.82 - 61.58); 12.7 (4.934 - 20.4); 1 (
25.2); 34.7611 (20.9314 - 48.5908); 25.712
H
View
Creatine
720
113 (0 - 654); 26 (5 - 95); 46 (9 - 135); 46 (
M
View
L-Tryptophan
35.78
13.52 (6.15 - 20.89); 5.263 (1.316 - 9.211);
H
View
L-Threonine
93.19
Study 1
Study 2
Study 3
Study 4
Study 5
Study 6
Study 7
Study 8
Study 9
Study 10
Study Reference
0
20
40
60
80
100
Concentration Range (umol/mmol_creatine)

Fig. 10.64 Single sample profiling analysis of a list of metabolites from urine

10.6.3 Quantitative Enrichment Analysis (QEA)

QEA is like SSP but directed to multiple comparisons rather than tailored to individual samples. QEA focuses on identifying whether certain categories, such as pathways or functional groups, are overrepresented or underrepresented in each dataset.

The input data should include the labels of the compared classes as well as the names and concentrations of the metabolites. Prior to the enrichment analysis, the data undergoes normalization and scaling. The results are presented in a manner like SSP, with the added capability for users to visualize the box plot of metabolite concentrations (Fig. 10.66).

10.7 Power Analysis

Power analysis is a statistical method employed to assess a study or experiment's ability to detect a true effect or difference, should one exist. Ideally, it is conducted before data collection to determine the necessary sample size based on the expected

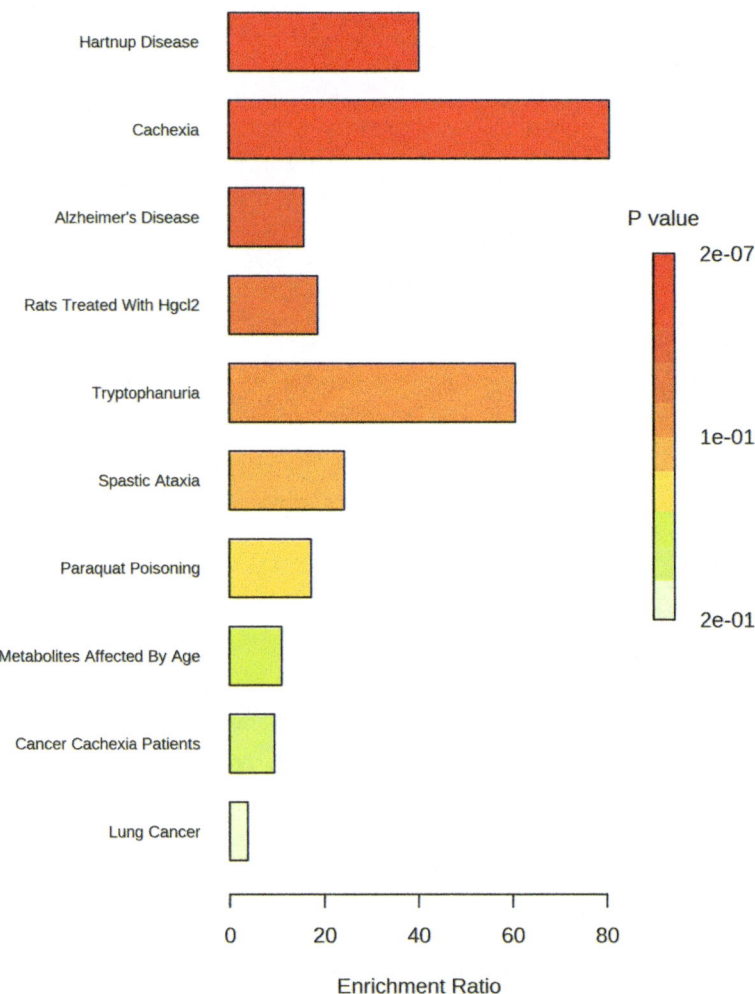

Fig. 10.65 Potential most significant diseases for a patient based on increased urine metabolites

effect size, significance level, and desired power. Nevertheless, performing power analysis after data collection enables an assessment of the achieved power based on the obtained effect size and sample size. A study with high power is more likely to detect an effect, while a study with low power may fail to do so even if the effect genuinely exists [121, 122].

10.7 Power Analysis

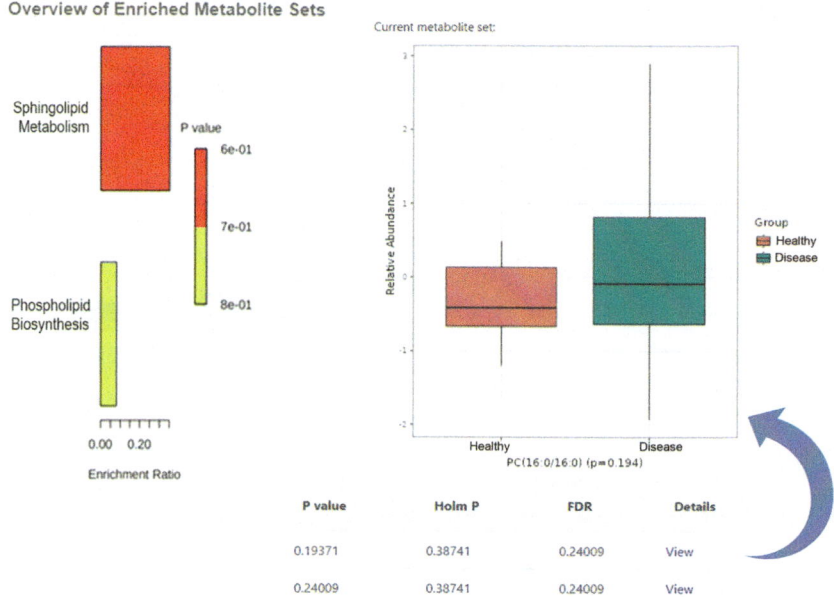

Fig. 10.66 Quantitative enrichment analysis. Bar plot showcasing the enriched pathways (left) and box plot of the metabolite concentrations within the analyzed cohort (right), utilizing the Small Molecule Database (SMPDB)

In the context of model building, sample sizes can be determined to test whether including additional or fewer parameters in a model will enhance its fit, as measured [123].

Power analysis encompasses three key components:

- **Effect size:** The magnitude of the difference or effect in the population. Larger effect sizes generally require smaller sample sizes to achieve the same level of power.
- **Significance level (α):** The probability of rejecting a true null hypothesis (Type I error rate). Commonly set at 0.05 (5%). Lowering the significance level increases the required sample size but also increases power.
- **Power (1-β (type 2 error)):** The probability of correctly rejecting a false null hypothesis. Usually, users aim for a power of 0.80 (80%) or higher.

Power analysis can address crucial questions such as "How powerful is my study?" and "How large of a sample do I need to obtain a reliable answer consistent with the objectives of my study?"

After selecting the module and uploading the data, clicking the "Submit" button initiates the integrity check of the data, followed by normalization and scaling procedures. Subsequently, clicking the "Proceed" button opens a new window containing some graphs (Fig. 10.67) where users can assess the normality of test statistics.

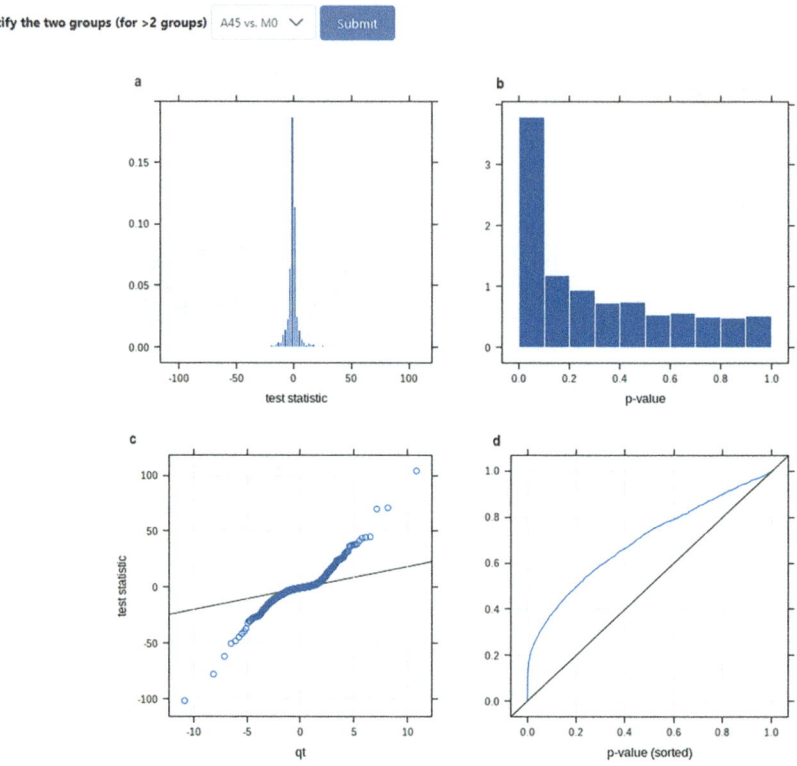

Fig. 10.67 Visual assessment of *t*-statistics normality (**a, c**), significance of the compounds (**b**), and data distribution

One way to visualize this is through the histogram displayed in the "a" graph of Fig. 10.67. The histogram of the statistics distribution should resemble a bell-shaped curve characteristic of a normal distribution [124–126] . If this is not achieved, the user must return to the normalization step and try different methods of normalization.

Another way of visualizing normality is through the Q-Q (Quantile-Quantile) plot (Fig. 10.67, c graph). In this graph, the quantiles of the test statistics are plotted against the quantiles of a theoretical normal distribution. If the points on the plot lie approximately along a straight line, the distribution is close to normal. An S-shaped curve may indicate skewness in the data. Discontinuities or steps in the plot may suggest departures from normality.

Users can assess the significance of specific compounds in the analyzed cohort by examining the "b" graph in Fig. 10.67. The histogram of the distribution of *p*-values should exhibit a left-skewed pattern, indicating that most compounds are associated with *p*-values close to zero. The observed left-skewness in the distribution of *p*-values indicates that these compounds are significantly different between the two conditions selected for the power analysis.

10.7 Power Analysis

The "d" graph in Fig. 10.67 is a P-P (Probability-Probability) plot utilized to evaluate whether the two samples originate from the same distribution. Various possibilities for interpretation include:

- If the points on the P-P plot fall along a straight line (the 45-degree line in the plot), it suggests that the two samples come from the same distribution.
- If the points consistently lie above the line, it suggests that the first sample has larger values than the second sample for a given cumulative probability. This indicates that the first sample has heavier tails.
- If the points lie below the line, it suggests that the data has lighter tails than the assumed distribution.
- If the P-P plot crosses the 45-degree line, it suggests that the distributional characteristics of the two samples differ at the point of intersection.
- An S-shaped curve in the P-P plot may indicate differences in the central portion of the distributions.

If the data conforms to normality, clicking the "Submit" button will display a new window with a default power curve (Fig. 10.68). However, users have the option to specify the false discovery rate (FDR) and the maximum sample size per group. Afterward, then can click on the "Submit" button to obtain a new curve.

A power curve typically plots power against varying sample sizes or effect sizes. An upward-sloping curve indicates how power increases with larger sample sizes, as shown in Fig. 10.68. Users can determine the sample size at which the power curve reaches a satisfactory level. In the given example, 40 samples in each group would be required for the study to have an 80% chance of concluding with a statistically significant effect.

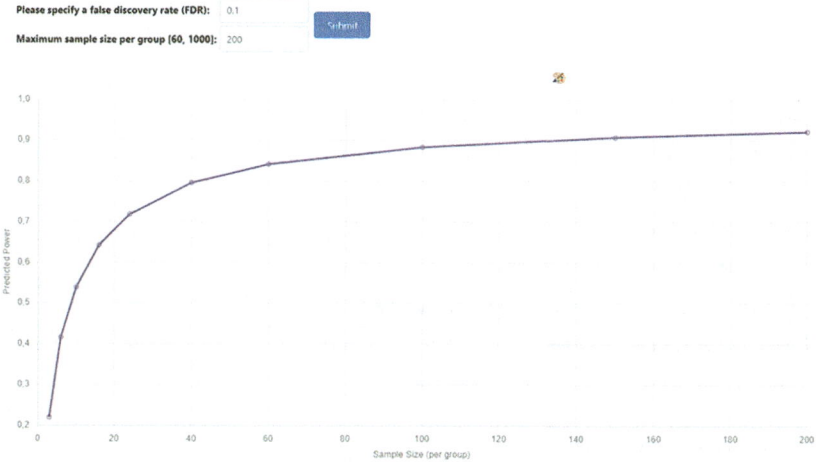

Fig. 10.68 Predicted power as a function of number of samples in each group

10.8 Biomarker Analysis

After selecting and quantifying the significant metabolites and identifying the altered pathways, the subsequent step involves validating the biomarkers to affirm their association with the specific condition or physiological process.

Biomarkers are assessed for their clinical utility, including their sensitivity, specificity, positive predictive value, and negative predictive value. These metrics help determine how well a biomarker can accurately identify the presence or absence of a particular condition. The gold standard method for obtaining these measures is through ROC analysis [55].

Clicking the "Biomarker Analysis" module opens the window to upload data, which follows the data integrity checking and normalization step as discussed in Sects. 10.3.2.1 and 10.3.3. In this step, it is possible to include the top 20 ratios to be assessed as potential biomarkers, as seen in Fig. 10.69.

Upon clicking the "Proceed" button, a new window will appear, presenting three available options for performing the biomarker analysis:

- **Classical Univariate ROC analysis (single biomarker analysis)**
- **Multivariate ROC curve-based exploratory analysis (multi-biomarker analysis):** involves the automated identification and evaluation of features using multivariate algorithms such as support vector machine (SVM), random forest, or PLS-DA.
- **ROC curve-based model evaluation:** offers users the flexibility to select any combination of biomarkers. Furthermore, users are allowed to upload a holdout unlabeled sample for testing the model.

Sample normalization

○ None
○ Sample-specific normalization (i.e. weight, volume)   Specify
○ Normalization by sum
⦿ Normalization by median
○ Normalization by a reference sample (PQN)   Specify
○ Normalization by a pooled sample from group (group PQN) Specify
○ Normalization by reference feature   Specify
○ Quantile normalization (suggested only for > 1000 features)

☑ **Compute and include metabolite ratios:** Top 20 ⌄

Fig. 10.69 Setting the type of normalization and determining the number of top ratios to be assessed as biomarkers

10.8 Biomarker Analysis

10.8.1 Classical Univariate ROC Analysis

Selecting this option will open a new window with the results (Fig. 10.70). Users can choose the method for calculating the optimal cutoff by selecting one of the two available options and clicking the "Update" button to visualize the result. The optimal cutoff point is the one that balances sensitivity and specificity, and the chosen method for its calculation often depends on the specific goals and requirements of the application.

- **Closest to top-left corner (Euclidean Distance):** This approach is applicable when dealing with a binary classification problem, and the model generates continuous probability scores for each instance. Converting these probability scores into binary predictions requires setting a cutoff threshold. Euclidean distance can be employed to identify the threshold that minimizes the overall distance between two points: the point (0, 1) representing perfect sensitivity and specificity, and the point (sensitivity, 1-specificity) on the ROC curve.
- **Farthest to diagonal line (Youden index):** This approach entails identifying the point with the furthest vertical distance from the diagonal line (representing random chance) with a slope of 1 in the coordinate (0, 1) direction. The Youden index (sensitivity + specificity − 1) is used to pinpoint the cutoff point that maximizes the difference between the true positive rate (sensitivity) and the false positive rate (1 − specificity) [55, 127].

Clicking "View" in the ROC curve column (Fig. 10.70) displays the ROC graph and a boxplot of the concentration of the selected feature (Fig. 10.71) under the analysed conditions. Clicking the "Details" button reveals a table containing values

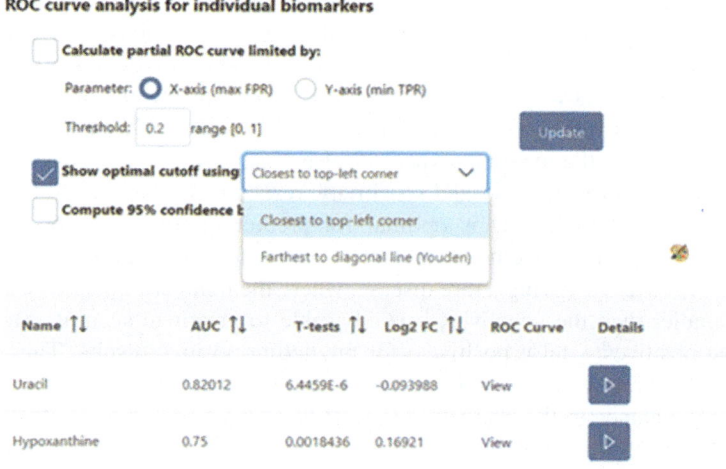

Fig. 10.70 Univariate ROC curve analysis

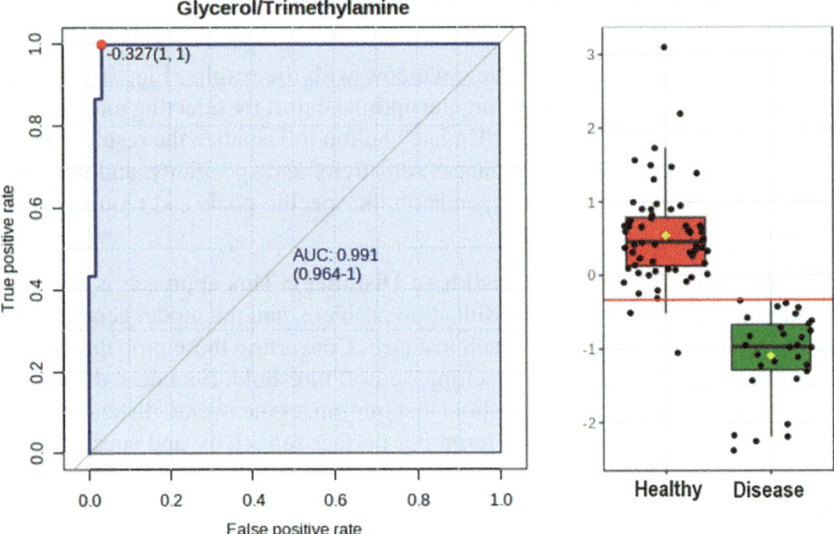

Fig. 10.71 ROC curve analysis of the glycerol/trimethylamine ratio (left) and box plots of the ratio concentrations within the analyzed cohort. The AUC is expressed with its confidence interval (95% CI), and the optimal cutoff (specificity, sensitivity) is indicated by the red notch closest to the top-left corner

for specificity, sensitivity, positive, and negative likelihood ratios based on the specified cutoff, as discussed before in Chap. 2, Sect. 2.4.3.

When models exhibit similar AUC values upon comparison, users have the option to construct a partial AUC at a specific threshold to discern sensitivity or specificity differences [127]. To utilize this attribute, users can click the "Calculate partial ROC curve" button (Fig. 10.70) and select the axis to be constrained. In the provided example in Sect. 2.4.3, users should opt for the Y-axis (representing the minimum expected true positive rate) and input a threshold value of 0.8 to constrain sensitivity at the point where the compared ROC curves intersect. To constrain specificity at 0.9, users must select the X-axis option (X-axis representing 1-specificity = 0.1, the maximal expected false positive ratio) and observe the new values obtained for specificity and sensitivity, respectively. The method exhibiting the highest sensitivity value is optimal for predicting true positives, while the method with greater specificity is better suited for predicting true negatives. However, in cases involving a rare disease, where the number of positives is significantly smaller than the negatives, it is advisable to prioritize accuracy (true positives/true positives + false positives) for predicting positive results. This measure avoids incorporating the number of negatives into the calculation, preventing a potential imbalance in the evaluation [55]. If there is no preference between sensitivity and specificity, or if both are equally important, then the most reasonable approach is to maximize them both.

10.8 Biomarker Analysis

10.8.2 Multivariate ROC Curve-Based Exploratory Analysis

Selecting the "Multivariate ROC curve" option will open a new window where the user can specify the classification method (Linear SVM, PLS-DA, or Random Forest) and choose the feature ranking method, as illustrated in Fig. 10.72. If the user opts for PLS-DA, they must specify the number of latent variables.

After clicking the "Submit" button, a new window will appear, presenting the results through various types of graphs (Fig. 10.73). In the "ROC View" tab, the default display is the ROC curve derived from all models averaged across all cross-validation runs, as depicted in Fig. 2.11. To view the ROC curve for a specific model, users can select the desired model in the "Select a model" tab. For a detailed examination of a single model, the confidence interval can be displayed by selecting the respective tab and clicking the "Update" button, as demonstrated in Fig. 10.73.

The "Prob. View" tab displays the predicted class probabilities plot, as detailed in Sect. 2.4.3, and illustrated in Fig. 2.8. Meanwhile, the "Pred. View" tab showcases the model's performance, specifically predictive accuracy, evaluated in relation to the number of predictors (Fig. 10.74). In the provided example, the optimal model was achieved with two predictors (highlighted by the red point in the graph), yielding an accuracy of 88.4%.

The "Sig. Features" tab presents a graph displaying features selected either by "mean importance measure" or "frequencies of being selected," as illustrated in (Fig. 10.75). Users have the option to specify the number of top features to be displayed in the graph by entering a value in the "Display features of top" box. Clicking on the spreadsheet icon will reveal a table containing the rank frequency and importance of the selected features.

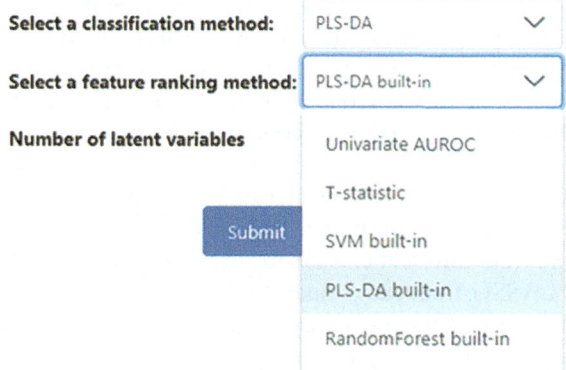

Fig. 10.72 Setting parameters for multivariate ROC analysis

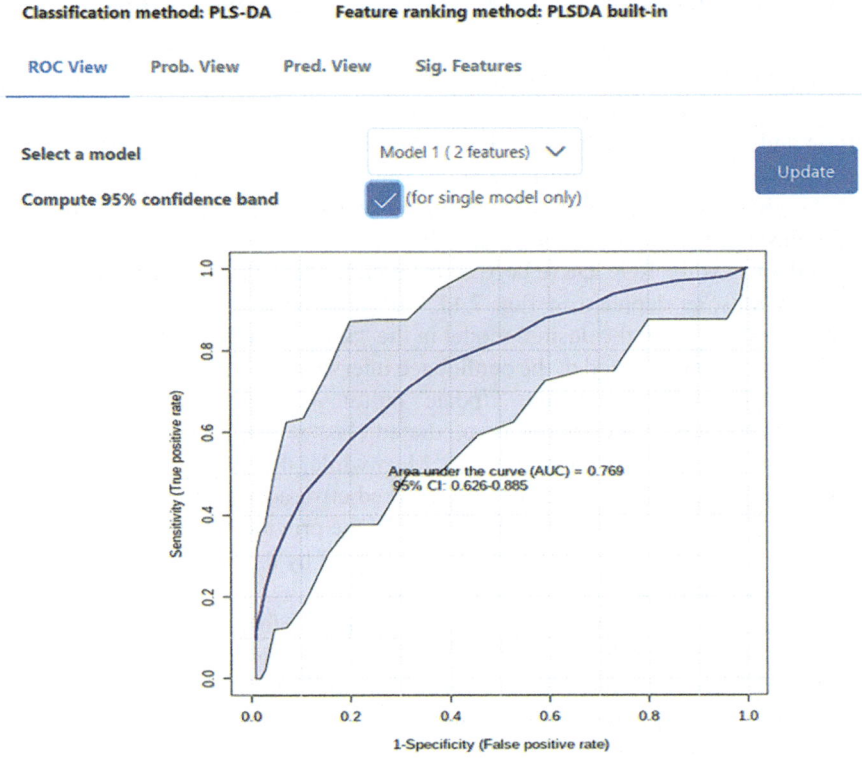

Fig. 10.73 ROC curve for a model considering 2 features, showing the AUC and its 95% confidence intervals

10.8.3 ROC Curve-Based Model Evaluation

Selecting this option will open a new window with three distinct actions. In the "Variable Selection" tab, users can manually choose the biomarkers that will constitute the model by clicking on the checkbox in the first column of the table. This selection is based on various criteria, including statistical measures (e.g., AUC, t-test, VIP), fold change, biological context, and the user's own judgment (Fig. 10.76).

The K-means clustering algorithm (refer to Sect. 10.4.3.2) categorizes features into groups with similar behavior. In the "KM cluster" column, users are prompted to select predictors from different clusters to reduce redundancy.

Clicking on "LASSO frequency" opens a window displaying a table with the selected variables (biomarkers) and their respective percentage of frequency in the LASSO modeling. This frequency indicates how often each variable is chosen or retained by LASSO across multiple runs or iterations. Higher LASSO frequency for

10.8 Biomarker Analysis

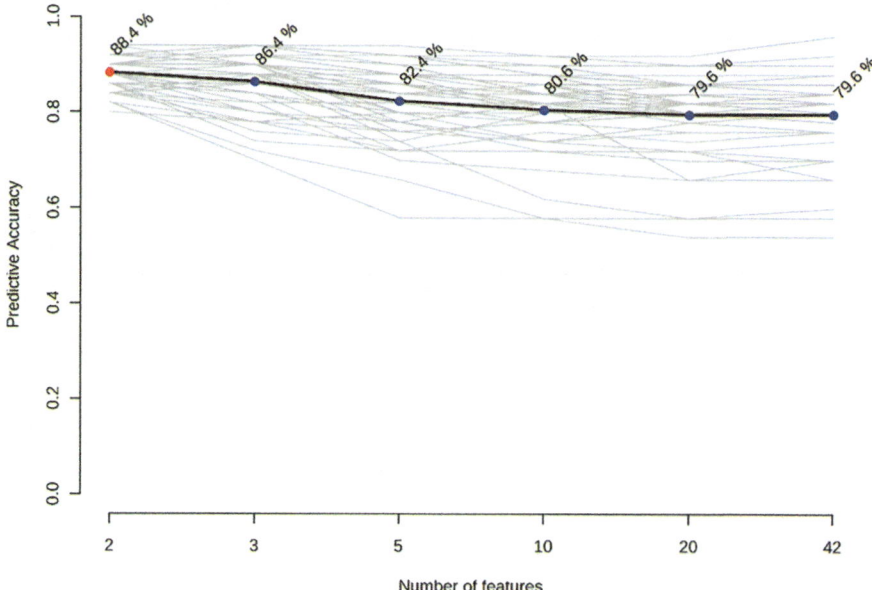

Fig. 10.74 Model performance (predictive accuracy) assessed as a function of the number of predictors. The optimal model is highlighted with a red notch

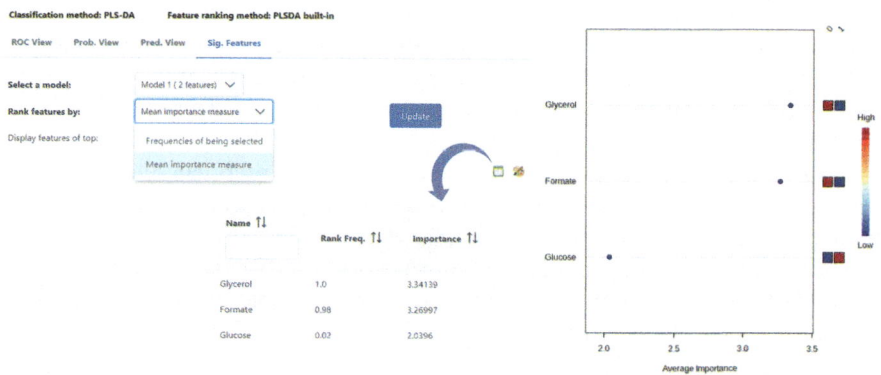

Fig. 10.75 Feature signature obtained based on the mean importance measure

a variable suggests that it is considered important by the algorithm and tends to be included in the selected subset of features.

LASSO, short for Least Absolute Shrinkage and Selection Operator, is an optimization algorithm widely used in scenarios where there are many features. It selects a subset of the most relevant features for the model, enhancing interpretability and potentially improving generalizability [128].

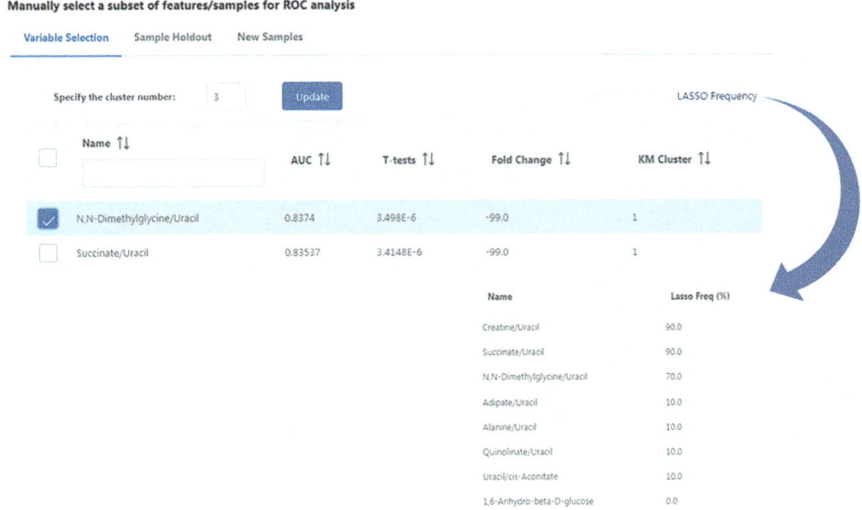

Fig. 10.76 Selecting the biomarkers for building a model

The key idea behind LASSO is to introduce a penalty term, known as L1 regularization, into the standard linear regression cost function. This penalty is proportional to the absolute values of the coefficients associated with the regression variables (the features). The L1 regularization encourages the algorithm to shrink some coefficients exactly to zero, effectively eliminating the corresponding features from the model. When all values are reduced to zero, it may indicate either unidentified errors or the presence of too many features (e.g., over 500 with a small sample size) for LASSO computation. Careful consideration of the model's output and validation procedures is essential in such cases.

Clicking the "Next" button opens a new window where users can select the learning algorithm, perform cross-validation, conduct a permutation test, and carry out predictions (testing) on a new set of samples (Fig. 10.77).

The box plot depicting predictive accuracy from cross-validation visually represents the dispersion of model performance across multiple folds or iterations. It summarizes some characteristics:

- **Median (Centerline of the box)**

The line inside the box represents the median predictive accuracy. It gives an idea of the central tendency of the accuracy values.

- **Box (Interquartile range—IQR)**

The box itself represents the IQR, which is the range between the first quartile (25th percentile) and the third quartile (75th percentile). It indicates the spread of the middle 50% of the accuracy values.

A larger box suggests greater variability in model performance across folds.

10.8 Biomarker Analysis

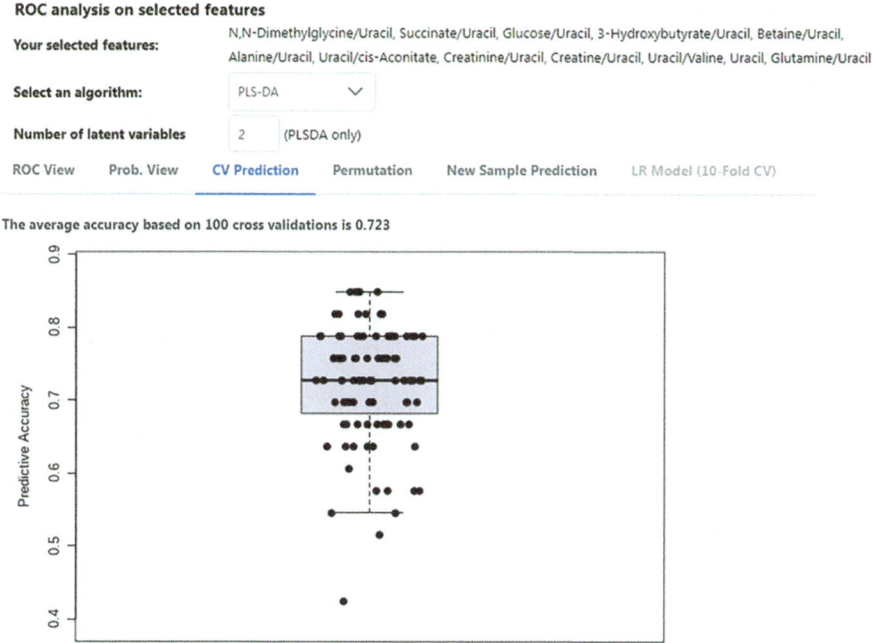

Fig. 10.77 Cross validation prediction assessed by average accuracy (72.3%) using 100 iterations

- **Whiskers**

The whiskers extend from the box to the data points that fall within a certain distance from the quartiles.

They show the range of accuracy values outside the IQR, providing an indication of the overall spread of the data or variability in predictive accuracy.

- **Outliers (Data points beyond whiskers)**

Individual data points beyond the whiskers are considered outliers. These are values that fall significantly outside the typical range and may represent extreme cases of model performance.

A narrow box and short whiskers suggest that the model's performance is relatively consistent across different folds or iterations of cross-validation. If the box is wide and the whiskers are long, it may indicate greater variability, and the model's performance could be less stable across folds.

When comparing multiple box plots, a higher median or a higher box might suggest a more accurate or stable model.

Clicking the "Permutation" tab, opens a new window to select the performance measure (predictive accuracy or AUC) and the number of permutations (Fig. 10.78).

A frequency graph resulting from a permutation test shows the distribution of a performance metric under the null hypothesis that there is no true effect or difference. It encapsulates some elements:

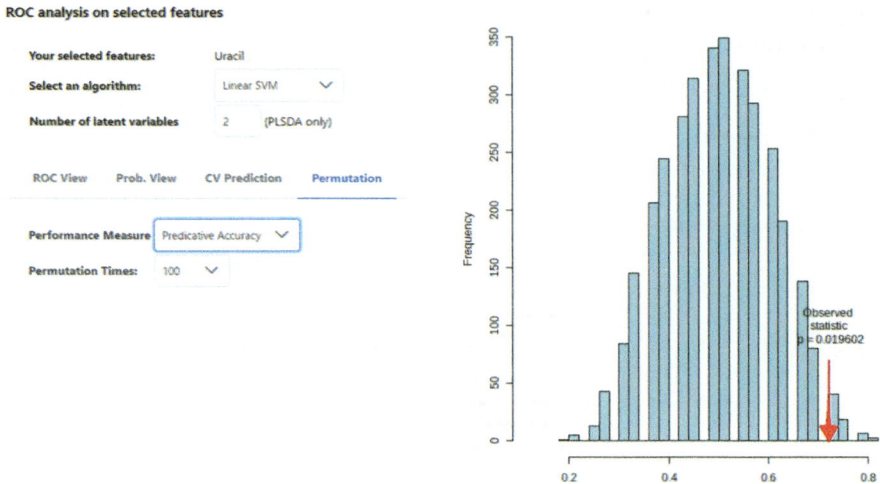

Fig. 10.78 Frequency graph from a permutation test using predictive accuracy and 100 permutations ($p = 0.0196$)

- **The axes**

The x-axis typically represents the values of the performance metric assessed. The y-axis represents the frequency or count of occurrences for each value on the x-axis.

- **The observed value (actual performance metric)**

A vertical line or marker (red arrow) on the graph indicates the observed value of the performance metric obtained from the actual data or model. This is the value to be compared with the distribution under the null hypothesis.

- **The null distribution**

The main purpose of the permutation test is to create a null distribution of the performance metric. To simulate the null distribution, the permutation test involves shuffling or randomly permuting the data labels or assignments multiple times. This creates a set of scenarios where the relationship between the variables is effectively destroyed, and any observed effect is solely a result of chance. The shape of the null distribution reflects the range of possible values the metric can take purely by random variation. If the null hypothesis is true, the metric should follow a certain pattern in terms of its distribution. This pattern is characterized by the central tendency (e.g., mean or median) and the variability (spread) of the metric when no true effect is present. A narrow or skewed null distribution may indicate that the metric is sensitive to the observed effect, making deviations from the null distribution more meaningful. A wider distribution suggests higher variability and a greater likelihood of observing a similar result by random chance.

10.8 Biomarker Analysis

- **Comparison with the null distribution**

  The observed value is compared with the null distribution. If the observed value falls within the central portion of the null distribution, it suggests that the result is consistent with what one would expect by random chance. If the observed value is in the tail or extreme end of the null distribution, it suggests that the result is unlikely to occur by random chance alone—that is the goal!

- ***p*-Value calculation**

  The *p*-value is often calculated as the proportion of values in the null distribution that are equal to or more extreme than the observed value. A smaller *p*-value indicates stronger evidence against the null hypothesis. To assess statistical significance, a significance level (e.g., 0.05) is chosen. If the observed value falls in the tail of the null distribution beyond the selected significance level, it is deemed statistically significant. This was the case in the given example (Fig. 10.78).

  Clicking the "New Sample Prediction" tab opens a new window (Fig. 10.79) that allows users to choose the learning algorithm for analyzing uploaded data without labels. Depending on the algorithm selected, the prediction probability can vary significantly.

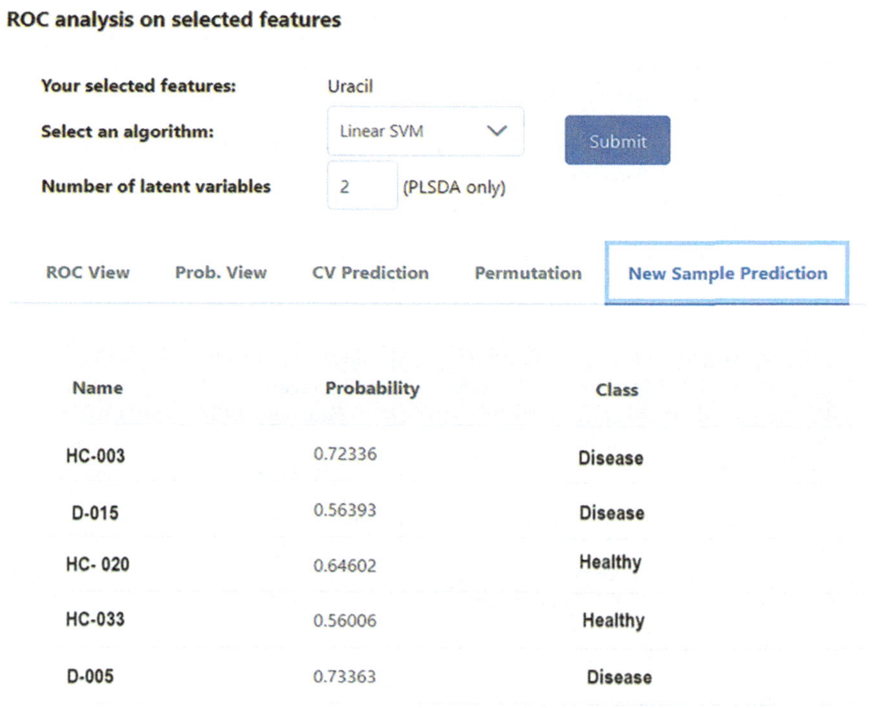

ROC analysis on selected features

Your selected features:	Uracil	
Select an algorithm:	Linear SVM	Submit
Number of latent variables	2 (PLSDA only)	

| ROC View | Prob. View | CV Prediction | Permutation | **New Sample Prediction** |

Name	Probability	Class
HC-003	0.72336	Disease
D-015	0.56393	Disease
HC-020	0.64602	Healthy
HC-033	0.56006	Healthy
D-005	0.73363	Disease

Fig. 10.79 Prediction of new samples ($n = 10$) using SVM to test the model

After reviewing the results and tallying the number of correctly and misclassified samples in the testing dataset, users can calculate sensitivity, specificity, positive predictive value, and negative predictive value. This allows for comparison with the values obtained in the training dataset and assesses its ability to generalize.

References

1. Danzi F, Pacchiana R, Mafficini A, Scupoli MT, Scarpa A, Donadelli M, Fiore A (2023) To metabolomics and beyond: a technological portfolio to investigate cancer metabolism. Signal Transduct Target Ther 8:137. https://doi.org/10.1038/s41392-023-01380-0
2. Wishart DS (2019) Metabolomics for investigating physiological and pathophysiological processes. Physiol Rev 99:1819–1875. https://doi.org/10.1152/physrev.00035.2018
3. Xia J (2017) Computational strategies for biological interpretation of metabolomics data. Adv Exp Med Biol 965:191–206. https://doi.org/10.1007/978-3-319-47656-8_8
4. Sussulini A (ed) (2017) Metabolomics: from fundamentals to clinical applications. Springer
5. Villas-Bôas SG, Rasmussen S, Lane GA (2005) Metabolomics or metabolite profiles? Trends Biotechnol 23:385–386. https://doi.org/10.1016/j.tibtech.2005.05.009
6. Podwojski K, Fritsch A, Chamrad DC et al (2009) Retention time alignment algorithms for LC/MS data must consider non-linear shifts. Bioinformatics 25:758–764. https://doi.org/10.1093/bioinformatics/btp052
7. MZmine 3. https://mzmine.github.io/. Accessed 1 Jan 2024
8. Tsugawa H, Cajka T, Kind T, Ma Y, Higgins B, Ikeda K, Kanazawa M, VanderGheynst J, Fiehn O, Arita M (2015) MS-DIAL: data-independent MS/MS deconvolution for comprehensive metabolome analysis. Nat Methods 12:523–526. https://doi.org/10.1038/nmeth.3393
9. Doshi M ProteoWizard. https://proteowizard.sourceforge.io. Accessed 1 Jan 2024
10. (2020) File conversion for SIRIUS 4 – MSconvert (ProteoWizard)
11. Holman JD, Tabb DL, Mallick P (2014) Employing ProteoWizard to convert raw mass spectrometry data. Curr Protoc Bioinformatics 46:13.24.1–13.24.9. https://doi.org/10.1002/0471250953.bi1324s46
12. File Conversion (Waters). https://ccms-ucsd.github.io/GNPSDocumentation/fileconversion_waters/. Accessed 1 Jan 2024
13. Smith R, Mathis AD, Ventura D, Prince JT (2014) Proteomics, lipidomics, metabolomics: a mass spectrometry tutorial from a computer scientist's point of view. BMC Bioinf 15(Suppl 7):S9
14. Schmid M, Rath D, Diebold U (2022) Why and how Savitzky-Golay filters should be replaced. ACS Meas Sci Au 2:185–196. https://doi.org/10.1186/1471-2105-15-S7-S9
15. Savitzky A, Golay MJE (1964) Smoothing and differentiation of data by simplified least squares procedures. Anal Chem 36:1627–1639. https://doi.org/10.1021/ac60214a047
16. Xia J, Sinelnikov IV, Han B, Wishart DS (2015) MetaboAnalyst 3.0 – making metabolomics more meaningful. Nucleic Acids Res 43:W251–W257. https://doi.org/10.1093/nar/gkv380
17. Pang Z, Zhou G, Ewald J, Chang L, Hacariz O, Basu N, Xia J (2022) Using MetaboAnalyst 5.0 for LC-HRMS spectra processing, multi-omics integration and covariate adjustment of global metabolomics data. Nat Protoc 17:1735–1761. https://doi.org/10.1038/s41596-022-00710-w
18. Wieder C, Frainay C, Poupin N, Rodríguez-Mier P, Vinson F, Cooke J, Lai RP, Bundy JG, Jourdan F, Ebbels T (2021) Pathway analysis in metabolomics: recommendations for the use of over-representation analysis. PLoS Comput Biol 17:e1009105. https://doi.org/10.1371/journal.pcbi.1009105
19. Burkov A (2019) The hundred-page machine learning bookISBN 1999579518, 9781999579517. Published by Andriy Burkov.

20. Vatcheva KP, Lee M, McCormick JB, Rahbar MH (2016) Multicollinearity in regression analyses conducted in epidemiologic studies. Epidemiology 6:227. https://doi.org/10.4172/2161-1165.1000227
21. Tautenhahn R, Böttcher C, Neumann S (2008) Highly sensitive feature detection for high resolution LC/MS. BMC Bioinf 9:504. https://doi.org/10.1186/1471-2105-9-504
22. Godzien J, Gil de la Fuente A, Otero A, Barbas C (2018) Metabolite annotation and identification. In: Comprehensive analytical chemistry. Elsevier, pp 415–445. https://doi.org/10.1016/bs.coac.2018.07.004
23. Sarker IH (2021) Machine learning: algorithms, real-world applications and research directions. SN Comput Sci 2:160. https://doi.org/10.1007/s42979-021-00592-x
24. Jang C, Chen L, Rabinowitz JD (2018) Metabolomics and isotope tracing. Cell 173:822–837. https://doi.org/10.1016/j.cell.2018.03.055
25. Smith R, Prince JT, Ventura D (2015) A coherent mathematical characterization of isotope trace extraction, isotopic envelope extraction, and LC-MS correspondence. BMC Bioinf 16(Suppl 7):S1. https://doi.org/10.1186/1471-2105-16-S7-S1
26. Barea-Sepúlveda M, Duarte H, Aliaño-González MJ, Romano A, Medronho B (2022) Total ion chromatogram and total ion mass spectrum as alternative tools for detection and discrimination (a review). Chemosensors (Basel) 10:465. https://doi.org/10.3390/chemosensors10110465
27. Adutwum LA, Abel RJ, Harynuk J (2018) Total ion spectra versus segmented total ion spectra as preprocessing tools for gas chromatography – mass spectrometry data. J Forensic Sci 63:1059–1068. https://doi.org/10.1111/1556-4029.13657
28. Renner G, Reuschenbach M (2023) Critical review on data processing algorithms in nontarget screening: challenges and opportunities to improve result comparability. Anal Bioanal Chem 415:4111–4123. https://doi.org/10.1007/s00216-023-04776-7
29. Smith CA, Want EJ, O'Maille G, Abagyan R, Siuzdak G (2006) XCMS: processing mass spectrometry data for metabolite profiling using nonlinear peak alignment, matching, and identification. Anal Chem 78:779–787. https://doi.org/10.1021/ac051437y
30. Li S, Siddiqa A, Thapa M, Chi Y, Zheng S (2023) Trackable and scalable LC-MS metabolomics data processing using asari. Nat Commun 14:4113. https://doi.org/10.1038/s41467-023-39889-1
31. Conley CJ, Smith R, Torgrip RJO, Taylor RM, Tautenhahn R, Prince JT (2014) Massifquant: open-source Kalman filter-based XC-MS isotope trace feature detection. Bioinformatics 30:2636–2643. https://doi.org/10.1093/bioinformatics/btu359
32. Pang Z, Chong J, Li S, Xia J (2020) MetaboAnalystR 3.0: toward an optimized workflow for global metabolomics. Metabolites. https://doi.org/10.3390/metabo10050186
33. (2023) SetPeakParam: set parameters for peak profiling and parameters optimization in xia-lab/OptiLCMS: optimized LC-MS spectra processing. https://rdrr.io/github/xia-lab/OptiLCMS/man/SetPeakParam.html. Accessed 1 Jan 2024
34. Mehl F, Gallart-Ayala H, Konz I, Teav T, Oikonomidi A, Peyratout G, van der Velpen V, Popp J, Ivanisevic J (2018) LC-HRMS data as a result of untargeted metabolomic profiling of human cerebrospinal fluid. Data Brief 21:1358–1362. https://doi.org/10.1016/j.dib.2018.10.113
35. Zhang Y, Liang H, Liu Y, Zhao M, Xu Q, Liu Z, Weng X (2021) Metabolomic analysis and identification of sperm freezability-related metabolites in boar seminal plasma. Animals (Basel) 11:1939. https://doi.org/10.3390/ani11071939
36. Gowda H, Ivanisevic J, Johnson CH et al (2014) Interactive XCMS online: simplifying advanced metabolomic data processing and subsequent statistical analyses. Anal Chem 86:6931–6939. https://doi.org/10.1021/ac500734c
37. Quintas G (2020) Targeted vs untargeted MS2 data-dependent acquisition for automated peak annotation in LC-MS metabolomics. Metabolites. https://doi.org/10.17632/FNZBXMKV83.1
38. Rafiei A, Sleno L (2015) Comparison of peak-picking workflows for untargeted liquid chromatography/high-resolution mass spectrometry metabolomics data analysis. Rapid Commun Mass Spectrom 29:119–127. https://doi.org/10.1002/rcm.7094

39. Kuhl C, Tautenhahn R, Böttcher C, Larson TR, Neumann S (2012) CAMERA: an integrated strategy for compound spectra extraction and annotation of liquid chromatography/mass spectrometry data sets. Anal Chem 84:283–289. https://doi.org/10.1021/ac202450g
40. Prince JT, Marcotte EM (2006) Chromatographic alignment of ESI-LC-MS proteomics data sets by ordered bijective interpolated warping. Anal Chem 78:6140–6152. https://doi.org/10.1021/ac0605344
41. Cleveland WS (1979) Robust locally weighted regression and smoothing scatterplots. J Am Stat Assoc 74:829. https://doi.org/10.1080/01621459.1979.10481038
42. Liu Q, Walker D, Uppal K, Liu Z, Ma C, Tran V, Li S, Jones DP, Yu T (2020) Addressing the batch effect issue for LC/MS metabolomics data in data preprocessing. Sci Rep 10:13856. https://doi.org/10.1038/s41598-020-70850-0
43. Di Guida R, Engel J, Allwood JW, Weber RJM, Jones MR, Sommer U, Viant MR, Dunn WB (2016) Non-targeted UHPLC-MS metabolomic data processing methods: a comparative investigation of normalisation, missing value imputation, transformation and scaling. Metabolomics 12:93. https://doi.org/10.1007/s11306-016-1030-9
44. Wei R, Wang J, Su M, Jia E, Chen S, Chen T, Ni Y (2018) Missing value imputation approach for mass spectrometry-based metabolomics data. Sci Rep 8:663. https://doi.org/10.1038/s41598-017-19120-0
45. Vinaixa M, Samino S, Saez I, Duran J, Guinovart JJ, Yanes O (2012) A guideline to univariate statistical analysis for LC/MS-based untargeted metabolomics-derived data. Metabolites 2:775–795. https://doi.org/10.3390/metabo2040775
46. Reisdorph NA, Walmsley S, Reisdorph R (2019) A perspective and framework for developing sample type specific databases for LC/MS-based clinical metabolomics. Metabolites 10:8. https://doi.org/10.3390/metabo10010008
47. (2023) SetAnnotationParam: Set annotation parameters in xia-lab/OptiLCMS: Optimized LC-MS Spectra Processing. https://rdrr.io/github/xia-lab/OptiLCMS/man/SetAnnotationParam.html. Accessed 1 Jan 2024
48. Stancliffe E, Schwaiger-Haber M, Sindelar M, Patti GJ (2021) DecoID improves identification rates in metabolomics through database-assisted MS/MS deconvolution. Nat Methods 18:779–787. https://doi.org/10.1038/s41592-021-01195-3
49. Li Y, Kind T, Folz J, Vaniya A, Mehta SS, Fiehn O (2021) Spectral entropy outperforms MS/MS dot product similarity for small-molecule compound identification. Nat Methods 18:1524–1531. https://doi.org/10.1038/s41592-021-01331-z
50. Blaženović I, Kind T, Ji J, Fiehn O (2018) Software tools and approaches for compound identification of LC-MS/MS data in metabolomics. Metabolites. https://doi.org/10.3390/metabo8020031
51. Masuya H, Makita Y, Kobayashi N et al (2011) The RIKEN integrated database of mammals. Nucleic Acids Res 39:D861–D870. https://doi.org/10.1093/nar/gkq1078
52. Sawada Y, Nakabayashi R, Yamada Y et al (2012) RIKEN tandem mass spectral database (ReSpect) for phytochemicals: a plant-specific MS/MS-based data resource and database. Phytochemistry 82:38–45. https://doi.org/10.1016/j.phytochem.2012.07.007
53. Lee Y, Hwang S, Seo M, Shin KB, Kim KH, Park GW, Kim JY, Yoo JS, No KT (2020) BMDMS-NP: a comprehensive ESI-MS/MS spectral library of natural compounds. Phytochemistry 177:112427. https://doi.org/10.1016/j.phytochem.2020.112427
54. Jeffryes JG, Colastani RL, Elbadawi-Sidhu M, Kind T, Niehaus TD, Broadbelt LJ, Hanson AD, Fiehn O, Tyo KEJ, Henry CS (2015) MINEs: open access databases of computationally predicted enzyme promiscuity products for untargeted metabolomics. J Cheminform 7:44. https://doi.org/10.1186/s13321-015-0087-1
55. Xia J, Broadhurst DI, Wilson M, Wishart DS (2013) Translational biomarker discovery in clinical metabolomics: an introductory tutorial. Metabolomics 9:280–299. https://doi.org/10.1007/s11306-012-0482-9
56. Gromski PS, Xu Y, Kotze HL, Correa E, Ellis DI, Armitage EG, Turner ML, Goodacre R (2014) Influence of missing values substitutes on multivariate analysis of metabolomics data. Metabolites 4:433–452. https://doi.org/10.3390/metabo4020433

57. Walach J, Filzmoser P, Hron K (2018) Data normalization and scaling: consequences for the analysis in omics sciences. In: Comprehensive analytical chemistry. Elsevier, pp 165–196. https://doi.org/10.1016/bs.coac.2018.06.004
58. Schiffman C, Petrick L, Perttula K, Yano Y, Carlsson H, Whitehead T, Metayer C, Hayes J, Rappaport S, Dudoit S (2019) Filtering procedures for untargeted LC-MS metabolomics data. BMC Bioinf 20:334. https://doi.org/10.1186/s12859-019-2871-9
59. Xia J, Wishart DS (2011) Web-based inference of biological patterns, functions and pathways from metabolomic data using metaboanalyst. Nat Protoc 6:743–760. https://doi.org/10.1038/nprot.2011.319
60. Ejigu BA, Valkenborg D, Baggerman G, Vanaerschot M, Witters E, Dujardin J-C, Burzykowski T, Berg M (2013) Evaluation of normalization methods to pave the way towards large-scale LC-MS-based metabolomics profiling experiments. OMICS 17:473–485. https://doi.org/10.1089/omi.2013.0010
61. Filzmoser P, Walczak B (2014) What can go wrong at the data normalization step for identification of biomarkers? J Chromatogr A 1362:194–205. https://doi.org/10.1016/j.chroma.2014.08.050
62. Kvalheim OM, Brakstad F, Liang Y (1994) Preprocessing of analytical profiles in the presence of homoscedastic or heteroscedastic noise. Anal Chem 66:43–51. https://doi.org/10.1021/ac00073a010
63. Dieterle F, Ross A, Schlotterbeck G, Senn H (2006) Probabilistic quotient normalization as robust method to account for dilution of complex biological mixtures. Application in 1H NMR metabonomics. Anal Chem 78:4281–4290. https://doi.org/10.1021/ac051632c
64. Correia GDS, Takis PG, Sands CJ et al (2022) H NMR signals from urine excreted protein are a source of bias in probabilistic quotient normalization. Anal Chem 94:6919–6923. https://doi.org/10.1021/acs.analchem.2c00466
65. Välikangas T, Suomi T, Elo LL (2018) A systematic evaluation of normalization methods in quantitative label-free proteomics. Brief Bioinform 19:1–11. https://doi.org/10.1093/bib/bbw095
66. Hicks SC, Irizarry RA (2015) Quantro: a data-driven approach to guide the choice of an appropriate normalization method. Genome Biol 16:117. https://doi.org/10.1186/s13059-015-0679-0
67. Karaman I (2017) Preprocessing and pretreatment of metabolomics data for statistical analysis. Adv Exp Med Biol 965:145–161. https://doi.org/10.1007/978-3-319-47656-8_6
68. Zhou B, Xiao JF, Tuli L, Ressom HW (2012) LC-MS-based metabolomics. Mol BioSyst 8:470–481. https://doi.org/10.1039/c1mb05350g
69. Liu X, Fang Y, Ma H, Zhang N, Li C (2023) Performance comparison of three scaling algorithms in NMR-based metabolomics analysis. Open Life Sci 18:20220556. https://doi.org/10.1515/biol-2022-0556
70. van den Berg RA, Hoefsloot HCJ, Westerhuis JA, Smilde AK, van der Werf MJ (2006) Centering, scaling, and transformations: improving the biological information content of metabolomics data. BMC Genomics 7:142. https://doi.org/10.1186/1471-2164-7-142
71. Smilde AK, van der Werf MJ, Bijlsma S, van der Werff-van der Vat BJC, Jellema RH (2005) Fusion of mass spectrometry-based metabolomics data. Anal Chem 77:6729–6736. https://doi.org/10.1021/ac051080y
72. Brown S, Tauler R, Walczak B (2020) Comprehensive chemometrics: chemical and biochemical data analysis. Elsevier
73. Varmuza K, Filzmoser P (2008) Introduction to multivariate statistical analysis in chemometrics. CRC Press. https://doi.org/10.1201/9781420059496
74. Jolliffe IT, Cadima J (2016) Principal component analysis: a review and recent developments. Philos Trans A Math Phys Eng Sci 374:20150202. https://doi.org/10.1098/rsta.2015.0202
75. Abdi H, Williams LJ (2010) Principal component analysis. Wiley Interdiscip Rev Comput Stat 2:433–459. https://doi.org/10.1002/wics.101
76. Gromski PS, Muhamadali H, Ellis DI, Xu Y, Correa E, Turner ML, Goodacre R (2015) A tutorial review: metabolomics and partial least squares-discriminant analysis – a marriage

of convenience or a shotgun wedding. Anal Chim Acta 879:10–23. https://doi.org/10.1016/j.aca.2015.02.012
77. James G, Witten D, Hastie T, Tibshirani R (2017) An introduction to statistical learning: with applications in R. Springer. Corr. 7th printing. ISBN-13 978-1461471370
78. Szymańska E, Saccenti E, Smilde AK, Westerhuis JA (2012) Double-check: validation of diagnostic statistics for PLS-DA models in metabolomics studies. Metabolomics 8:3–16. https://doi.org/10.1007/s11306-011-0330-3
79. Ghosh T, Zhang W, Ghosh D, Kechris K (2020) Predictive modeling for metabolomics data. Methods Mol Biol 2104:313–336. https://doi.org/10.1007/978-1-0716-0239-3_16
80. Bevilacqua M, Bro R (2020) Can we trust score plots? Metabolites 10:278. https://doi.org/10.3390/metabo10070278
81. Galindo-Prieto B, Eriksson L, Trygg J (2014) Variable influence on projection (VIP) for orthogonal projections to latent structures (OPLS). J Chemom 28:623–632. https://doi.org/10.1002/cem.2627
82. Worley B, Powers R (2013) Multivariate analysis in metabolomics. Curr Metabolomics 1:92–107. https://doi.org/10.2174/2213235X11301010092
83. Lindgren F, Hansen B, Karcher W, Sjöström M, Eriksson L (1996) Model validation by permutation tests: applications to variable selection. J Chemom 10:521–532. https://doi.org/10.1002/(SICI)1099-128X(199609)10:5/6<521::AID-CEM448>3.0.CO;2-J
84. Fattuoni C, Palmas F, Noto A et al (2016) Primary HCMV infection in pregnancy from classic data towards metabolomics: an exploratory analysis. Clin Chim Acta 460:23–32. https://doi.org/10.1016/j.cca.2016.06.005
85. Ojala M, Garriga GC (2010) Permutation tests for studying classifier performance. JMLR 11:1833–1886. https://www.jmlr.org/papers/volume11/ojala10a/ojala10a.pdf. Accessed 2 Jan 2024
86. Mehmood T, Liland KH, Snipen L, Sæbø S (2012) A review of variable selection methods in partial least squares regression. Chemom Intell Lab Syst 118:62–69. https://doi.org/10.1016/j.chemolab.2012.07.010
87. Sorochan Armstrong MD, de la Mata AP, Harynuk JJ (2022) Review of variable selection methods for discriminant-type problems in chemometrics. Front Anal Sci. https://doi.org/10.3389/frans.2022.867938
88. Nagpal A, Jatain A, Gaur D (2013) Review based on data clustering algorithms. In: 2013 IEEE conference on information and communication technologies. https://doi.org/10.1109/cict.2013.6558109
89. Rasyid LA, Andayani S (2018) Review on clustering algorithms based on data type: towards the method for data combined of numeric-fuzzy linguistics. J Phys Conf Ser 1097:012082. https://doi.org/10.1088/1742-6596/1097/1/012082
90. Harrington P (2012) Machine learning in action. Simon and Schuster
91. Bridges CC Jr (1966) Hierarchical cluster analysis. Psychol Rep 18:851–854. https://doi.org/10.2466/pr0.1966.18.3.851
92. Xia J, Psychogios N, Young N, Wishart DS (2009) MetaboAnalyst: a web server for metabolomic data analysis and interpretation. Nucleic Acids Res 37:W652–W660. https://doi.org/10.1093/nar/gkp356
93. Stanimirova I, Daszykowski M (2018) Exploratory analysis of metabolomic data. In: Comprehensive analytical chemistry. Elsevier, pp 227–264
94. Moseley B, Vassilvitskii S, Wang Y (2021) Hierarchical clustering in general metric spaces using approximate nearest neighbors. In: Proceedings of the 24th international conference on artificial intelligence and statistics (AISTATS), San Diego, California, USA. PMLR: volume 130. https://proceedings.mlr.press/v130/moseley21a/moseley21a.pdf. Accessed 2 Jan 2024
95. Gal J, Bailleux C, Chardin D et al (2020) Comparison of unsupervised machine-learning methods to identify metabolomic signatures in patients with localized breast cancer. Comput Struct Biotechnol J 18:1509–1524. https://doi.org/10.1016/j.csbj.2020.05.021
96. Webb AR, Copsey KD, Cawley G (2011) Statistical pattern recognition, 3rd edn. Wiley. https://doi.org/10.1002/9781119952954

97. Breiman L (2001) Random forests. Mach Learn 45:5–32. https://doi.org/10.102 3/A:1010933404324
98. Grissa D, Pétéra M, Brandolini M, Napoli A, Comte B, Pujos-Guillot E (2016) Feature selection methods for early predictive biomarker discovery using untargeted metabolomic data. Front Mol Biosci 3:30. https://doi.org/10.3389/fmolb.2016.00030
99. Chen T, Cao Y, Zhang Y, Liu J, Bao Y, Wang C, Jia W, Zhao A (2013) Random Forest in clinical metabolomics for phenotypic discrimination and biomarker selection. Evid Based Complement Alternat Med 2013:298183. https://doi.org/10.1155/2013/298183
100. James G, Witten D, Hastie T, Tibshirani R (2021) Resampling methods. In: Springer texts in statistics. Springer US, New York, pp 197–223. https://doi.org/10.1007/978-1-0716-1418-1_5
101. Rustam Z, Kharis SAA (2020) Comparison of support vector machine recursive feature elimination and Kernel function as feature selection using support vector machine for lung cancer classification. J Phys Conf Ser 1442:012027. https://doi.org/10.1088/1742-6596/1442/1/012027
102. Zhang X, Lu X, Shi Q, Xu X-Q, Leung H-CE, Harris LN, Iglehart JD, Miron A, Liu JS, Wong WH (2006) Recursive SVM feature selection and sample classification for mass-spectrometry and microarray data. BMC Bioinf 7:197. https://doi.org/10.1186/1471-2105-7-197
103. Hastie T, Tibshirani R, Friedman J (2013) The elements of statistical learning: data mining, inference, and prediction. Springer Science & Business Media
104. Kohavi R (1995) A study of cross validation and bootstrap for accuracy estimation and model selection. https://ai.stanford.edu/~ronnyk/accEst.pdf. Accessed 2 Jan 2024
105. Nakatsu RT (2023) Validation of machine learning ridge regression models using Monte Carlo, bootstrap, and variations in cross-validation. J Intell Syst. https://doi.org/10.1515/jisys-2022-0224
106. Steyerberg EW, Harrell FE Jr, Borsboom GJ, Eijkemans MJ, Vergouwe Y, Habbema JD (2001) Internal validation of predictive models: efficiency of some procedures for logistic regression analysis. J Clin Epidemiol 54:774–781. https://doi.org/10.1016/s0895-4356(01)00341-9
107. Molinaro AM, Simon R, Pfeiffer RM (2005) Prediction error estimation: a comparison of resampling methods. Bioinformatics 21:3301–3307. https://doi.org/10.1093/bioinformatics/bti499
108. Ortmayr K, Charwat V, Kasper C, Hann S, Koellensperger G (2016) Uncertainty budgeting in fold change determination and implications for non-targeted metabolomics studies in model systems. Analyst 142:80–90. https://doi.org/10.1039/C6AN01342B
109. Aguinis H, Vassar M, Wayant C (2021) On reporting and interpreting statistical significance and p values in medical research. BMJ Evid Based Med 26:39–42. https://doi.org/10.1136/bmjebm-2019-111264
110. Hayter AJ (1986) The maximum familywise error rate of fisher's least significant difference test. J Am Stat Assoc 81:1000. https://doi.org/10.2307/2289074
111. Proschan M (1997) Conditional power with Fisher's least significant difference procedure. Biometrika 84:197–208. https://www.jstor.org/stable/2337566
112. Nanda A, Mohapatra DBB, Mahapatra APK, Mahapatra APK, Mahapatra APK (2021) Multiple comparison test by Tukey's honestly significant difference (HSD): do the confident level control type I error. Int J Stat Appl Math 6:59–65. https://doi.org/10.22271/maths.2021.v6.i1a.636
113. Rosato A, Tenori L, Cascante M, De Atauri Carulla PR, Martins Dos Santos VAP, Saccenti E (2018) From correlation to causation: analysis of metabolomics data using systems biology approaches. Metabolomics 14:37. https://doi.org/10.1007/s11306-018-1335-y
114. Schober P, Boer C, Schwarte LA (2018) Correlation coefficients: appropriate use and interpretation. Anesth Analg 126:1763–1768. https://doi.org/10.1213/ANE.0000000000002864
115. Taylor R (1990) Interpretation of the correlation coefficient: a basic review. J Diagn Med Sonogr 6:35–39. https://doi.org/10.1177/875647939000600106
116. Xia J, Wishart DS (2010) MSEA: a web-based tool to identify biologically meaningful patterns in quantitative metabolomic data. Nucleic Acids Res 38:W71–W77. https://doi.org/10.1093/nar/gkq329

117. Marco-Ramell A, Palau-Rodriguez M, Alay A, Tulipani S, Urpi-Sarda M, Sanchez-Pla A, Andres-Lacueva C (2018) Evaluation and comparison of bioinformatic tools for the enrichment analysis of metabolomics data. BMC Bioinf 19:1. https://doi.org/10.1186/s12859-017-2006-0
118. Lu Y, Pang Z, Xia J (2023) Comprehensive investigation of pathway enrichment methods for functional interpretation of LC-MS global metabolomics data. Brief Bioinform 24:bbac553. https://doi.org/10.1093/bib/bbac553
119. Picart-Armada S, Fernández-Albert F, Vinaixa M, Rodríguez MA, Aivio S, Stracker TH, Yanes O, Perera-Lluna A (2017) Null diffusion-based enrichment for metabolomics data. PLoS One 12:e0189012. https://doi.org/10.1371/journal.pone.0189012
120. Wieder C, Lai RPJ, Ebbels TMD (2022) Single sample pathway analysis in metabolomics: performance evaluation and application. BMC Bioinf 23:481. https://doi.org/10.1186/s12859-022-05005-1
121. Balthazart J, McCormick C (2022) Statistical rules versus biological reasoning: some apparent conflicts and how to solve them. Horm Behav 137:104938. https://doi.org/10.1016/j.yhbeh.2021.104938
122. Hickey GL, Grant SW, Dunning J, Siepe M (2018) Statistical primer: sample size and power calculations – why, when and how? Eur J Cardiothorac Surg 54:4–9. https://doi.org/10.1093/ejcts/ezy169
123. Lovell DP (2020) Null hypothesis significance testing and effect sizes: can we "effect" everything … or … anything? Curr Opin Pharmacol 51:68–77. https://doi.org/10.1016/j.coph.2019.12.001
124. Chambers JM, Cleveland WS, Kleiner B, Tukey PA (2018) Graphical methods for data analysis. https://doi.org/10.1201/9781351072304
125. Cleveland W (1994) Elements Graphing Data – Ed2. AT&T Bell Laboratories. ISBN 9780963488411
126. DuToit SHC, Steyn AGW, Stumpf RH (2012) Graphical exploratory data analysis. Springer Science & Business Media
127. Nahm FS (2022) Receiver operating characteristic curve: overview and practical use for clinicians. Korean J Anesthesiol 75:25–36. https://doi.org/10.4097/kja.21209
128. Tibshirani R (1996) Regression shrinkage and selection via the Lasso. J R Statist Soc B 58:267–288. http://www.jstor.org/stable/2346178

Index

A
Area under the curve (AUC), 26, 29–31, 38, 39, 48, 66, 68, 69, 79, 80, 84, 88–90, 97–110, 112, 117, 120, 121, 126, 127, 130–132, 137, 140–143, 147, 176, 234, 236, 239

B
Biomarker, 2, 13, 38, 78, 96, 122, 125, 135, 147, 184
Biomarker analysis, 157, 160, 205, 232–242
Biomarker classification, 14–15
Biomarker statistical analysis, 22–31
Brain derived neurotrophic factor (BDNF), 3, 16, 57, 78, 125–133

C
Cannabinoid hypothesis, 5
Cognitive impairment (Cimp), 1, 56, 66, 67, 125, 136
Cortisol, 6, 7, 17, 65, 77, 78, 115–117, 120, 121

D
Diagnostic models, 15–23, 26–28, 38, 67, 69, 70, 89, 90, 96, 97, 110, 112, 126, 136, 143, 157–242
Dopamine hypothesis, 3
Dopaminergic, 3–5, 7, 18, 116, 125, 136, 142
Dysbiosis, 7, 19, 54, 95–97, 109, 148

E
Endocannabinoid (eCB), 5, 143
Epigenetic biomarker, 37–38

G
GABAergic, 4, 18, 55, 136, 142
Genetic biomarker, 37–38
Genetic hypothesis, 2
Glial cell line-derived neurotrophic factor (GDNF), 131, 133
Glutamate hypothesis, 4
Glutamatergic, 6, 18, 54, 55, 67, 135, 136, 142, 143
Gut microbiota, 7, 54, 61, 67, 68, 95–110, 112, 147, 148
Gut microbiota-brain axis, 19, 95–112

H
Hormone regulation, 115
HPA axis, 5, 7, 13, 115, 121

I
Immune biomarkers, 15
Immune dysfunction, 95
Immunoinflammatory biomarker, 91
Inflammation, 6, 7, 16, 44, 48, 67, 77, 95, 136
Inflammatory biomarkers, 85, 89, 122

L
Learning algorithm, 20–22, 27, 39, 69, 70, 188, 205, 238, 241

M

Machine learning, 15, 188, 199, 208
Machine learning-based diagnostic model, 15–23
MetaboAnalyst, 70–72, 110, 111, 157–242
Metabolic biomarker, 38–72, 136, 147
Metabolomics, 14, 40, 41, 45, 47, 51, 53, 55, 65, 98, 100, 101, 136, 148, 149, 154, 157–161, 174, 175, 183, 188, 196, 208, 212–214, 219, 220, 222, 223
Metabolomic workflow, 157
Microbiome, 54, 95, 96
Multivariate analysis, 22, 24, 26, 30, 89
Multivariate analysis of biomarker, 22, 26

N

Nerve growth factor (NGF), 3, 16, 66, 125–127, 131, 133
Neurobiological hypothesis, 3–5
Neurochemical hypothesis, 3–5
Neurodevelopmental hypothesis, 125
Neuroendocrine biomarker, 116–121
Neuronal plasticity, 125
Neurotransmitter system, 115–116
Neurotrophin, 3, 125, 126
Neurotrophins hypothesis, 3
Norepinephrine hypothesis, 4–5

O

Oxidative-stress-related biomarkers, 77–91

P

Partial AUC, 29, 30, 234
Pathophysiology of schizophrenia, 1–8
Pathway analysis, 154, 160, 184
Power analysis, 153, 160, 227–231
Proteomic biomarker, 37–38

R

ROC analysis, 38–72, 78–90, 96–110, 116–121, 126–132, 136–143, 232–235
ROC curve analysis, 26–30, 39–66, 80–88, 121, 126, 136, 147, 233, 234

S

Schizophrenia spectrum disorder (SSD), 1–8, 13, 14, 16, 17, 19, 22, 26, 37–72, 77–91, 95–112, 115–122, 125–133, 135–143, 147, 148
Serotonergic, 4, 5, 18, 125, 136
Serotonin hypothesis, 4
Stress-related biomarker, 91
Stress response, 4, 7, 115, 116
Study design, 149–152

U

Univariate analysis, 22, 142, 212–222

The manufacturer's authorised representative in the EU is Springer Nature Customer Service Centre GmbH, Europaplatz 3, 69115 Heidelberg, Germany. If you have any concerns regarding our products, please contact ProductSafety@springernature.com

Printed and bound by CPI Group (UK) Ltd, Croydon, CR0 4YY

25/03/2026

02078171-0007